中国科学院植物研究所系统与进化植物学国家重点实验室
学术著作出版基金资助出版

THALICTRUM (RANUNCULACEAE) IN CHINA
中国唐松草属植物

Wen-Tsai Wang
王文采　著图
孙英宝　绘图

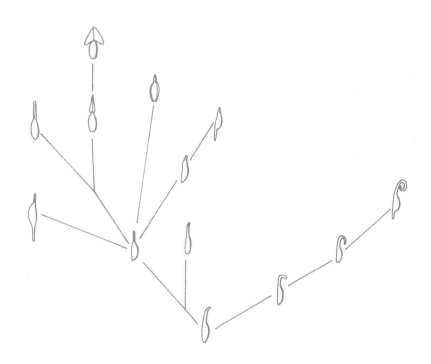

图书在版编目(CIP)数据

中国唐松草属植物 / 王文采 著. —北京:北京大学出版社,2018.9

ISBN 978-7-301-29817-6

Ⅰ.①中… Ⅱ.①王… Ⅲ.①双子叶植—植物志—中国 Ⅳ.①Q949.72

中国版本图书馆 CIP 数据核字(2018)第 191440 号

书　　名	中国唐松草属植物 ZHONGGUO TANGSONGCAOSHU ZHIWU
著作责任者	王文采　著　孙英宝　绘图
责任编辑	陈　静
标准书号	ISBN 978-7-301-29817-6
出版发行	北京大学出版社
地　　址	北京市海淀区成府路 205 号　100871
网　　址	http://www.pup.cn　新浪微博:@北京大学出版社
电子信箱	zpup@pup.cn
电　　话	邮购部 62752015　发行部 62750672　编辑部 62707542
印刷者	北京中科印刷有限公司
经销者	新华书店 787 毫米×1092 毫米　16 开本　24.5 印张　465 千字 2018 年 9 月第 1 版　2018 年 9 第 1 次印刷
定　　价	128.00 元

未经许可,不得以任何方式复制或抄袭本书之部分或全部内容。
版权所有,侵权必究
举报电话:010-62752024　电子信箱:fd@pup.pku.edu.cn
图书如有印装质量问题,请与出版部联系,电话:010-62756370

谨以此书庆祝

中国科学院植物研究所成立九十周年

目 录

前言 ··· 1
一、分类学研究简史 ·· 1
   A. 关于种的数目 ·· 1
   B. 六分类系统简介 ··· 3
      1. De Candolle 系统 ·· 3
      2. Lecoyer 系统 ·· 3
      3. Prantl 系统 ·· 4
      4. Boivin 系统 ··· 6
      5. Tamura 系统 ··· 9
      6. 根据分子系统学研究给出的谱系图 ··· 14

二、形态特征分析 ·· 17
   A. 关于唐松草属在毛茛科中的系统位置 ··· 17
   B. 扁果草族 Isopyreae 的形态特征 ··· 17
   C. 七形态特征的分析 ·· 19
      1. 茎的高度 ··· 19
      2. 对单叶特征的认识 ·· 20
      3. 花序 ··· 20
      4. 萼片 ··· 20
      5. 雄蕊 ··· 23
      6. 心皮 ··· 25
      7. 瘦果 ··· 27

三、分类系统 ·· 28
   A. 王文采，王蜀秀分类系统简介 ·· 28
   B. 本书的分类系统简介 ·· 31

四、地理分布 ·· 35
   A. 唐松草属地理分布简介 ··· 35
   B. 中国唐松草属地理分布 ··· 35

    1. 概况 ·· 35
    2. 广域分布种 ·· 35
    3. 广布的和稍广布的特有种 ·· 38
    4. 狭域分布种 ·· 41
    5. 五原始种 ·· 42
    6. 分布中心 ·· 42
    7. 姊妹群 ··· 44

五、经济用途 ·· **51**
  A. 药用 ··· 51
  B. 观赏 ··· 52

六、分类学处理 ·· **56**
  唐松草属 ·· 56
  分亚属、组、系检索表 ·· 58
  分种、变种、变型检索表 ··· 62
  亚属 1. 唐松草亚属 Subgen. Thalictrum ··· 80
    组 1. 叉枝唐松草组 Sect. Leptostigma B. Boivin ··· 81
      系 1. 叉枝唐松草系 Ser. Saniculiformia W. T. Wang & S. H. Wang ············ 81
        1. 叉枝唐松草 T. saniculiforme *DC.* ·· 81
        2. 小花唐松草 T. minutiflorum *W. T. Wang* ·· 84
        3. 绢毛唐松草 T. brevisericeum *W. T. Wang & S. H. Wang* ······················· 86
      系 2. 糙叶唐松草系 Ser. Scabrifolia *W. T. Wang & S. H. Wang* ··················· 90
        4. 糙叶唐松草 T. scabrifolium *Franch.* ··· 90
        5. 鹤庆唐松草 T. leve (*Franch.*) *W. T. Wang* ··· 92
        6. 大关唐松草 T. daguanense *W. T. Wang* ··· 94
        7. 华东唐松草 T. fortunei *S. Moore* ··· 96
        8. 多枝唐松草 T. ramosum *B. Boivin* ··· 99
        9. 新宁唐松草 T. xinningense *W. T. Wang* ·· 102
        10. 密叶唐松草 T. myriophyllum *Ohwi* ·· 104
        11. 云南唐松草 T. yunnanense *W. T. Wang* ··· 106

  12. 爪哇唐松草 T. javanicum *Bl.* ·· 109
  13. 微毛唐松草 T. lecoyeri *Franch.* ·· 112
  14. 玉山唐松草 T. sessile *Hayata* ··· 116
  15. 峨嵋唐松草 T. omeiense *W. T. Wang & S. H. Wang* ······················ 118
  16. 巨齿唐松草 T. grandidentatum *W. T. Wang & S. H. Wang* ················ 120
  17. 大叶唐松草 T. faberi *Ulbr.* ·· 122
  18. 粗壮唐松草 T. robustum *Maxim.* ·· 125
 系 3. 小喙唐松草系 Ser. Rostellata *W. T. Wang & S. H. Wang* ··············· 128
  19. 小喙唐松草 T. rostellatum *Hook. f. & Thoms.* ···························· 128
  20. 菲律宾唐松草 T. philippinense *C. B. Rob.* ································ 131
  21. 察隅唐松草 T. chayuense *W. T. Wang* ····································· 133
  22. 纺锤唐松草 T. fusiforme *W. T. Wang* ······································ 135
  23. 澜沧唐松草 T. lancangense *Y. Y. Qian* ····································· 137
  24. 弯柱唐松草 T. uncinulatum *Franch.* ·· 139
  25. 长喙唐松草 T. macrorhynchum *Franch.* ···································· 142
  26. 南湖唐松草 T. rubescens *Ohwi* ·· 145
  27. 台湾唐松草 T. urbainii *Hayata* ·· 147
 系 4. 思茅唐松草系 Ser. Simaoensia *W. T. Wang* ······························· 149
  28. 思茅唐松草 T. simaoense *W. T. Wang* ····································· 150
  29. 短蕊唐松草 T. brachyandrum *W. T. Wang* ································ 152
 系 5. 钩柱唐松草系 Ser. Uncata *W. T. Wang & S. H. Wang* ··················· 154
  30. 钩柱唐松草 T. uncatum *Maxim.* ·· 154
  31. 狭序唐松草 T. atriplex *Finet & Gagnep.* ··································· 157
  32. 螺柱唐松草 T. spiristylum *W. T. Wang* ···································· 160
组 2. 钻柱唐松草组 Sect. Dolichostylus *W. T. Wang* ······························· 162
  33. 钻柱唐松草 T. tenuisubulatum *W. T. Wang* ······························· 162
组 3. 宽柱唐松草组 Sect. Platystylus *W. T. Wang* ·································· 164
  34. 宽柱唐松草 T. latistylum *W. T. Wang* ····································· 165
组 4. 瓣蕊唐松草组 Sect. Erythrandra *B. Boivin* ··································· 167
 系 1. 细柱唐松草系 Ser. Megalostigmata *W. T. Wang* ························· 167

35. 细柱唐松草 T. megalostigma (*B. Boivin*) *W. T. Wang* ·················· 167
系 2. 贝加尔唐松草系 Ser. Baicalensia (*Tamura*) *W. T. Wang & S. H. Wang* ·········· 170
 36. 贝加尔唐松草 T. baicalense *Turcz.* ························· 170
系 3. 瓣蕊唐松草系 Ser. Petaloidea (*Prantl*) *W. T. Wang & S. H. Wang* ·········· 174
 37. 瓣蕊唐松草 T. petaloideum *L.* ··························· 175
系 4. 网脉唐松草系 Ser. Reticulata *W. T. Wang & S. H. Wang* ················ 179
 38. 网脉唐松草 T. reticulatum *Franch.* ······················· 180
 39. 武夷唐松草 T. wuyishanicum *W. T. Wang & S. H. Wang* ············ 182
系 5. 西南唐松草系 Ser. Fargesiana *W. T. Wang & S. H. Wang* ··············· 184
 40. 西南唐松草 T. fargesii *Franch.* ex *Finet & Gagnep* ··············· 184
 41. 兴山唐松草 T. xingshanicum *G. F. Tao* ····················· 188
 42. 丽江唐松草 T. wangii *B. Boivin* ························· 190
 43. 察瓦龙唐松草 T. tsawarungense *W. T. Wang & S. H. Wang* ·········· 192
 44. 矮唐松草 T. pumilum *Ulbr.* ··························· 194
组 5. 散花唐松草组 Sect. Omalophysa *Turcz.* ex *Fisch. & Mey.* ················ 196
 45. 毛蕊唐松草 T. lasiogynum *W. T. Wang* ····················· 197
 46. 长柄唐松草 T. przewalskii *Maxim.* ························ 199
 47. 散花唐松草 T. sparsiflorum *Turcz.* ························ 203
组 6. 尖叶唐松草组 Sect. Physocarpum *DC.* ························· 205
系 1. 尖叶唐松草系 Ser. Clavata *W. T. Wang & S. H. Wang* ················· 206
 48. 阴地唐松草 T. umbricola *Ulbr.* ··························· 206
 49. 稀蕊唐松草 T. oligandrum *Maxim.* ························ 209
 50. 岳西唐松草 T. yuexiense *W. T. Wang* ······················ 211
 51. 尖叶唐松草 T. acutifolium (*Hand.-Mazz.*) *B. Boivin* ················ 213
 52. 盾叶唐松草 T. ichangense *Lecoy.* ex *Oliv.* ···················· 217
 53. 深山唐松草 T. tuberiferum *Maxim.* ······················· 221
 54. 花唐松草 T. filamentosum *Maxim.* ······················· 223
 55. 小果唐松草 T. microgynum *Lecoy.* ex *Oliv.* ··················· 225
系 2. 拟盾叶唐松草系 Ser. Pseudoichangensia *W. T. Wang* ··················· 228
 56. 拟盾叶唐松草 T. pseudoichangense *Q. E. Yang & G. Zhu* ············ 228

组 7. 唐松草组 Sect. Tripterium *DC.* ···230
    57. 唐松草 T. aquilegifolium *L.* var. sibiricum *Regel & Tiling* ·························230

组 8. 芸香叶唐松草组 Sect. Rutifolia (*Prantl*) *W. T. Wang* ····································234
    系 1. 小金唐松草系 Ser. Xiaojinensia *W. T. Wang* ···234
        58. 小金唐松草 T. xiaojinense *W. T. Wang* ··235
        59. 六脉萼唐松草 T. sexnervisepalum *W. T. Wang* ···237
    系 2. 芸香叶唐松草系 Ser. Rutifolia *W. T. Wang & S. H. Wang* ····························239
        60. 芸香叶唐松草 T. rutifolium *Hook. f. & Thoms.* ···239
    系 3. 长柱唐松草系 Ser. Sinomacrostigmata *W. T. Wang* ·····································242
        61. 长柱唐松草 T. sinomacrostigma *W. T. Wang* ···243
    系 4. 白茎唐松草系 Ser. Leuconota *W. T. Wang* ··245
        62. 吉隆唐松草 T. jilongense *W. T. Wang* ···245
        63. 白茎唐松草 T. leuconotum *Franch.* ···247

组 9. 帚枝唐松草组 Sect. Platystachys *W. T. Wang* ···250
    系 1. 帚枝唐松草系 Ser. Virgata *W. T. Wang & S. H. Wang* ································250
        64. 帚枝唐松草 T. virgatum *Hook. f. & Thoms.* ···251
        65. 希陶唐松草 T. tsaii *W. T. Wang* ···254
        66. 丝叶唐松草 T. foeniculaceum *Bunge* ··256
        67. 大花唐松草 T. grandiflorum *Maxim.* ···259
        68. 攀枝花唐松草 T. panzhihuaense *W. T. Wang* ··261
        69. 粘唐松草 T. viscosum *C. Y. Wu ex W. T. Wang & S. H. Wang* ···························263
        70. 金丝马尾连 T. glandulosissimum (*Finet & Gagnep.*) *W. T. Wang &*
            S. H. Wang ··265
    系 2. 圆叶唐松草系 Ser. Indivisa (*DC.*) *W. T. Wang* ··268
        71. 圆叶唐松草 T. rotundifolium *DC.* ···269
        72. 心基唐松草 T. punduanum *Wall.* var. hirtellum *W. T. Wang* ···············271
    系 3. 偏翅唐松草系 Ser. Violacea *W. T. Wang & S. H. Wang* ································273
        73. 错那唐松草 T. cuonaense *W. T. Wang* ···274
        74. 珠芽唐松草 T. chelidonii *DC.* ···276
        75. 美花唐松草 T. callianthum *W. T. Wang* ···278

76. 美丽唐松草 T. reniforme *Wall.* ···280
77. 堇花唐松草 T. diffusiflorum *Marq. & Shaw* ···282
78. 偏翅唐松草 T. delavayi *Franch.* ··285

组 10. 二翅唐松草组 Sect. Dipterium *Tamura* ···291
    79. 二翅唐松草 T. elegans *Wall.* ···291

组 11. 腺毛唐松草组 Sect. Thalictrum DC. ··294
  系 1. 川鄂唐松草系 Ser. Osmundifolia *W. T. Wang* ··295
    80. 毛发唐松草 T. tricopus *Franch.* ···295
    81. 河南唐松草 T. honanense *W. T. Wang & S. H. Wang* ·······································297
    82. 川鄂唐松草 T. osmundifolium *Finet & Gagnep.* ···300
    83. 亚东唐松草 T. yadongense *W. T. Wang* ···302
    84. 陕西唐松草 T. shensiense *W. T. Wang & S. H. Wang* ······································304
    85. 滇川唐松草 T. finetii *B. Boivin* ···306

  系 2. 腺毛唐松草系 Ser. Thalictrum ··309
    86. 星毛唐松草 T. cirrhosum *Lévl.* ···309
    87. 札达唐松草 T. zhadaense *W. T. Wang* ···312
    88. 藏南唐松草 T. austrotibeticum *J. Y. Li, L. Xie & L. Q. Li* ·······························314
    89. 多叶唐松草 T. foliolosum *DC.* ··316
    90. 高原唐松草 T. cultratum *Wall.* ··319
    91. 腺毛唐松草 T. foetidum *L.* ··322
    92. 亚欧唐松草 T. minus *L.* ··326
    93. 细唐松草 T. tenue *Franch.* ···335
    94. 紫堇叶唐松草 T. isopyroides *C. A. Mey.* ···337
    95. 箭头唐松草 T. simplex *L.* ··339
    96. 黄唐松草 T. flavum *L.* ···345
    97. 展枝唐松草 T. squarrosum *Steph. ex Willd.* ··347

组 12. 石砾唐松草组 Sect. Schlagintweitella (*Ulbr.*) *W. T. Wang & S. H. Wang* ···········351
  系 1. 石砾唐松草系 Ser. Squamifera *W. T. Wang & S. H. Wang* ······························351
    98. 石砾唐松草 T. squamiferum *Lecoy.* ···351
  系 2. 高山唐松草系 Ser. Alpina (*Tamura*) *W. T. Wang & S. H. Wang* ······················355

99. 高山唐松草 T. alpinum *L.* ……………………………………………………… 355

亚属 2. 单性唐松草亚属 Subgen. Lecoyerium *B. Boivin* …………………………… 361

100. 鞭柱唐松草 T. smithii *B. Boivin* ……………………………………………… 361

101. 定结唐松草 T. dingjieense *W. T. Wang* …………………………………… 365

不了解的种 …………………………………………………………………………… 367

1. Thalictrum oshimae *Masamune* ……………………………………………… 367

2. Thalictrum ussuriense *Luferov* ……………………………………………… 367

中名索引 ……………………………………………………………………………… **368**

拉丁学名索引 ………………………………………………………………………… **372**

前　言

1959年冬，"中国植物志编纂委员会"在北京成立，中国科学院植物研究所领导派我参加会议的记录工作。在会议临近结束时，编委们讨论了各科的编写计划。这时，江苏植物研究所所长裴鉴教授召集陈封怀教授和我讨论毛茛科六大属的编写任务，决定由裴老承担铁线莲属和毛茛属，陈老承担乌头属和翠雀花属，我承担唐松草属和银莲花属，并决定将所有向兄弟所借用的六属植物标本均集中到江苏植物研究所。

那年秋季，王蜀秀先生大学毕业，分配到植物研究所，我遂邀请她合作编写唐松草属志。我们先根据中国科学院植物研究所植物标本馆(PE)收藏的唐松草属植物标本写出该属志初稿，又于1961年到江苏植物研究所研究了该所和自各兄弟所借到的该属植物标本，对初稿进行了补充，共鉴定出67种，划分为2亚属、4组、17系。经过对形态学的研究，认识到在唐松草属，雄蕊花丝呈条形和心皮的柱头不明显是两个重要的原始特征，将具这两个特征的叉枝唐松草 *Thalictrum saniculiforme* DC. 放在全属植物系统排列的首位。大概就在这时，陈老来信辞去乌头属和翠雀花属的编写任务。1968年，裴老不幸过早逝世。此后，毛茛科志的编写工作就由我来主持。

1973年在广州召开的三志会议上，铁线莲属的编写任务决定由江苏植物研究所的张美珍、凌萍萍两位先生和四川大学的方明渊先生一同承担，毛茛属的编写任务由沈阳林业土壤研究所承担。但会后不久沈阳林业土壤研究所就辞去了毛茛属的编写任务。植物研究所领导找到禾本科专家刘亮先生来担任毛茛属的编写工作。此外，早在1962年，药学家肖培根先生承担了扁果草族 Trib. Isopyreae 的编写工作，在研究此族的过程中他发表了分布于东亚的新属——人字果属 *Dichocarpum*，为毛茛科分类研究做出重要贡献。

大约在20世纪70年代初期，中国毛茛科志稿全部完成，我根据瑞典植物学家 L. A. Langlet 的毛茛科细胞分类学重要著作，将染色体大型、基数主要为8、果实为蓇葖果的属都归到金莲花亚科 Subfam. Helleboroideae；将染色体小型、基数为7或9、果实为蓇葖果或瘦果的属都归到唐松草亚科 Subfam. Thalictroideae；将染色体大型、基数主要为8、果实为瘦果的属都归到毛茛亚科 Subfam. Ranunculoideae。至于 Paeonia 芍药属，由于《中国植物志》采用 A. Engler & L. Diels 编著的 *Syllabus der Pflanzenfamilien* 第11版（1936）的被子植物分类系统，则被放在毛茛科中，但在形态描述之后说明 Paeonia 在形态学、解剖学、孢粉学、细胞学、胚胎学和植物化学等方面存在一系列特征与毛茛科不同，所以，应承认 Paeoniaceae 芍药科的建立。这样，将芍药亚科、金莲花亚科和唐松草亚科放在《中国植物志》第二十七卷中，于1979年出版；将毛茛亚科放在第二十八卷中，于1980年出版。

1988年，中美合作的英文版 *Flora of China* 编委会成立，编写工作启动。在此志的毛茛科编写工作方面，我承担了翠雀花属、铁线莲属和毛茛属，其他属由傅德志、李良千两位先生承担。在1996年我访问美国，与美国合作者进行翠雀花属和苦苣苔科的定稿工作。在密苏里植物园，我向该志的项目共同主任（Project co-director）朱光华先生建议，应将Paeonia自毛茛科移出，接受Paeoniaceae。之后，朱先生向诸编委发函，征求相关意见。不久，朱先生告诉我，诸编委对上述建议没有意见。这样，在该志的vol.6中，Paeoniaceae被放在毛茛科之前。就在某一天，朱先生向我询问唐松草属的分类方法，我当即做了介绍，当时完全不了解他是该志此属的作者之一。一直到该志于2001年出版，我得到此书后才了解唐松草属的作者署名是Fu Dezhi; Zhu Guanghua。再看此书的唐松草属志，种数增加到76种，但对这76种未进行分类，而是按照英文字母顺序加以排列，这样做就使读者难于了解诸种间的亲缘关系和诸种的演化水平，这不符合分类学研究的要求。后来，我向傅德志先生谈到这个问题，他告诉我，此属稿由朱先生一人编写，他并未参与编写工作。

在 Flora of China vol.6 于2001年出版之后到2017年年初，中国唐松草属又发现17新种，在2017年10月又发现3新种和2新变种，现在中国此属增加到99种，20变种。在新种中，宽柱唐松草 T. latistylum 和钻柱唐松草 T. tenuisubulatum 二种很独特。前者的花柱是一长圆形小薄片，看不到柱头；后者的花柱是一狭条形膜质薄片，顶端钩状弯曲，也看不到柱头。此外，在西藏南部发现了唐松草属进化群单性唐松草亚属 Subgen. Lecoyerium 在亚洲分布的第二种——定结唐松草 T. dingjieense。此种与第一种鞭柱唐松草 T. smithii 有甚为相近的亲缘关系。在亚洲分布的这两种均是单性唐松草亚属的原始种，这对此亚属的系统发育研究有重要意义。种数增加，一些新的形态特征的出现，就需要对中国唐松草属进行一次分类学修订。

2016年艺术家孙英宝先生在为唐松草属新种绘图时，我谈到20世纪40年代胡先骕教授编著《中国森林植物图志》，写出桦木科专著，对每一种都绘制一图版，在中国这种图谱性的著作甚是罕见。我没想到，孙先生听后建议我编一本唐松草属图谱，他愿承担绘图工作。对2016年发现的11新种，我写了两篇文章，已于2017年先后在《广西植物》上发表。孙先生为新种绘出11幅精彩图版，使我的文章倍增光彩。听他表示愿承担图谱绘图工作，我大喜过望，遂接受他的建议，决定编写这部图谱性著作《中国唐松草属植物》，并完成中国唐松草属新的一次修订。

在此，我对孙英宝先生的热心大力支持谨表示衷心的感谢；此外，对班勤、马欣堂两位先生在提供标本等方面，杨志荣博士在借用模式标本等方面，韩芳桥先生在提供文献方面给予的热心帮助表示衷心的感谢；还对杨宗愈博士惠允抄用他的唐松草属植物绘图，以及哈佛大学植物标本馆（GH）、纽约植物标本馆（NY）、江苏植物研究所植物标本馆（NAS）惠借珍贵植物标本，表示衷心的感谢。在2017年12月，本书全稿完成，交北京大学出版社，该社编辑陈静审阅全稿，对校样认真校对，对此作者表示衷心的感谢。

中国唐松草属植物丰富、复杂，而我的研究工作不深入，不仔细，在本书中定存在各种错误，衷心欢迎读者给予批评指正。

王文采

2018 年 7 月 12 日

一、分类学研究简史

A. 关于种的数目

描述中国唐松草属第一个种的是俄国植物学家 A. Bunge，1833 年他根据自己可能采自北京的一号标本发表 1 种 *Thalictrum foeniculaceum*。1842 年，不平等条约《南京条约》签订，五口通商，国门大开，很快大批欧洲国家的植物采集人员拥入中国各省区，采走大量植物标本。在 19 世纪下半叶，根据这些标本，中国唐松草属有 20 新种发表。英国植物学家 S. Moore 于 1878 年发表 1 种（*T. fortunei*），俄国植物学家 C. J. Maximowicz 从 1859 年到 1890 年发表了 7 种（*T. filamentosum*，*T. tuberiferum*，*T. przewalskii*，*T. robustum*，*T. uncatum*，*T. oligandrum* 和 *T. grandiflorum*）。比利时植物学家 C. J. Lecoyer 于 1880 年发表了 1 种（*T. squamiferum*）。法国植物学家 A. Franchet 从 1883 年到 1894 年发表了 9 种（*T. tenue*，*T. uncinulatum*，*T. scabrifolium*，*T. reticulatum*，*T. delavayi*，*T. lecoyeri*，*T. leuconotum* 和 *T. macrorhynchum*）；他在 1889 年发表的变种 *T. scabrifolium* var. *leve*，被我在 1993 年提升到种级。英国植物学家 D. Oliver 在 1886 年和 1888 年共发表了 2 种（*T. microgynum* 和 *T. ichangense*）。

从 20 世纪初到 1945 年又有 18 种发表。两位法国植物学家 E. A. Finet 和 F. Gagnepain 于 1903 年发表 3 种（*T. atriplex*，*T. fargesii* 和 *T. osmundifolium*），于 1906 年发表 1 种（*T. macrostigma*）；另一位法国植物学家 H. Leveillé 于 1909 年发表 1 种（*T. cirrhosum*）（他在 20 世纪初期还发表唐松草属 7 种，但后来均被归并）。日本植物学家 B. Hayata 于 1911 年和 1913 年发表台湾 2 种（*T. urbainii* 和 *T. sessile*）；另一日本植物学家 J. Ohwi 于 1933 年又发表台湾 2 种（*T. myriophyllum* 和 *T. rubescens*）。两位英国植物学家 C. V. B. Marquand 和 H. K. Airy-Shaw 于 1929 年发表 1 种（*T. diffusiflorum*）。德国植物学家 E. Ulbrich 于 1925 年发表 2 种（*T. faberi* 和 *T. umbricola*）；此外，他在 1929 年发表了一个重要变种（*T. alpinum* var. *elatum*）。美国植物学家 B. Boivin 在 1944 年发表了一变种（*T. baicalense* Turcz. var. *megalostigma*，我在 1980 年将此变种提升到种级），同时将奥地利植物学家 H. Handel-Mazzetti 于 1926 年发表的变种 *T. clavatum* DC. var. *acutifolium* Hand.-Mazz. 正确地提升到种级，他于 1945 年又发表了 4 种（*T. ramosum*，*T. wanglii*，*T. finetii* 和 *T. smithii*）。

上述两位法国植物学家 Finet 和 Gagnepain 于 1903 年发表一篇关于东亚毛茛科、

木兰科等科的重要分类学研究论文，在此文中列出东亚唐松草属植物51种，按照花丝规则或不规则，花柱明显或不明显，以及瘦果是否具柄这几个相当重要的形态特征，这51种被划分为8群，但对这8群都未予命名。在51种中，在中国有分布的有39种（包括此二位学者在此文中发表的3新种），这39种是中国唐松草属在1903年了解到的第一个名录。这39种加上20世纪上半叶发表的上述其他14种为53种，即为中国唐松草属在1945年的种的数目。

1961年，我（W. T. Wang）和王蜀秀（S. H. Wang）同志承担《中国植物志》唐松草属的编写任务，研究了不少标本馆的此属植物标本，发现8新种（*T. shensiense, T. brevisericeum, T. omeiense, T. wuyishanicum, T. grandidentatum, T. tsawarungense, T. honanense, T. viscosum*），并将变种 *T. foetidum* L. var. *glandulosissimum* Finet & Gagnep.（1903）提升到种级。此外，还鉴定出 *T. saniculiforme* DC.，*T. elegans* Wall.，*T. isopyroides* C.A. Mey. 和 *T. flavum* L. 在中国的新分布。这样，在1979年出版的《中国植物志》27卷中的唐松草属的种数增到67种。

后来，在1982年和1994年我共发表了4种（*T. chayuense, T. fusiforme, T. tenuisubulatum* 和 *T. yunnanense*）。在1996年，我与朱光华（G. Zhu）博士合作发表1种（*T. simaoense*）。1980年，陶光复发表1种（*T. xingshanicum* G. F. Tao）。1997年，钱义咏发表1种（*T. lancangense* Y. Y. Qian）。此外，于1994年发现了 *T. rotundifolium* DC. 在西藏的分布。这样，在2001年出版的 *Flora of China* 第6卷由傅德志和朱光华编著的唐松草属增到76种。

2001年之后到2017年年初又发表了17种。2004年，杨亲二和朱光华发表1种（*T. pseudoichangense* Q. E. Yang & G. Zhu）。2015年，李进宇、谢磊、李良千合作发表1种（*T. austrotibeticum* Z. Y. Li, L. Xie & L. Q. Li）。从2013年到2016年，我发表了4种（*T. callianthum, T. yuoxiense, T. cuonaense* 和 *T. panzhihuaense*）；在2017年发表了11种（*T. minutiflorum, T. xinningense, T. brachyandrum, T. lasiogynum, T. xiaojinense, T. daguanense, T. latistylum, T. sexnervisepalum, T. jilongense, T. spiristylum* 和 *T. zhadaense*）。此外，我将过去被归并的 *T. lecoyeri* Franch.，*T. sinomacrostigma* W. T. Wang（*T. macrostigma* Finet & Gagnep. 1906, non Edgew. 1851）和 *T. sessile* Hayata 恢复其等独立种地位，并将变种 *T. baicalense* Turcz. var. *macrostigma* B. Boivin 提升到种级地位。由于未看到 *T. pumilum* Ulbr.（1913）的标本，此种未收入到《中国植物志》第27卷和 *Flora of China* 第6卷的唐松草属志中；在2017年8月，有幸从纽约植物

园标本馆借到由 E. E. Maire 采自云南东川的此种的 isosyntype，此种得以收入本书中，被置于 43. *T. tsawarungense* 之后。此外，在 2017 年 10 月，我又发现了 3 新种和 2 新变种（*T. lecoyeri* Franch. var. *debilistylum*, *T. tsaii*, *T. punduanum* Wall. var. *hirtellum*, *T. yadongense*, *T. dingjieense*。*T. dingjieense* 的花两性，柱头细长，伸出花外，是单性唐松草 Subgen. *Lecoyerium* 分布在亚洲的第二个原始种，此种的发现，对此亚属系统发育的研究有重要意义），这些新分类群就放在本书中发表，这样，中国唐松草属现在的种和变种的数目增到 99 种，20 变种。

B. 六分类系统简介

1. De Candolle 系统

瑞士大植物分类学家 A. P. de Candolle（1817，1824）首次对唐松草属植物进行分类，根据瘦果形态特征，他将当时了解的世界该属植物约 50 种划分为 3 组：Sect. *Tripterium*, Sect. *Physocarpum* 和 Sect. *Euthalictrum*。又根据花的性别、根、叶、萼片等形态特征将包含约 40 种的 Sect. *Euthalictrum* 划分为 4 小群：§1. *Heterogama*, §2. *Genuina*, §3. *Indivisa* 和 §4. *Grumosa*。

2. Lecoyer 系统

比利时植物学家 C. J. Lecoyer 于 1885 年编写发表了唐松草属专著，对收载的当时的 69 种进行分类，并划分为 2 组和 4 亚组。

Sect. 1. *Macrogynes*　雌蕊在花期长度超过萼片而伸出。

　Sons-sect. A. *Anemalocarpes*　瘦果形状不规则，扁平或强度扁压；缝线与侧脉的形状不同。有分布于墨西哥和秘鲁的 *T. lanatum* Lecoy., *T. pubigerum* Benth., *T. longistylum* DC., *T. podocarpum* H. B. K. 和 *T. gibbosum* Lecoy., 以及分布于欧洲西南部的 *T. macrocarpum* Gren. 等 14 种。

　Sons-sect. B. *Homalocarpes*　瘦果卵球形，近卵球形，纺锤形或近纺锤形；缝线与侧脉同样弯曲。有分布于北美的 *T. dioicum* L., *T. debile* Buckl., *T. revolutum* DC. 和分布于非洲南部的 *T. rhynchocarpum* Dill. & Rich 等 6 种。

Sect. 2. *Microgynes*　雌蕊在花期长度不超过萼片，内藏。

　Sons-sect. A. *Longistamines*　雄蕊在花期长度超过萼片，伸出。

　　A. *Claviformes*　雄蕊花丝与花药等宽，或比花药宽。有包括分布于中国的 *T. fortunei* S. Moore, *T. javanicum* Bl., *T. uncinulatum* Franch., *T. baicalense* Turcz.,

T. petaloideum L., *T. przewalskii* Maxim., *T. sparsiflorum* Turcz., *T. tuberiferum* Maxim., *T. filamentosum* Maxim. 等 17 种。

B. *Filiformes* 雄蕊花丝全长的直径相同。

α. 瘦果扁平或强度扁压。有包括分布于中国的 *T. chelidonii* DC., *T. reniforme* Wall., *T. elegans* Wall., *T. cultratum* Wall., *T. foetidum* L., *T. tenue* Franch., *T. squamiferum* Lecoy. 等 8 种。

β. 瘦果卵球形，近卵球形，纺锤形或近纺锤形。有包括分布于中国的 *T. saniculiforme* DC., *T. rutifolium* Hook. f. & Thoms., *T. virgatum* Hook. f. & Thoms., *T. rotundifolium* DC., *T. foliolosum* DC., *T. isopyroides* C. A. Mey., *T. minus* L., *T. squarrosum* Steph., *T. alpinum* L. 等 17 种。

Sons-sect. B. *Brevistamines* 在花期，雄蕊的长度短于萼片，内藏。有包括分布于中国的 *T. rostellatum* Hook. f. & Thoms. 和 *T. foeniculaceum* Bunge 等 6 种。

本系统的 Sect. 1. *Macrogynes* 具单性花（少数种具两性花），即为 Boivin 系统的 Subgen. 2. *Lecoyerium*（见下）；第二组 Sect. 2 *Microgynes* 具两性花，即为 Boivin 系统的 Subgen. 1. *Thalictrum*（见下）。

3. Prantl 系统

德国毛茛科专家 K. Prantl 于 1887 年制定的唐松草属分类系统根据瘦果背部和腹部的凸起或凹入情况将此属植物划分为 2 群，在每群再以花序、花被、花丝、花柱、果实等特征进行小群划分。

Sect. 1. *Camptonota* 果实背部较强凸起，腹部则稍微凸起或甚至凹入。22 种，这些种被划分为 3 小群。

1. *Rotundifolia* 花组成稀疏圆锥花序；果实多数，几无柄，具向后弯曲的花柱和狭柱头。叶或小叶近圆形。8 种，其中 *T. rotundifolium* DC.、*T. dalzellii* Hook. 和 *T. saniculiforme* DC. 产喜马拉雅山区，*T. javanicum* Bl. 产印度、斯里兰卡和印尼爪哇岛。

2. *Rutifolia* 花组成简单总状花序；果实前部弯曲，花柱向内弯曲。只 1 种，*T. rutifolium* Hook. f. & Thoms. 产喜马拉雅山区。

3. *Petaloidea* 花由于增宽的花丝或花瓣状花被而明显；果实数目稍多，无柄或具柄，通常具短花柱；花序聚伞状。13 种，这些种被再划分为 3 小群。

a. 具花瓣状花丝。

α. 果实具柄。

★果实具翅，具1条侧脉。*T. aquilegifolium* L. 产欧洲及俄罗斯（西伯利亚）、日本。

★★果实无翅，具数条侧脉。*T. tuberiferum* Maxim. 产日本，中国东北部；*T. clavatum* DC., 产北美洲东部。

β. 果实无柄，具数条侧脉。*T. petaloideum* L. 产中亚、东亚；*T. filamentosum* Maxim., 产中国东北部，*T. tibeticum* Franch. 产中国西藏，等4种。

b. 花被花瓣状，花丝狭窄。

α. 果实具柄，具明显向后弯曲的花柱：*T. pedunculatum* Edgew. 产喜马拉雅山区西部。

β. 果实无柄，具短花柱：*T. virgatum* Hook. f. & Thoms., *T. foeniculaceum* Bunge 等5种。

Sect. 2. *Camptogastra*① 果实腹部比背部较强凸起。

4. *Sparsiflora* 花序聚伞状；果实膨胀，具数条侧脉；花柱与子房等长。

a. 雄蕊与花被等长，上部增宽。*T. sparsiflorum* Turcz. 产东亚、北美洲东部；*T. przewalskii* Maxim. 产中国西部；*T. baicalense* Turcz. 产东亚。

b. 雄蕊比花被短，狭窄。*T. rostellatum* Hook. f. & Thoms. 产喜马拉雅山区。

5. *Macrocarpa* 花序较大；果实膨胀，有数条侧脉；花柱比子房长。*T. rhychocarpum* Dill. & Rich. 产埃塞俄比亚；*T. macrocarpum* Gren. 和 *T. calabricum* Spreng. 产欧洲南部。

6. *Platycarpa* 花序总状或圆锥状；花两性；果实有短或长柄，具4条或较多条侧脉；花柱和柱头一起才与子房等长。

a. 果实具4条侧脉。*T. elegans* Wall. 产喜马拉雅山区。

b. 果实具较多侧脉。*T. cultratum* Wall., *T. pauciflorum* Royle, *T. chelidonii* DC., *T. reniforme* Wall. 均产喜马拉雅山区；*T. squamiferum* Lecoy. 产中国西藏；*T. tenue* Franch. 产蒙古，中国北部。

7. *Podocarpa* 花序圆锥状；花杂性，雌雄同株或雌雄异株；果实具短柄，具1或数条侧脉；具柱头的花柱与子房等长，并超过花被。*T. fendleri* Engelm., *T. polycarpum* Wats., *T. wrightii* A. Gray. 产北美洲西部；*T. gibbosum* Lecoy., *T. pubigerum* Benth., *T. peltatum* DC., *T. lanatum* Lecoy. 产墨西哥；*T. longistylum* DC., *T. rutidocarpum* DC., *T.*

---

① Prantl 未给出此组的种的数目。

podocarpum Kunth, *T. veciculosum* Lecoy. 产南美洲。

8. *Dioica* 花序圆锥状；花杂性，雌雄同株或雌雄异株；果实无柄或具短柄，膨胀，具数条直侧脉；具柱头的花柱比子房长，并超过花被。*T. dioicum* L., *T. dasycarpum* Fish., Mey. & Lall. 广布北美；*T. debile* Buckl., *T. corynellum* DC., *T. revolutum* DC. 产北美洲东部。

9. *Flexuosa* 花序总状或圆锥状；花两性；果实膨胀，具数条直侧脉；柱头无柄，宽与子房等长。

    a. 子房具柄；花组成简单总状花序。*T. alpinum* L. 产中亚、欧洲北部、北美加拿大等地。

    b. 子房无柄；花组成圆锥花序。约 20 个难于区别的种[①]，产中亚、东亚、欧洲以及印度热带地区和南非。

4. Boivin 系统

    Lecoyer 系统将唐松草属正确地划分为具两性花和具单性花的两大群，但是他命名的组（section）名和亚组（subsection）名均未用拉丁文，不符合后来订立的《国际植物命名法规》关于各级分类群的名称均须用拉丁文的规定。此外，Lecoyer 缺乏生物演化的认识，将具单性花的进化群 Sect. *Macrogynes* 放在具两性花的原始群 Sect. *Microgynes* 之前，居于分类系统的首位。Prantl 系统则将具单性花的两群 *Podocarpa* 和 *Dioica* 置于瘦果腹部较强凸起的 Sect. *Camptocastra* 中，未认识到具单性花的群是唐松草属的进化群。上述两系统对唐松草属单性群处理的缺点在近六十年之后才由一位美国植物学家进行了相当完美的弥补。1944 年，美国唐松草属专家 B. Boivin 发表了一篇长达 181 页的关于唐松草属分类学研究的文章，他研究了大量采自北美洲、南美洲以及不少采自亚洲、非洲、欧洲的标本，将唐松草属划分为具两性花的第一亚属 Subgen. 1. *Thalictrum* 和具单性花由他建立的第二亚属 Subgen. 2. *Lecoyerium* Boivin。他对第二亚属进行了深入的研究，建立了 4 新组和 10 新亚组，描述了产北美洲的 14 新种，产南美洲的 7 新种，和产非洲的 4 新种。他建立的 Subgen. *Lecoyerium* 分类系统被后来 Tamura 建立的唐松草属分类系统完全接受，只是 Sect. *Cincinneria* Boivin 被处理为 Sect. *Macrogynes* 的异名（见下）。此外，对 Subgen. *Thalictrum* 的分类学处理，由于研究的标本等方面不充分，因此不够全面，不够仔细，但还是划分出 7 组，承认 19 世

[①] Prantl 这里提到的难于区别的种是指 *T. minus* L., *T. flavum* L., *T. foetidum* L., *T. simplex* L. 等种。

纪早就建立的3组：Sect. *Physocarpum* DC., Sect. *Omalophysa* Turcz. 和 Sect. *Tripterium* DC.（在 Lecoyer 和 Prantl 的二系统中，此3组均未被接受），并描述了3新组：Sect. *Homothalictrum*, Sect. *Leptostigma* 和 Sect. *Erythrandra*，将 De Candolle 描述的一亚组提升为组级分类群：Sect. *Genuina* (DC.) Boivin。后来，在 Tamura 的唐松草属分类系统中都被归并，Sect. *Homothalictrum* 和 Sect. *Genuina*（DC.）Boivin 分别作为 Sect. *Thalictrum* 和 Subsect. *Thalictrum* 的异名，sect. *Leptostigma* Boivin 被处理为 Sect. *Comptonota* Prantl 的异名，其他1新组则被承认。

上面我说 B.Boivin 对 Subgen. *Thalictrum* 的研究不够仔细，是由于看到两个分类学处理不正确的情况：在 Boivin 系统中，新组 Sect. *Leptostigma* 的模式是 *T. saniculiforme* DC., 给出的特征集要（diagnosis）是"stigmatibus angustissimis nec alatis"，但在列出的此组其他种有 *T. dalzelii* Hook., *T. rotundifolium* DC. 和 *T. virgatum* Hook. f. & Thoms., 这三种的宿存花柱很短，长 0.2–0.8 mm，直，在腹面有狭长圆形或长圆形的明显柱头，与 *T. saniculiforme*（宿存花柱长 2.2 mm，顶端钩状弯曲，腹面的柱头极小，不容易看出来）区别甚大，与上述特征集要不符。另一新组 Sect. *Erythrandra* 的模式是 *T. petaloideum* L., 给出的特征集要是"Filamenta antherarum apice clavata, alba vel rufescentia. Carpella sessilia vel breviter stipitata, costata, in flata. Stigma nunqum alatum nec sagittatum."。在列出的此组其他种中有 *T. fortunei* S. Moore 和 *T. javanicum* Bl., 这两种的花柱较长，顶端钩状弯曲，腹面的柱头很小，难于看出（这些特征与 *T. saniculiforme* DC. 很相似），而与 *T. petaloideum* L.（花柱短，直，腹面有明显柱头）不同，说明与 *T. petaloideum* 不是隶属同一组的种。另外，还列出有 *T. thibeticum* Franch., 这个种的萼片紫色，花丝丝形，子房沿腹、背缝线均有狭翅，与 *T. petaloideum*（萼片白色，花丝上部倒披针形，下部丝形，子房无翅）大不相同，是与后者在亲缘关系上距离相当远的种。

Boivin 在文中给出用拉丁文写的区分唐松草属亚属、组和亚组的检索表，在下面作者译出的此表中，可了解他对此属分类学处理的方法，在其中 Subgen. *Lecoyerium* 附上一些代表种，供参考。

1. 花为具备花；柱头在花期短于萼片···Subgen. *Thalictrum*
　2. 柱头无柄，具2翅，呈正三角形；花丝丝形，顶端近花药处稍增粗；成熟心皮无柄，具肋
　　3. 成熟心皮不扁···Sect. *Homothalictrum* Boivin
　　3. 成熟心皮扁·· Sect. *Genuina* (DC.) Boivin
　2. 柱头无翅，不呈箭头形，如具狭翅，则呈条形。
　　4. 花丝丝形，常在近花药处稍增粗；柱头有时拳卷状弯曲；瘦果常具短柄，具肋···············

.... Sect. *Leptostigma* Boivin
 4. 花丝多少呈棒状，顶端近花药处稍缢缩，白色或淡红色，稀紫色；花药球形，可演变成长圆状披针形；柱头无翅，不呈箭头形，不拳卷状弯曲。
  5. 果实膨胀或扁，无翅。
   6. 果实无柄或近无柄，不扁，具肋；柱头位花柱腹面 ············· Sect. *Erythrandra* Boivin
   6. 果实具柄，多少扁，脉隆起。
    7. 果实多少扁，背脉较凸起或似腹脉弯曲；柱头位花柱腹面或顶端；花梗直；茎生叶常不存在 ·· Sect. *Physocarpum* DC.
    7. 果实强度扁，腹脉凸起；茎生叶存在，为3-5回三出复叶；柱头细圆柱状 ··· Sect. *Omalophysa* Turcz.
  5. 瘦果具长柄，具3-4翅；柱头位于花柱腹面 ························ Sect. *Tripterium* DC.
1. 花雌雄异株或杂性，稀为具备花；花柱于花期在长度上超过萼片 ········· Subgen. *Lecoyerium* Boivin
 8. 花杂性同株，有时为具备花；柱头丝形，稀具狭翅；萼片不二形。
  9. 子房少数，单生或成对生，稀3-4；花梗果期很长；花为具备花（10种，*T. rhynchocarpum* Dill. & Rich., *T. aduncum* Boivin, *T. innitens* Boivin 等5种产非洲，*T. steyermarkii* Stand. 产危地马拉，*T. cincinatum* Boivin 和 *T. steinbachii* Boivin 产玻利维亚，*T. macrocarpum* Gren. 产法国西南部） ··· Sect. *Cincinneria* Boivin.
  9. 子房多数；花梗长达6 mm；花杂性同株，稀为具备花；柱头细圆柱形（*T. johnstonii* 除外）
   10. 小叶不呈盾形 ·· Sect. *Camptogastrum* Prantl
    11. 子房10，或更少；成熟心皮扁。
     12. 心皮腹脉不呈浅囊形（9种，*T. galeotii* Lecoy., *T. grandifolium* Wats., *T. henandeeii* Tausch ex Presl, *T. viridulum* Boivin 等6种产中美洲，*T. decipiens* Boivin, *T. inuncans* Boivin 等3种产南美洲） ·· Subsect. *Simplicia* Boivin
     12. 心皮腹脉呈浅囊形（19种，*T. pachuense* Rose, *T. standleyi* Stey., *T. lanatum* Lecoy., *T. johnstonii* Stand. & Stey., *T. strigillosum* Hemsl. *T. gibbosum* Lecoy., *T. pubigerum* Benth., *T nelsonii* Boivin, *T. subpubescens* Rose 等15种产中美，*T. vesiculosum* Lecoy., *T. rutidocarpum* DC., *T. longistylum* HBK ex DC., *T. lasiostylum* Presl 产南美洲） ··· Subsect. *Gibbosa* Boivin
    11. 子房18-36；成熟心皮腹面呈浅囊形；柱头稍呈棒状（1种，*T. venturii* Boivin 产南美洲）
     ·· Subsect. *Venturiana* Boivin
   10. 小叶呈盾形 ··· Sect. *Pelteria* Boivin
    13. 小叶较小，小叶柄在边缘处着生（3种，*T. lankesteri* Stand., *T. torresii* Stand. & Boivin, *T. guatemalense* C. DC. & Rose 产中美洲） ······························ Subsect. *Subpeltata* Boivin
    13. 小叶大，小叶柄在叶片1/3处着生（7种，*T. peltatum* DC., *T. pringlei* Wats., *T. jaliseanum* Rose, *T. treleassii* Boivin, *T. roseanum* Boivin 等产中美洲） ········ Subsect. *Eupeltata* Boivin
 8. 花雌雄异株或杂性异柱；花柱短，多少具2翅；萼片二形，雌花的萼片较小。
  14. 花丝有颜色，丝形，花雌雄异株，稀杂性；小叶常3浅裂，裂片有圆齿 ······· Sect. *Heterogamia* Boivin
   15. 心皮直，不扁，腹面对称，具较粗脉；根为须根或具块根，无匍匐茎；植株无毛。
    16. 块根存在。

17. 心皮的脉不弯曲；叶为三出复叶（3 种，*T. texanum* (A. Gray) Small, *T. debile* Buck., *T. arkansanum* Boivin 产美国东南部）·················· Subsect. *Debilia* Boivin

17. 心皮的脉弯曲（3 种，*T. pudicum* Stand., *T. pinnatum* Wats., *T. madrense* Rose 产墨西哥）·················· Subsect. *Sinuosa* Boivin

16. 根为须根；小叶柄具关节；心皮的脉不弯曲（1 种，*T. dioicum* L. 产美国和加拿大）·················· Subsect. *Dioica* (Prantl) Boivin

15. 心皮弯曲或扁；植株有时被柔毛，多少具匍匐茎。

18. 心皮壁厚，坚实，不扁，向内弯曲，具肋，背脉较腹脉凸起（2 种，*T. venulosum* Trel., *T. confine* Fernald 产美国）·················· Subsect. *Incurvata* Boivin

18. 心皮壁有时膜质，多少扁，直或基部多少向后弯曲，或很扁，脉明显。

19. 心皮稍扁，具柄，直，腹脉比背脉较凸起（2 种，*T. coriaceum* (Britt.) Small, *T. steeleanum* Boivin 产美国）·················· Subsect. *Clavocarpa* Boivin

19. 心皮近无柄，如具柄，柄向后弯曲，或腹脉比背脉较凸起。

20. 心皮的厚度达到或超过宽度之半（2 种，*T. occidentale* A. Gray, *T. nigromontanum* Boivin 产美国）·················· Subsect. *Compressa* Boivin

20. 心皮的厚度不到宽度之半（2 种，*T. polycarpum* (Torrey) Wats., *T. fendleri* Engelm. ex A. Gray 产美国，墨西哥）·················· Subsect. *Laminaria* Boivin

14. 花丝白色，常多少呈棒状；花杂性异株；小叶不分裂，或顶端 3 浅裂，裂片全缘（5 种，*T. revolutum* DC., *T. polycarpum* Muhl ex Spreng., *T. dasycarpum* Fisch. & Lall., *T. macrostylum* Small & Hell., *T. subrotundum* Boivin 产美国）·················· Sect. *Leucocoma* (Greene) Boivin

5. Tamura 系统

日本植物学家田村道夫（M. Tamura）一生精心致力于毛茛科植物分类学的研究，对此科分类系统的建立做出重要贡献。对唐松草属，他也进行过长期研究，在 1995 年，他最后一次对此属的分类系统进行了修订，现在下面予以介绍。他像 Boivin 一样将此属具两性花的群作为第一亚属，将具单性花的群作为第二亚属，但他对第一亚属 Subgen. *Thalictrum* 进行了全面、深入的研究，将此亚属划分为 9 组，12 亚组。并在形态描述、地理分布之后对此亚属各群的演化水平，也可以说是对其系统发育进行了说明："在 Subgen. *Thalictrum* 中，Sect. *Comptonotum* subsect. *Indivisa* 的种具单叶、丝形花丝和无柄，不扁压的瘦果，是最原始的群。Sect. *Comptonotum* 的其他亚组（subsections）以及 Sect. *Thalictrum* 紧随其后。Sect. *Platycarpa* 由于其瘦果强度扁压，Sect. *Dipterium* 由于其下垂、具翅的瘦果，Sect. *Erythrandra* 由于其变宽的花丝，Sect. *Physocarpum* 由于其变宽的花丝和强度扁的瘦果，以及 Sect. *Tripterium* 由于其变宽的花丝和下垂具翅的瘦果而显得特化，而被认为是 Subgen. *Thalictrum* 的进化群。"关于 Subgen. *Lecoyerium* 的分类，如前所述，他近完全地接受了 Boivin 的分类方法。

1. Subgen. 1. *Thalictrum*　花两性；柱头在花期通常不超出萼片。约125种，广布欧洲、亚洲、非洲东北部和北美洲。

　　Sect. 1. *Camptonatum* Prantl　花丝丝形；瘦果扁或不扁，腹缝线凸起、直或稍凹，背缝线凸起；宿存花柱多少伸长，钩状弯曲或近直；柱头不变宽或稍变宽。20种，产东亚。

　　　　Subsect. 1. *Indivisa* DC.　叶为单叶或1-2回三出复叶；瘦果无柄，不扁，背、腹缝线均凸起，具肋；稍向后弯曲。3种：*T. rotundifolium* DC., *T. dalzellii* Hook., *T. punduanum* Wall, 产喜马拉雅山脉和印度东北部。

　　　　Subsect. 2. *Saniculiformia* (W. T. Wang & S. H. Wang) Tamura　瘦果多数，稍扁，背腹缝线均凸起，宿存花柱伸长，与瘦果等长，拳卷状。1种，*T. saniculiforme* DC., 产喜马拉雅地区和中国西南部。

　　　　Subsect. 3. *Rutifolia* Prantl　花序似总状花序；瘦果具柄，多少扁压，腹缝线近直；花柱向内弯曲。2种：*T. rutifolia* Hook. f. & Thoms., *T. leuconotum* Franch.

　　　　Subsect. 4. *Grumosa* DC.　块根有时存在；瘦果具柄或无柄，不扁，背、腹缝线均凸起；宿存花柱短，直。3种：*T. tuberosum* L. 产南欧, *T. orientale* Boiss. 产南欧、中东, *T. foeniculaceum* Bunge 产中国北部。

　　　　Subsect. 5. *Violacea* (W. T. Wang & S. H. Wang) Tamura　花序圆锥状；萼片大，紫色，粉红色，稀白色；瘦果具柄，稍扁，腹缝线凸起或近直，有条纹；花柱向后弯曲或直。7种：*T. delavayi* Franch., *T. chelidonii* DC., *T. virgatum* Hook. f. & Thoms., *T. grandiflorum* Maxim. 等。

　　　　Subsect. 6. *Purpurea* Tamura　与上亚组相近，区别为瘦果具粗纵肋。2种：*T. grandisepalum* Lévl. 产朝鲜, *T. rochebrantanum* Franch. & Sav. 产日本。

　　Sect. 2. *Erythrandra* Boivin　花丝棒状；瘦果无柄或具短柄，不扁或稍扁，背、腹缝线均稍凸起；花柱弯曲或直；柱头不变宽。35种，广布东亚。

　　　　Subsect. 1. *Actaeifolia* Tamura　瘦果无柄或具短柄，狭卵球形，有粗纵肋；花柱长，钩状。约25种：*T. actaeifolium* Sieb. & Zucc., *T. scabrifolium* Franch., *T. javanicum* Bl., *T. sessile* Hayata, *T. omeiense* W. T. Wang & S. H. Wang. *T. ramosum* Boivin, *T. robustum* Maxim. 等，广布自喜马拉雅山区起；东达朝鲜、日本，南达斯里兰卡、印度尼西亚。

Subsect. 2. *Baicalensia* Tamura　瘦果无柄或有短柄，卵球形或倒卵球形，直或稍向后弯曲；柱头短，头状，或伸长。6 种：*T. baicalense* Turcz., *T. reticulatum* Franch., *T. fargesii* Franch. ex Finet & Gagnep., *T. wangii* Boivin 等，产俄罗斯（西伯利亚）、中国、朝鲜、日本。

Subsect. 3. *Petaloidea* Prantl　瘦果无柄或具短柄，卵球形，具粗肋；宿存花柱长，钩状；柱头多少增长，无翅。4 种：*T. petaloideum* L., *T. calabricum* Spreng., *T. uncinatum* Rehmann 和 *T. sachalinense* Lecoy., 产南、东欧，中亚，俄罗斯（西伯利亚），中国北部和东北，朝鲜，日本。

Sect. 3. *Thalictrum*　花丝丝形；瘦果无柄，稀具柄，不扁，背、腹缝线均凸起，具粗肋；宿存花柱不伸长；柱头多少增宽，或具翅，三角形，心形或卵形。约 30 种，广布欧、亚、美洲和非洲东北部。

Subsect. 1. *Thalictrum*　花序多花，圆锥状；瘦果无柄。约 25 种：*T. foeticum* L., *T. minus* L., *T. simplex* L., *T. flavum* L., *T. lucidum* L., *T. foliolosum* DC., *T. isopyroides* C. A. Mey., *T. sultanabadense* Stapf 等，广布亚洲、欧洲和非洲北部；*T. simperianum* Hochst. ex Schweinf. 和另 2 种产非洲南部热带地区。

Subsect. 2. *Alpina* Tamura　叶基生；花序总状；瘦果无柄或具柄。1 种：*T. alpina* L., 分布于环北极地区，南达欧洲西南部、高加索地区和喜马拉雅山区。

Subsect. 3. *Squamifera* (W. T. Wang & S. H. Wang) Tamura　茎上部具叶，下部具鳞片；花单生叶腋；瘦果无柄。1 种：*T. squamiferum* Lecoy. 产中国西南部、喜马拉雅山区。

Sect. 4. *Platycarpa* (Prantl) Tamura　花丝丝形；瘦果具柄，扁，腹缝线凸起，背缝线稍凸起或不凸，具条纹；宿存花柱伸长，向外钩状弯曲；柱头增宽或稍增宽。4 种：*T. cultratum* Wall., *T. platycarpum* Hook. f. & Thoms., *T. rostellatum* Hook. f. & Thoms. 和 *T. uncatum* Maxim. 产中国西南部和喜马拉雅山区。

Sect. 5. *Dipterium* Tamura　花丝丝形；瘦果具细长柄，扁平，沿腹、背缝线具狭翅，每侧有 1 条纵脉；柱头不增宽。1 种：*T. elegans* Wall. ex Royle. 产喜马拉雅山区和中国西南部。

Sect. 6. *Anemonella* (Spach) Tamura　花序伞状，具 1 或少数花；萼片 4–10，大；花丝丝形；瘦果无柄，不扁，背、腹缝线均凸起，具粗肋；宿存花柱短；柱头宽。1 种：*T. thalictroides* (L.) Eames & Boivin 产北美东部。

Sect. 7. *Physocarpum* DC. 花丝棒状；瘦果具柄，扁，背缝线凸，腹缝线稍凸或凹；宿存花柱短或长，柱头不增宽，稍伸长或头状。约 29 种，分布于东亚和北美东部；*T. ichangense* Lecoy. ex Oliv., *T. acutifolium* (Hand.-Mazz.) Boivin 等产中国；*T. tuberiferum* Maxim. 产中国东北部、俄罗斯（西伯利亚）和日本；*T. integrilobum* Maxim., *T. watnabei* Yatabe 和 *T. toyamae* Ohwi 产日本；*T. clavatum* DC. 和 *T. mirabile* Small 产美国东部。

Sect. 8. *Omalophysa* Turcz. ex Fisch. & C. A. Mey. 花丝上部稍增宽；瘦果具柄，腹缝线明显凸，背缝线稍凸或稍凹；宿存花柱长，直。2 种：*T. sparsiplorum* Turcz. ex Fisch. & C. A. Mey 产亚洲东北部和北美西部，*T. przewalskii* Maxim. 产中国四川西部至内蒙古西南部。

Sect. 9. *Tripterium* DC. 花丝上部增宽；瘦果具细长柄，倒卵球形，具 1 纵脉，沿脉具翅；宿存花柱直；柱头短，不增宽。1 种：*T. aquilegifolium* L. 广布欧洲和亚洲。

2. Subgen. 2. *Lecoyerium* Boivin 花单性，稀两性；柱头在花期长度超出萼片。约 70 种，分布于北、南美洲，非洲，欧洲南部，中国西南部有 1 种。

Sect. 1. *Macrogynes* Lecoy. 植物两性；花序圆锥状；花丝丝形；花有 1-4 枚心皮，柱头丝形；瘦果具柄，多少扁。约 10 种：*T. macrocarpum* Gren. 产欧洲西南部；*T. rhynchocarpum* Dill. & A. Rich. 和其他 5 种产南非；*T. steyermarkii* Standl. 产危地马拉。

Sect. 2. *Camptogastrum* Prantl 小叶不盾状；植物雌雄同株或两性；花丝丝形；瘦果扁。约 30 种，广布中美洲和南美洲。

Subsect. 1. *Simplicia* Boivin 花有 10 枚以下的心皮；瘦果腹缝线凸，背缝线不凸，侧脉不波状或稍波状。10 种：*T. galeottii* Lecoy., *T. hernandezii* Tausch ex Presl，产中美洲；*T. decipiens* Boivin，产南美洲。

Subsect. 2. *Venturiana* Boivin 植物两性；花具 15 枚以上心皮；瘦果膨胀，背、腹缝线均凸，侧脉波状。1 种：*T. venturii* Boivin，产南美洲。

Subsect. 3. *Gibbosa* Boivin 植物杂性同株；花有 10 枚以下心皮；瘦果腹缝线凸，背缝线稍凸或凹。侧脉波状或网状。约 20 种：*T. gibbosum* Lecoy., *T. podocarpum* Benth., *T. lanatum* Lecoy., *T. podocarpum* HBK ex DC., 和 *T. longistylum* DC.，产墨西哥和秘鲁。

Sect. 3. *Pelteria* Boivin 植物杂性同株或两性；小叶盾状；花序圆锥状；花丝丝形；柱

头丝形，不增宽；瘦果扁，侧脉波状或直，约 10 种，产中美洲。

  Subsect. 1. *Subpeltata* Boivin 小叶较小，稍盾状，小叶柄在近边缘处着生。3 种：*T. lankesteri* Stand., *T. torrestii* Stand. & Boivin, *T. guatemalense* C. DC. & Rose. 产中美洲。

  Subsect. 2. *Eupeltata* Boivin 小叶大，明显盾状。7 种：*T. peltatum* DC., *T. pringlei* Wats. 及其他 5 种产墨西哥。

Sect. 4. *Heterogamia* (DC.) Boivin 植物雌雄异株，稀杂性；萼片二形，雄花萼片比雌花萼片大；花丝黄色或淡紫色，丝形；柱头短，多少具 2 翅。15 种分布于北美洲，1 种产中国。

  Subsect. 1. *Dioica* Prantl 根为须根；瘦果无柄，直立，不扁，侧脉直。2 种：*T. dioicum* L. 产北美洲，*T. smithii* Boivin 产中国西南部。

  Subsect. 2. *Debilia* Boivin 近 *Dioica* Prant 1，但根为块根。3 种：*T. debile* Buckley 及另 2 种产美国东南部。

  Subsect. 3. *Sinuosa* Boivin 近前 2 亚组，但瘦果侧脉弯曲。3 种：*T. pinnatum* Wats. 及另 2 种产墨西哥。

  Subsect. 4. *Clavocarpa* Boivin 瘦果具柄，直立，不扁或稍扁，侧脉不或稍弯曲。2 种：*T. coriaceum* (Britt.) Small 和 *T. steeleanum* Boivin，产北美洲。

  Subsect. 5. *Incurvata* Boivin 瘦果无柄，向内弯曲，不扁，侧脉不明显。2 种：*T. venulosum* Trel. 和 *T. confine* Fernald，产北美洲。

  Subsect. 6. *Compressa* Boivin 瘦果近无柄或具短柄，直立或向后弯曲，扁，其厚度为宽度之半，侧脉不弯曲。2 种：*T. occidentale* A. Gray, *T. nigromontanum* Boivin，产北美洲西部。

  Subsect. 7. *Laminaria* Boivin 瘦果近无柄，直立，强度扁，其厚度小于宽度之半，侧脉稍弯曲。2 种：*T. fendleri* A. Gray 和 *T. polycarpum* (Torrey) Wats. 产北美洲西部。

Sect. 5. *Leucocoma* Greene. 植物杂性异株；花序圆锥状或伞房状；萼片二形；花丝白色，丝形或棒状；柱头短，多少具 2 翅；瘦果无柄或具短柄。5 种：*T. polygamum* Muhl., *T. revolutum* DC. 及另 2 种的花丝棒状，*T. dasycarpum* Fisch. & Lall. 的花丝丝形，均产北美洲。

6. 根据分子系统学研究给出的谱系图

药学专家郝大程和肖培根（2016）进行了唐松草属的分子系统学研究，给出二谱系图，现抄录于下：

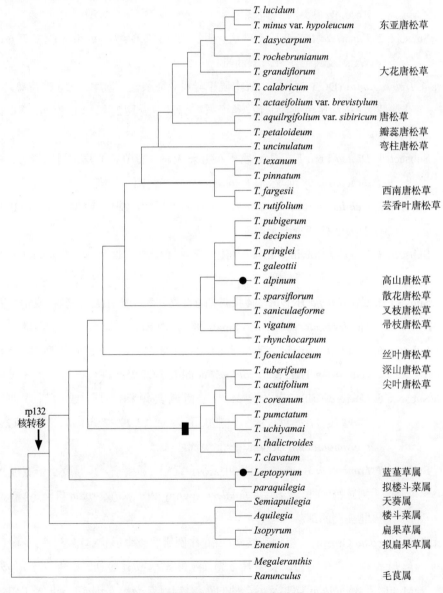

唐松草亚科 37 种系统发育关系

（用 ML 法和 5 个 cp 标记构建，即 rbcL, ndhF, ndhA 内含子, trnL 内含子, trnL-F 基因间隔序列。圆表示 rpl32 从质体完全丢失，矩形表示此演化支共有的一个 indel 事件）

一、分类学研究简史

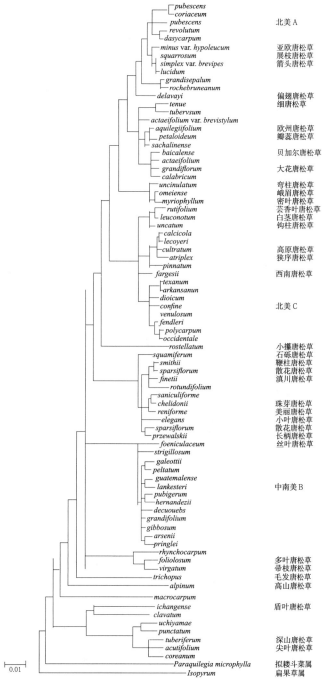

唐松草属药用植物的亲缘关系——基于叶绿体分子标记 rpl32-trnL+ndhA 内含子 +rbcL+trnL-trnF 的最大似然树（标尺长度代表每位点核苷酸替换数目）

我对分子系统学了解不深，因此，没有能力提出任何评论意见，但是看到了这两幅谱系图，发现与根据形态学建立的分类系统之间存在的一些甚大区别：(1) 二谱系图的基部群中的 *T. acutifolium*，*T. ichangense*，*T. coreanum*，*T. clavatum* DC.，*T. tuberiferum* 和 *T. microgynum* 诸种的雄蕊花丝上部倒披针形，下部丝形，心皮基部具细柄，花柱强度变短，腹面具密集柱头组织，这些形态特征甚为特化，说明诸种隶属的尖叶唐松草组 Sect. *Physocarpum* 是唐松草亚属 Subgen. *Thalictrum* 的一个进化群，不应是原始群。(2) 由 Subgen. *Thalictrum* 的腺毛唐松草组 Sect. *Thalictrum* 演化出的单性唐松草亚属 Subgen. *Lecoyerium*（花单性，稀两性；柱头细长，超出萼片而伸出花外）是唐松草属进化的单系分类群（monophyletic taxon），但在二谱系图中被分成互相分离的多个部分。此外，一些隶属唐松草亚属 Subgen. *Thalictrum*（花两性；柱头不细长，内藏）的种与隶属 Subgen. *Lecoyerium* 的种聚在一起，如 (a) 将特产东亚的 *T. virgatum* 和 *T. foliolosum*（*T. virgatum* 的雄蕊花丝狭条形，花柱无翅，*T. foliolosum* 的花丝丝形，柱头具 2 条侧生翅，此二种的亲缘关系疏远）与特产非洲的 *T. rhynchocarpum* Dill & Rich. 聚在一起；(b) 将特产东亚的 *T. lecoyeri*，*T. atriplex* 和 *T. cultratum*（前 2 种的心皮花柱伸长，呈钩状弯曲，柱头不明显，与具直而较短花柱，柱头具 2 侧生翅的 *T. cultratum* 在亲缘关系上甚为疏远）与特产墨西哥的 *T. pinnatum* S. Wats. 聚在一起；(c) 将隶属 Subgen. *Thalictrum* sect. *Omalophysa* 的 *T. sparsiflorum* 与隶属 Subgen. Lecoy erium 的 *T. smithii* 聚在一起。从上述情况就会误认为 Subgen. *Lecoyerium* 是一个不自然的复系分类群（polyphyletic taxon）。(3) 由 Sect. *Thalictrum* 演化出的一对姊妹群 *T. squamiferum* 和 *T. alpinum* 的花部构造，瘦果形态均相同，花的发育方向均是向顶发育，二种的亲缘关系密切，但在谱系图见上页"唐松草属药用植物的亲缘关系"图中被分开距离甚大。(4) 特产中国北部的 *T. tenue*（植株不具块根，花小，萼片长 3–3.6 mm）与特产欧洲南部的 *T. tuberosum* L.（植株具块根，花大，萼片长 15 mm）被聚在一起，此二种的亲缘关系疏远。

《中国植物志》第 27 卷中的中国唐松草属志是由中国植物学家编写的中国唐松草属的第一次分类学修订（revision），将收载的 67 种划分为 2 亚属，4 组，17 系（王文采，王蜀秀，1979），在本书下面的第三节将对此分类系统加以介绍。*Flora of China* vol.6 中的"中国唐松草属志"是由中国植物学家编写的中国唐松草属的第二次分类学修订，对收载的 76 种未进行分类，而是按照种加词第一个字母的顺序将 76 种加以排列（Fu & Zhu, 2001），这样的处理就使读者难于了解各种间的亲缘关系，以及各个种的演化水平等情况。

二、形态特征分析

A. 关于唐松草属的毛茛科中的系统位置

为了进行种的排列，需要了解各个种的演化水平，一些重要形态特征的演化趋势，以及唐松草亚科的系统位置。瑞典系统学家 O. Langlet（1933）撰写的关于毛茛科细胞分类学论文不长，虽只有 20 页，却根据染色体的形态、大小和数目大致显示出毛茛科系统发育的情况：此种大多数属，包括原始属驴蹄草属 *Caltha*，鸡爪草属 *Calathodes* 和金莲花属 *Trollius* 的染色体均为大型，基数 $x = 8$，这些属形成了毛茛科的演化主干。自其演化出具小型，$x = 7$ 染色体的唐松草亚科 subfam. Thalictroideae；具小型，$x = 9$ 染色体的黄连亚科 Subfam. Coptidoideae 二分枝（Okada & Tamura, 1979；张芝玉，1982；Fu, 1990；Tamura, 1995）。因此，近年来出版的一些著作（如 Takhtajan, 2009; W. Wang et al., 2009）将 Coptidoideae 和 Thalictroideae 放在毛茛科分类系统的原始位置上，就与上述毛茛科系统发育情况不相符合了。

具 $x = 7$ 染色体的唐松草亚科包括二群，一个是扁果草族 Trib. Isopyreae，其特征是子房具 2 颗以上的胚珠，受精后形成开裂、具细横脉的蓇葖果，其花粉具三沟；另一群是具单属的唐松草族 Trib. Thalictreae，其特征是子房具 1 颗胚珠，受精后形成不开裂、每侧具 1-3 条纵脉或纵肋的瘦果，其花粉具散孔（Xi, 1973; Fu, 1990; Nowicke & Shvarla, 1995）。从上述区别特征可了解 Isopyreae 较原始，Thalictreae 较进化，二群的区别明显，Langlet（1932）指出 Thalictreae 由 Isopyreae 演化而出，完全正确，所以由上述可确定组成单属族 Thalictreae 的唐松草属 *Thalictrum* 是 Thalictroideae 的进化群。

B. 扁果草族 Isopyreae 的形态特征

唐松草属由 Isopyreae 演化而出，要了解唐松草属形态特征的演化趋势，就有必要对其祖先群的形态特征进行了解。

扁果草族 Isopyreae 包含 9 属，约 120 种，分布于北温带；为多年生草本植物，具根状茎，稀为一年生草本植物；茎高 10-25 cm，稀达 40-80 cm（楼斗菜属 *Aquilegia*）；叶为 1-3 回三出复叶，稀为 3 全裂的单叶（尾囊草属 *Urophysa*）；花序为少花的单歧或二歧聚伞花序，或花单朵顶生；花两性；萼片 5，花瓣状，具 3（-7）条

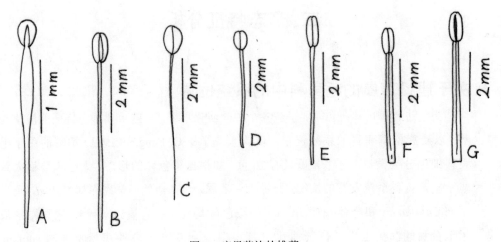

图 1　扁果草族的雄蕊
Figure 1　Stamens of Trib. Isopyreae

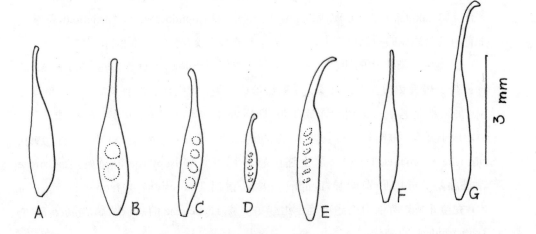

图 2　扁果草族的心皮
Figure 2　Carpels of Trib. Isopyreae

A. 拟扁果草 *Enemion raddeanum* Regel（据李冀云1124）

B. 东北扁果草 *Isopyrum manshuricum* Kom.（据王庭芬229）

C. 扁果草 *Isopyrum anemonoides* Kar. & Kir.（据关克俭2080）

D. 蓝堇草 *Leptopyrum fumarioides* (L.) Reichb.（据邵经荣87004）

E. 拟耧斗菜 *Paraquilegia microphylla* (Royle) Drumm. & Hutch.（据俞德浚6491）

F. 天葵 *Semiaquilegia adoxoides* (DC.) Makino（据236组0002）

G. 无距耧斗菜 *Aquilegia ecalcarata* Maxim.（据杨光辉55353）

二、形态特征分析

基生脉；花瓣小，稀大（*Aquilegia*），为分泌器官，稀不存在（拟扁果草属 *Enemion*）；雄蕊（8-）10-多数，花丝狭条形或丝形，稀上部稍变宽，花药宽椭圆形或长圆形，顶端钝（图 1）；心皮（1-）2-10（-20），子房具 2 至数颗、稀多颗胚珠，无柄，花柱通常直，比子房短，稀伸长（*Aquilegia*），腹面具不明显的柱头（图 2）；果实为具细横脉的蓇葖果。（Foster & Gifford, 1959; Hsiao, 1979; Tamura, 1995）；本族植物的雄蕊花丝和心皮花柱的形态与毛茛科原始群金莲花族 Trib. Trollieae 的花丝和花柱的形态十分相似。（W. T. Wang, 1979）

C. 七形态特征的分析

1. 茎的高度

在唐松草属，可看到茎由低向高的演化趋势，以及由高向低的演化趋势。

中国此属大多数种茎的高度达 40-60（-80）cm，而一些种的茎较高：高达 100 cm 的有爪哇唐松草 *T. javanicum*、细柱唐松草 *T. megalostigma*、大叶唐松草 *T. faberi*、小金唐松草 *T. xiaojinense*、珠芽唐松草 *T. chelidonii*、堇花唐松草 *T. diffusiflorum*、川鄂唐松草 *T. osmundifolium*、腺毛唐松草 *T. foetidum*、箭头唐松草 *T. simplex* 和展枝唐松草 *T. squarrosum*；高达 120 cm 的有长柄唐松草 *T. przewalskii*、毛发唐松草 *T. trichopus* 和高原唐松草 *T. cultratum*；高达 150 cm 的有粗壮唐松草 *T. robustum*、唐松草 *T. aquilegifolium* var. *sibiriculm*、白茎唐松草 *T. leuconotum*、错那唐松草 *T. cuonaense*、美花唐松草 *T. callianthum*、美丽唐松草 *T. reniforme*、河南唐松草 *T. honanense*、东亚唐松草 *T. minus* var. *hypoleucum* 和黄唐松草 *T. flavum*；高达 200 cm 的有偏翅唐松草 *T. delavayi*、川滇唐松草 *T. finetii*、藏南唐松草 *T. austrotibeticum* 和多叶唐松草 *T. foliolosum*。

在唐松草属进化群单性唐松草亚属 Subgen. *Lecoyerium*，茎高度进一步增大，其多数种的茎高达 100-150 cm，一些种的茎更高，其中，高达 200 cm 的有 *T. pubigerum* Benth. 和 *T. jaliscanum* Rose（二种均产墨西哥），*T. inuncans* Boivin（产玻利维亚）和 *T. rhynchocarpum* Dill. & Rich. 和 *T. implexum* Boivin（二种均产非洲）；高达 250 cm 的有 *T. gibbosum* Lecoy.（产墨西哥）；高达 300 cm 的有 *T. standleyi* Steyer.（产危地马拉），*T. steinbachii* Boivin 和 *T. cincinnatum* Boivin（二种均产玻利维亚）。（Boivin, 1944）

另一方面，由具较高茎的唐松草亚属 Subgen. *Thalictrum* 的进化群腺毛唐松草组 Sect. *Thalictrum* 演化出的石砾唐松草组 Sect. *Schlagintweitella* 的一对姊妹群，石砾唐松草 *T. squamiferum* 的茎长 6–20 cm，下部在石块中，以上部分渐升。另一种高山唐松草 *T. alpinum* var. *alpinum* 的植株通常高 6–10 cm。

2. 对单叶特征的认识

特产中国云南西北部的糙叶唐松草 *T. scabrifolium* 和产喜马拉雅山区的 *T. dalzellii* Hook 和 *T. punduanum* 的复叶小叶呈近圆形或圆肾形，不明显浅裂或不分裂，这种形态在唐松草属中甚为罕见。也产喜马拉雅山区的圆叶唐松草 *T. rotundifolium* DC. 的叶为单叶，形状与 *T. scabrifolium* 的小叶甚为相似，也呈圆肾形。（图版 1）上述叶的形状特征说明在花的构造方面这四种有相近的亲缘关系。*T. scabrifolium* 的心皮花柱大部直，只顶端稍向后弯曲，在腹面具不明显的柱头组织，这种情况与扁果草族 Isopyreae 的心皮极为相似，显示为唐松草属演化水平较低的原始种之一（见下文）。上述产喜马拉雅山区的三种的心皮花柱腹面具明显的狭长圆形或狭条形柱头，这说明此三种的演化水平显然比 *T. sabrifolium* 的演化水平高。由此，我推测产喜马拉雅山区的三种可能由特产云南西北部的 *T. scabrifolium* 演化出来的；也就是说 *T. rotundifolium* 的单叶是由复叶经过强烈退化而形成的。这里的单叶是一个进化特征，而不是原始特征。

3. 花序

扁果草族 Isopyreae 的花序主要是少花的单歧聚伞花序，唐松草属中的南湖唐松草 *T. rubescens*、华东唐松草 *T. fortunei*、多枝唐松草 *T. ramosum* 等种的花序继承了上述花序特征，是唐松草属花序的原始形态。这种花序的分枝和花数目增加就导致多花复单歧聚伞花序（如在瓣蕊唐松草 *T. petaloideum*）和多花的聚伞圆锥花序（如在腺毛唐松草组 Sect. *Thalictrum*）的形成。这时，突然发生了一个不小的变化，由上述离顶发育的多花花序突然演化出花向顶发育的少花短总状花序（高山唐松草 *T. alpinum*）和沿茎向顶发育、单生叶腋的花（石砾唐松草 *T. squamiferum*）（图 3）。Takhtajan（1991）介绍在毛茛科乌头属 *Aconitum* 和飞燕草属 *Consolida* 中一些种的花序为聚伞花序，在另一些种的花序转变为总状花序，但未提到唐松草属。

4. 萼片

唐松草属的花被只有一层，即萼片，不具花瓣。在中外有关唐松草属的植物分类学著作（尤其是植物志）中，关于萼片的描述大多很简单，给出的字数不多。B. Boivin（1944）在他关于唐松草属的论文中首先对营养和生殖器官的形态特征的分类

二、形态特征分析　　　　　　　　　　　　　　　21

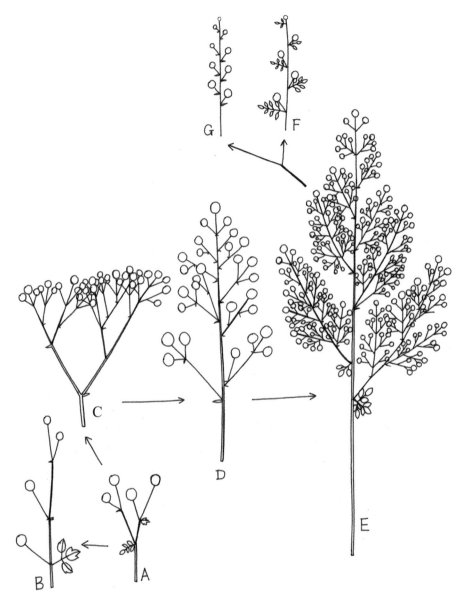

图 3　唐松草属花序类型推测的演化图

A. 少花单歧聚伞花序，B. 少花蝎尾状聚伞花序，C. 多花伞房状复单歧聚伞花序，D. 少花聚伞圆锥花序，E. 密集多花聚伞圆锥花序，F. 花单生上部茎生叶腋部，以向顶发育次序开放，G. 总状花序。

Figure 3　Presumed evolutionary development of inflorescence types in the genus *Thalictrum*.

A. few-flowered monochasium, B. few-flowered scorpioid cyme, C. many-flowered corymbiform compound monochasium, D. sparse-flowered thyrse, E. densely many-flowered thyrse, F. flowers singly and acropetally opening along the axils of upper cauline leaves, G. raceme.

学价值进行评估，在讨论到萼片时他说"They are of little taxonomic importance"。我在研究中国唐松草属分类学之后，关于萼片的分类学价值我看到3点值得注意：

(1) 数目　在扁果草族 Isopyreae，花的萼片数目是5枚，没有变化。在唐松草属，花的萼片数目是3–10枚（Tamura, 1995），主要是4枚（Boivin, 1944）。中国此属花的萼片数目，绝大多数种为4枚，8种（台湾唐松草 T. urbainii、钩柱唐松草 T. uncatum、帚枝唐松草 T. virgatum、偏翅唐松草 T. delavayi、堇花唐松草 T. diffusiflorum、川鄂唐松草 T. osmundifolium、川滇唐松草 T. finetii、腺毛唐松草 T. foetidum）为4–5枚，1种（盾叶唐松草 T. ichangense）为5枚，1种（小喙唐松草 T. rostellatum）为3枚，1变种（大花台湾唐松草 T. urbainii var. majus）为6枚。上述情况说明唐松草属从 Isopyreae 演化出来之后，在萼片数目上发生了不小的变化。

(2) 大小和颜色　唐松草属多数种的萼片较小，长2–5 mm，呈白色或黄绿色，只有大花唐松草 T. grandiforum 和偏翅唐松草系 Ser. Violacea 的种的萼片较大，长达20–30 mm，呈紫色、堇色或粉红色，这些进化特征的出现可能是适应虫媒传粉的结果；同时，上述多数种所具的基生脉的数目增加，每条脉较多二歧状分枝。

(3) 基生脉　在扁果草族 Isopyrea，多数属的萼片有3条平行、不分枝的基生纵脉，只在进化属耧斗菜属 Aquilegia，每一萼片的基生脉多达7条，并且每条脉1–3回二歧状分枝。在中国唐松草属植物中可看到关于萼片基生脉的以下7种情况：(a) 多数种的萼片有3条基生脉，这里又可区分出2种情况，有36种的萼片的3条脉不分枝，另32种的萼片有1–3条脉发生二歧状分枝；(b) 3种（螺柱唐松草 T. spiristylum、花唐松草 T. filamentosum、小果唐松草 T. microgynum）的萼片只有1条基生脉，此外，狭序唐松草 T. atriplex 的萼片有1或2条基生脉；(c) 1种（稀蕊唐松草 T. oligandrum）的萼片有2条基生脉；(d) 1种（金丝马尾连 T. glandulosissimum）的萼片有4条基生脉；(e) 8种（钻柱唐松草 T. tenuisubulatum、长柄唐松草 T. przewalskii、散花唐松草 T. sparsiflorum、丝叶唐松草 T. foeniculaceum、圆叶唐松草 T. rotundifolium、珠芽唐松草 T. chelidonii、美花唐松草 T. callianthum 和偏翅唐松草 T. delavayi）的萼片具5条基生脉；(f) 2种（六脉萼唐松草 T. sexnervisepalum 和大花唐松草 T. grandiflorum）的萼片具6条基生脉；(g) 1种（堇花唐松草 T. diffusiflorum）的萼片具7条基生脉。从上述情况推测萼片具3条不分枝的基生脉可能是原始的状态。此外，脉的数目可以作为区分种的特征之一，例如在长柄唐松草组 Sect. *Omalophysa*，毛蕊唐松草 *T. lasiogynum* 的萼片具 3 条基生脉，其他 2 种，长柄唐松草 *T. przewalskii* 和散花唐松草 *T. sparsiflorum* 的萼片均具 5 条基生脉，这里，萼片数目是一个很好的区别特征。

5. 雄蕊

(1) 花丝 在叉枝唐松草组 Sect. *Leptostigma* 中可以看到唐松草属雄蕊花丝演化的情况：叉枝唐松草 *T. saniculiforme*，小花唐松草 *T. minutiflorum* 和绢毛唐松草 *T. brevisericeum* 的花丝呈条形或狭条形；新宁唐松草 *T. xinningense* 和大叶唐松草 *T. faberi* 的花丝上部条形，比花药窄，与上述叉枝唐松草等种的花丝相似，但下部变细，近似丝形，也就是说花丝开始分化成上部和下部两部分；在多枝唐松草 *T. ramosum*，花丝上部呈狭倒披针形，仍比花药窄，下部呈丝形，在华东唐松草 *T. fortunei* 和峨眉唐松草 *T. omeiense*，花丝上部呈倒披针形，其宽度稍超过花药，下部丝形；到了瓣蕊唐松草 *T. petaloideum* 等种，花丝上部呈倒披针形，其宽度明显超过花药，下部呈丝形，这时，花的花丝起到了花瓣的作用，可能是适应虫媒传粉的结果。同组的思茅唐松草 *T. simaoense* 和短蕊唐松草 *T. brachyandrum* 的花丝均呈丝形。从上述情况推测唐松草属的花丝呈条形或狭条形可能是原始的形态，由这种形态的花丝演化出丝形和分化成上、下两部分的花丝。在唐松草亚属的进化群，腺毛唐松草组 Sect. *Thalictrum*，花丝呈丝形，且伸长并下垂，可起到促进花丝散布花粉的作用；同时，心皮的柱头生出二侧生宽翅，这些形态构造的出现可能是适应风媒传粉的结果。

(2) 花药 扁果草组 Isopyreae 的花药多呈宽椭圆形，顶端钝，像这样的花药在中国唐松草属中甚为稀见，只见于少数种如叉枝唐松草 *T. saniculiforme*、武夷唐松草 *T. wuyishanicum*、阴地唐松草 *T. umbricola*、花唐松草 *T. filamentosum* 等种。中国此属多数种的花药均呈长圆形，顶端钝；由此演变为狭长圆形，顶端钝（如在丽江唐松草 *T. wangli* 和毛发唐松草 *T. trichopus*）；再演变为狭长圆形，在顶端出现短尖头（如白茎唐松草 *T. leuconotum*、大花唐松草 *T. grandiflorum*、美丽唐松草 *T. reniforme*、高山唐松草 *T. alpinum*，以及腺毛唐松草组 Sect. *Thalictrum* 的不少种）；最后演变到条形，顶端具短尖头（如 Sect. *Thalictrum* 的高原唐松草 *T. cultratum*、陕西唐松草 *T. shensiense*、东亚唐松草 *T. minus* var. *hypoleucum*、腺毛箭头唐松草 *T. simplex* var. *glandulosum* 和直梗高山唐松草 *T. alpinum* var. *elatum*）。

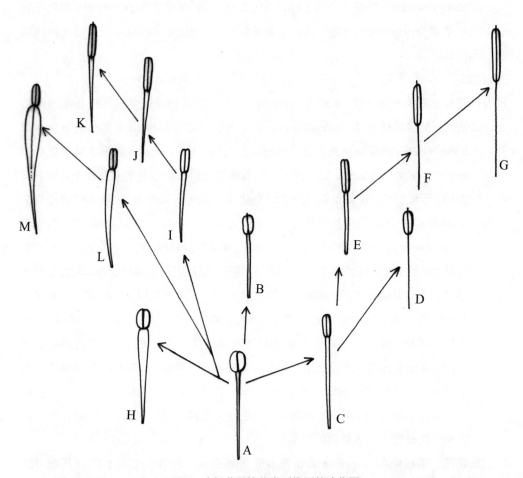

图 4 唐松草属雄蕊类型推测的演化图

A. 扁果草族一雄蕊。B—G. 花丝规则，由条形或狭条形演变为丝形；花药由宽椭圆形（叉枝唐松草）演变为长圆形、狭长圆形，甚至条形。H—M. 花丝多少分化成上、下在形状和大小上不同的二部分。（H. 武夷唐松草，I. 狭序唐松草，J. 小喙唐松草，K. 钩柱唐松草，L. 峨嵋唐松草，长喙唐松草，贝加尔唐松草等种，M: 瓣蕊唐松草，台湾唐松草，深山唐松草等种）

Figure 4 Presumed evolutionary development of stamen types in the genus *Thalictrum*

A. a stamen of Trib. Isopyreae. B—G. filaments regular, from linear or narrow-linear changing to filiform; anthers from broad-elliptic (in *T. saniculiforme*) changing to oblong, narrow-oblong and linear at length. H—M. filaments more or less differentiated into upper and lower two parts different in shape and size (H, *T. wuyishanicum*; I. *T, atriplex*; J, *T. rostellatum*; K, *T. uncatum*; L, *T. omeiense*, *T. macrorhynchum*, *T. baicalense* et al.; M, *T. petaloideum*, *T. urbainii*, *T. tuberiferum* et al.)

图 4 显示了唐松草属雄蕊的演化情况。

6. 心皮（单雌蕊）

毛茛科的原始属驴蹄草属 *Caltha*，鸡爪草属 *Calathodes* 和金莲花属 *Trollius* 的心皮的顶端具不长的直花柱，花柱腹面的柱头不明显。（W. T. Wang, 1979）扁果草族 Isopyreae 诸属的心皮仍保持了上述三属心皮的形态特征，只在二进化属——尾囊草属 *Urophysa* 和耧斗菜属 *Aquilegia*，花柱伸长，其长度常超过子房。（Hsiao, 1979）（图 2）

在中国唐松草属，叉枝唐松草组 Sect. *Leptostigma* 具有的 31 种的心皮仍保持上述三属心皮的形态特征，但是花柱或多或少伸长，并在顶端向后弯曲，多呈钩状或拳卷状，只有 2 种，糙叶唐松草 *T. scabrifolium* 和鹤庆唐松草 *T. leve* 的心皮花柱大部分直，只在顶端稍向后弯曲（图 5: A-D；图版 4-5），这种近似扁果草族 Isopyreae 心皮形态的情况说明此 2 种是唐松草属的原始类型。此组的微毛唐松草 *T. lecoyeri* 的心皮花柱呈钩状或拳卷状弯曲，在果期变坚硬，使宿存花柱像一个金属的钩，据此推测这可能是适应动物传播果实的结果。另有 2 种，钻柱唐松草 *T. tenuisubulatum* 和宽柱唐松草 *T. latistylum* 的心皮柱头强烈退化而消失；前者，心皮花柱狭长，长于子房，是呈狭条形的一条薄膜，顶端钩状弯曲，看不到柱头；后者，心皮花柱是一长圆形的小薄片，也看不到柱头。（图版 33-34）中国此属其他 70 种的心皮在花柱腹面或顶端出现了明显的柱头，后者呈狭长圆形、狭条形、椭圆形等形状（图 5: E-M）。在芸香叶唐松草组 Sect. *Rutifolia* 的吉隆唐松草 *T. jilongense* 和白茎唐松草 *T. leuconotum*（图 5: K），以及进化群腺毛唐松草组 Sect. *Thalictrum* 的多数种（图 5: L-M），柱头全部与花柱贴生，并向两侧生出两条狭或宽的膜质翅。长柄唐松草组 Sect. *Omalophysa* 的心皮具细长柄和稍长的花柱（图 5: J）。尖叶唐松草组 Sect. *Physocarpum* 和唐松草组 Sect. *Tripterium* 的心皮也具细长柄，但花柱变短（图 5: H）。网脉唐松草系 Ser. *Reticulata* 二种的花柱也强烈缩短，但心皮无柄（图 5: G），与上述二组的心皮比较，心皮无柄的状态应较为原始。偏翅唐松草 *T. delavayi* 的子房沿腹、背缝线均具一条狭翅，是唐松草属中一罕见的进化形态特征，中国此属其他种的心皮子房均无翅。上述柱头和子房具翅、心皮具柄和花柱缩短或伸长等现象显示出唐松草属心皮形态具有丰富的多样性，同时也说明这些形态都是进化的特征。

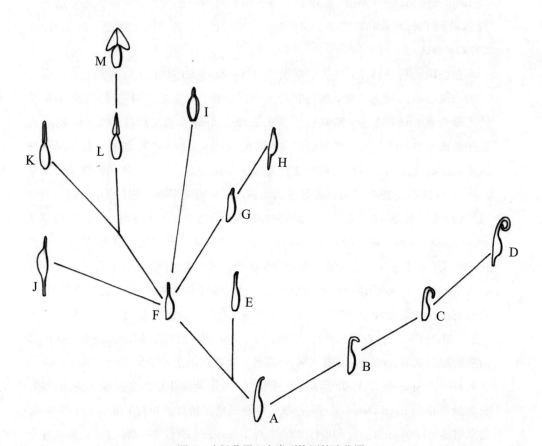

图 5 唐松草属心皮类型推测的演化图
A-D. 花柱腹面具不明显的柱头；E—M: 花柱腹面具明显的柱头。（见"6. 心皮"一文）
Figure 5 Presumed evolutionary development of carpel types in the genus *Thalictrum*
A-D: style ventrally with an inconspicuous stigma; E—M: style ventrally with a conspicuous stigma. (see text of 6. Carpel)

7. 瘦果

唐松草属的果实为含1种子、不开裂的瘦果，通常稍两侧扁，腹、背缝线轻度弧形凸起，果体呈狭椭圆形或纺锤形，在每侧具（2–）3（–4）条细纵肋。有时腹、背缝线均较强度凸起，果体呈宽椭圆体形或扁球形，如在狭序唐松草 *T. atriplex*，细柱唐松草 *T. megalostigma*，贝加尔唐松草 *T. baicalense*。有时腹缝线近直，背缝线弧形凸起，如小喙唐松草 *T. rostellatum*，钩柱唐松草 *T. uncatum*，瓣蕊唐松草 *T. petaloideum*，澜沧唐松草 *T. lancangense*，高原唐松草 *T. cultratum*。有时腹缝线较强度凸起，背缝线近直，果体呈斜倒卵形，如长柄唐松草 *T. przewalskii*，散花唐松草 *T. sparsiflorum*，芸香叶唐松草 *T. rutifolium*，美丽唐松草 *T. reniforme*，堇花唐松草 *T. diffusiflorum*，川滇唐松草 *T. finetii*，或果体呈新月形，如丽江唐松草 *T. wangii*；或果体呈斜卵形，如毛发唐松草 *T. trichopus*。上述的 *T. uncatum*，*T. wangii*，*T. przewalskii*，*T. reniforme*，*T. diffusiflorum*，*T. finetii*，*T. cultratum* 等种的瘦果均强烈两侧扁，果体扁平。中国此属稍超半数的种的瘦果无柄；有约46种的瘦果具柄，其中，多数种的果柄长0.5–1.2 mm，少数种的果柄较长，如芸香叶唐松草 *T. rutifolium* 和白茎唐松草 *T. leuconotum* 的果柄长达2 mm，尖叶唐松草 *T. acutifolium* 的果柄长达2.5 mm，阴地唐松草 *T. umbricola* 的果柄长达3 mm，散花唐松草 *T. sparsiflorum* 和稀蕊唐松草 *T. oligandrum* 的果柄长达3.5 mm，长柄唐松草 *T. przewalskii* 和唐松草 *T. aquilegifolium* var. *sibiricum* 的果柄长达5 mm。中国此属绝大多数种的瘦果无翅，只有少数种的瘦果具翅：川滇唐松草 *T. finetii* 的瘦果沿背缝线上部具1条狭翅；偏翅唐松草 *T. delavayi* 的瘦果沿腹背缝线各具1条狭翅；二翅唐松草 *T. elegans* 的瘦果扁平，具长4 mm 的果柄，每侧具1条纵脉，沿腹、背缝线各具1条狭翅；唐松草 *T. aquilegifolium* var. *sibiricum* 的瘦果具4条纵翅。上述果体扁平、具柄、具翅的形态构造当均是进化的特征。

三、分类系统

A. 王文采，王蜀秀分类系统简介

在20世纪60年代，我与王蜀秀同志承担了《中国植物志》唐松草属的编写任务，那时除 PE 以外，还研究了 NAS, SCIB, KUN, WUG 等兄弟所收藏的唐松草属植物标本。在研究标本的过程中我们对此属的形态特征进行了分析，认识到雄蕊花丝呈条形（线形），以及心皮的柱头不明显是唐松草属的两个重要原始特征，也就是说花丝呈丝形或分化成上、下两部，以及心皮具明显柱头等状况是进化特征。然后根据上述认识对当时中国唐松草属鉴定出的67种，做出了如《中国植物志》27卷（1979）中"唐松草属分类系统一览表"，分类如下：

亚属 1. 唐松草亚属 Subgen. **Thalictrum**　花两性。雌蕊不伸出花外。（种 1–66）
　组 1. 叉枝唐松草组 Sect. **Leptostigma**　花离顶发育。花丝线形或倒披针形。花柱通常长，拳卷，沿腹面上部有不明显的柱头组织，未形成明显的柱头。（种 1–18）
　　系 1. 叉枝唐松草系 Ser. **Saniculaeformia**　花序有少数花。花丝线形。（种 1）
　　系 2. 糙叶唐松草系 Ser. **Scabrifolia**　花序有少数或多数花，呈伞房状或圆锥状。花丝倒披针形或棒形，下部变细。叶片多为纸质，脉网通常明显隆起。（种 2–15）
　　系 3. 小喙唐松草系 Ser. **Rostellata**　花序有少数花。花丝上部狭倒披针形，其他部分狭线形。叶片草质，脉近平，脉网不明显。（种 16）
　　系 4. 钩柱唐松草系 Ser. **Uncata**　复单歧聚伞花序像总状花序。花丝近丝形或上部棒形。叶片草质，脉网不明显。（种 17–18）
　组 2. 唐松草组 Sect. **Tripterium**　花离顶发育。花丝倒披针形或棒形。花柱不拳卷，通常较短，在腹面形成较明显的无翅的柱头。（种 19–37）
　　系 1. 西南唐松草系 Ser. **Fargesiana**　花柱明显，上部成柱头。瘦果有短柄，卵球形或狭卵球形，稍两侧扁，无翅。叶片草质。（种 19–21）
　　系 2. 网脉唐松草系 Ser. **Reticulata**　花柱短，腹面全部生柱头组织。瘦果无柄，卵球形或纺锤形，稍两侧扁，无翅。叶片纸质或厚纸质，脉网明显。（种 22–23）
　　系 3. 贝加尔唐松草系 Ser. **Baicalensia**　花柱明显，上部形成柱头。瘦果有很短的柄，扁球形，无翅。叶片草质。花序短，圆锥状。（种 24）
　　系 4. 瓣蕊唐松草系 Ser. **Petaloidea**　花柱明显，上部形成柱头。瘦果无柄，卵球形，无翅。叶片草质。花序呈伞房状，大，有多数花。（种 25）
　　系 5. 唐松草系 Ser. **Triptera**　花柱明显，上部形成柱头。瘦果有细长柄和3–4条纵翅。叶片草质。花序呈伞房状，有多数花。（种 26）
　　系 6. 散花唐松草系 Ser. **Sparaiflora**　花柱明显，上部形成柱头。瘦果有细心皮柄，近扁平，无翅。

叶片草质。花序圆锥状。（种 27–28）

系 7. 尖叶唐松草系 Ser. **Clavata**　花柱极短，腹面全部密布柱头组织。瘦果有细长柄，近扁平，无翅。叶片通常草质。花序呈伞房状或伞状。（种 29–37）

组 3. 腺毛唐松草组 Sect. **Thalictrum**　花离顶发育。花丝狭线形或丝形。花柱不拳卷，腹面全部或上部形成无翅或有翅的柱头。（种 38–64）

系 1. 芸香叶唐松草系 Ser. **Rutifolia**　花序像总状花序。柱头无翅或有狭翅。（种 38–39）

系 2. 帚枝唐松草系 Ser. **Virgata**　花序伞房状，有少数花。花小，少有较大；萼片白色或淡黄绿色，少有粉红色。柱头无翅或有 2 狭翅。（种 40–45）

系 3. 偏翅唐松草系 Ser. **Violacea**　花序圆锥状。花常较大；萼片堇色或紫色。柱头无翅。（种 46–49）

系 4. 腺毛唐松草系 Ser. **Flexuosa**　花序圆锥状。花小；萼片通常淡黄绿色。柱头无翅，或有 2 狭或宽翅。（种 50–64）

组 4. 石砾唐松草组 Sect. **Schlagintweitella**　花向顶发育，小。萼片淡黄绿色。花丝丝形。花柱直，腹面形成有 2 宽翅的正三角形柱头。（种 65–66）

系 1. 石砾唐松草系 Ser. **Squamifera**　基生叶不存在；茎下部叶鳞片状；中部和上部叶正常，为三至四回复叶。花单生于茎上部叶腋。（种 65）

系 2. 高山唐松草系 Ser. **Alpina**　叶全部基生，正常发育。花序总状；苞片小，狭卵形，不分裂。（种 66）

亚属 2. 美洲唐松草亚属 Subgen. **Lecoyerium**　花单性，雌雄异株，稀两性。雌蕊在花期通常伸出花外。（种 67）

在"唐松草属分类系统一览表"之下，《中国植物志》还给出唐松草属各亚属、组、系的系统发育亲缘关系图，现附之于后（图 6），供读者参考。

此演化亲缘关系图将花柱腹面不具明显柱头的叉枝唐松草组 Sect. *Leptostigma* 作为唐松草属的原始群，居于分类系统的开始位置，认为组成单种系 Ser. *Saniculiformia* 的叉枝唐松草 *Thalictrum saniculiforme* 是此属的原始种，还认为自腺毛唐松草系 Ser. *Flexuosa* 演化出石砾唐松草组 Sect. *Schlagintweitella* 和单性唐松草亚属（美洲唐松草亚属）Subgen. *Lecoyerium*，现在，我对这些观点仍保持不变。

药学专家朱敏和肖培根（1991）对中国唐松草属的化学成分进行了深入研究，发现各生物碱结构类型在不同的属下分类群中所占比例则不相同，并按照上述 1979 年唐松草属分类系统的属下分类群给出以下介绍：（1）在唐松草亚属 Subgen. *Thalictrum* 中，叉枝唐松草组 Sect. *Leptostigma* 中双分子化合物占比例很大，其中阿朴菲-苄基异喹啉（简称 ABI）型又比双苄基异喹啉（简称 BBI）型更多。如在峨嵋唐松草 *T. omeiense* 中所得双分子化合物均为 ABI 型；而在大叶唐松草 *T. faberi* 中分离的 20 个

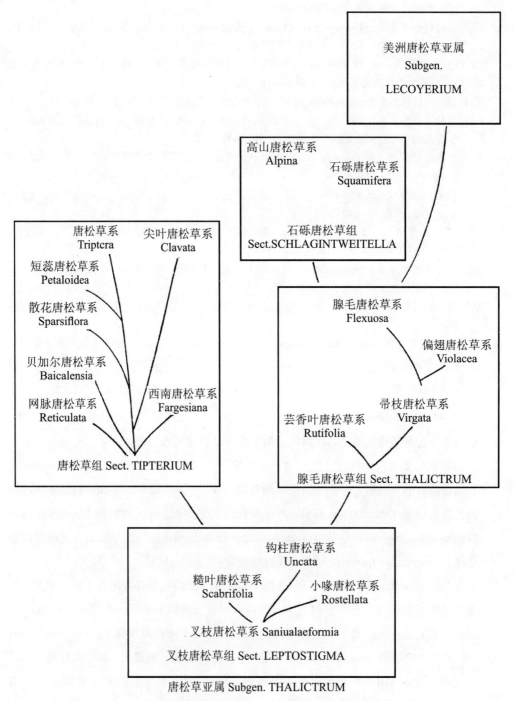

图 6 唐松草各亚属、组、系的系统发育亲缘关系图

双分子化合物中 12 个为 ABI 型, 8 个为 BBI 型。(2) 唐松草组 Sect. *Tripterium* 所含生物碱的类型以阿朴菲为主, 其特殊性表现在本组植物含有大量非生物碱化合物, 如唐松草 *T. aquilegifolium* var. *sibiricum* 只分离出两个阿朴菲碱, 其他为有机酸、黄酮、氰甙等化合物。再如在尖叶唐松草 *T. acutifolium* 中分离出的 8 个化合物中只有 3 个生物碱, 其他多为长链碳氢化合物。这个组的植物由于含生物碱较少, 作药用的种也较少。(3) 腺毛唐松草组 Sect. *Thalictrum* 是生物碱较集中分布的类群, 但在组内较原始的系中, 生物碱的含量并不高, 如系 1 的白茎唐松草 *T. leuconotum* 的生物碱含量低于 0.001%。在较进化的类群中, 生物碱的含量较高, 出现一些总碱达 2% 以上的高含量种。(4) 石砾唐松草组 Sect. *Schlagintweitella* 所含生物碱与 Sect. *Thalictrum* 相似, 以 BBI 的数目为多。(5) 单性唐松草亚属 Subgen. *Lecoyerium* 在中国产一种, 其生物碱含量测定结果与 Sect. *Thalictrum* 相仿。根据上述情况二专家得出重要结论: 植物中所含生物碱的种类及含量与植物的进化有一定关系, 在较原始的组中, 成分的类型和含量相对较少, 而在较进化的组中, 各种结构类型得到较大发展。此外还指出: 从化学系统学角度看,《中国植物志》中所安排的属内各组植物间的关系是适宜的。

B. 本书的分类系统简介

在 2016 年秋季我开始编写本书, 又将中科院植物研究所收藏的唐松草属标本进行研究, 将全部种的花进行解剖, 并对此属的重要形态特征进行观察和分析, 写出第二节。其中对叶(在 20 世纪 60 年代, 还未发现具单叶的圆叶唐松草 *T. rotundifolium* 在中国的分布)和萼片两项, 我均未曾考虑, 而对其他器官如雄蕊、心皮、瘦果等形态特征演化趋势的认识, 和现在得到的认识大致相同, 仍然认为雄蕊具条形花丝, 以及心皮具不明显柱头是唐松草属的二个重要原始特征, 因此, 本书做出的唐松草属诸种的系统排列与 1979 年《中国植物志》做出的基本相同, 或说是大同小异。小异在于以下 7 点:(1) 发现绢毛唐松草 *T. brevisericeum* 的花丝虽有些变异, 是呈狭条形的, 过去误认为其花丝分化成上、下两部分, 在本书中将其与新发现的小花唐松草 *T. minutiflorum* 一同补充到叉枝唐松草系 Ser. *Saniculiformia* 中。(2) 1993 年, 我有机会看到糙叶唐松草 *T. scabrifolium* Franch. 的模式标本, 才知道在《中国植物志》(1979) 中我鉴定的 *T. scabrifolium* 有误。此种和鹤庆唐松草 *T. leve* 的心皮花柱大部分直, 只在顶端稍向后弯曲, 其腹面柱头不明显, 因此其心皮与扁果草族 Isopyreae 的心皮甚为相似, 被认为是唐松草属的原始种。在本书中, 将二种置于糙叶唐松草系 Ser. *Scabrifolia* 的开始位置。（3）菲律宾唐松草 *T. philippinense* 和台湾唐松草 *T. urbainii* 的心皮花柱顶端钩状弯曲，腹面具不明显柱头，大概由于二种的心皮具细柄的缘故，在《中国植物志》（1979）中被误置于尖叶唐松草组 Sect. *Physocarpum* 中，在本书中，将此二种移至 Sect. *Leptostigma* 的小喙唐松草系 Ser. *Rostellata* 中。（4）近年来发现的二新种，思茅唐松草 *T. simaoense* 和短蕊唐松草 *T. brachyandrum* 的心皮花柱钩状弯曲，柱头不明显，同时雄蕊花丝均呈丝形，本书将二种组成一新系 Ser. *Simaoensia*，置于叉枝唐松草组 Sect. *Leptostigma* 中。（5）近年发现的 2 心皮无柱头的新种，钻柱唐松草 *T. tenuisubulatum*（花柱伸长，狭条形，顶端拳卷状弯曲）和宽柱唐松草 *T. latistylum*（花柱为一长圆形薄膜），此 2 种的花柱形态独特，互相区别甚大，本书根据此二新种分别建立二新组：Sect. *Dolichostylus* 和 Sect. *Platystylus*。（6）*T. baicalense* Turcz. var. *megalostigma* Boivin 的雄蕊花丝呈狭条形，心皮柱头也呈狭条形，顶端稍向后弯曲，而 *T. baicalense* 的花丝的上部狭倒披针形，下部丝形，柱头狭椭圆形，位于花柱腹面顶端，此外二者的花序也有区别，据此，我已于 1980 年将 var. *megalostigma* 升为种级群，在本书中，根据此种成立一新系，细柱唐松草系 Ser. *Megalostigmata*，置于贝加尔唐松草系 Ser. *Baicalensia* 之前。（7）在《中国植物志》（1979）中，几位植物学家建立的 Sect. *Physocarpum* DC. 和 Sect. *Omalophysa* Turcz. ex Fisch. & Mey. 未被采用，将 Sect. *Tripterium* DC. 的内容过分扩大，本书改正了这些不恰当处理，将 3 组恢复。但是，这导致组数目的增加。

下面是本书的"唐松草属分类系统一览表"：

亚属 1. 唐松草亚属 Subgen. *Thalictrum*　花两性；柱头不呈鞭形，内藏。
　组 1. 叉枝唐松草组 Sect. *Leptostigma*　花离顶发育；心皮花柱多少伸长，顶端向后弯曲，多呈钩状或拳卷状，在腹面具不明显柱头。
　　系 1. 叉枝唐松草系 Ser. *Saniculiformia*　花序伞房状；花丝条形或狭条形。（种 1-3）
　　系 2. 糙叶唐松草系 Ser. *Scabrifolia*　花序伞房状；花丝上部狭倒披针形或倒披针形，下部狭条形或丝形，心皮无柄，子房比花柱长。（种 4-18）
　　系 3. 小喙唐松草系 Ser. *Rostellata*　近上系，但心皮具柄。（种 19-27）
　　系 4. 思茅唐松草系 Ser. *Simaoensia*　近上二系，但花丝丝形。（种 28-29）
　　系 5. 钩柱唐松草系 Ser. *Uncata*. 花序狭长，似总状花序；花丝狭条形、丝形或上部棒状（种 30-32）
　组 2. 钻柱唐松草组 Sect. *Dolichostylus*　心皮无柱头；花柱钻状狭条形，比子房长，顶端钩状弯曲。（种 33）
　组 3. 宽柱唐松草组 Sect. *Platystylus*　心皮无柱头；花柱为一长圆形膜质片，比子房短。（种 34）

组 4. 瓣蕊唐松草组 Sect. *Erythrandra* 花离顶发育；花序伞房状；花丝分化为上、下两部分，稀呈狭条形（Ser. *Megalostigmata*），花柱直；柱头明显。

系 1. 细柱唐松草系 Ser. *Megalostigmata* 花丝狭条形；柱头狭条形；瘦果扁球形。（种 35）

系 2. 贝加尔唐松草系 Ser. *Baicalensia* 花柱腹面顶端具椭圆体形柱头；瘦果具短柄，扁球形。（种 36）

系 3. 瓣蕊唐松草系 Ser. *Petaloidea* 花柱腹面具狭条形柱头，瘦果无柄，狭卵形，具粗纵肋。（种 37）

系 4. 网脉唐松草系 Ser. *Reticulata* 近上系，但花柱极短。（种 38-39）

系 5. 西南唐松草系 Ser. *Fargesia* 花丝分化成上、下两部分，稀呈狭条形（*T. tsawarungense*）；花柱直，柱头狭条形或狭椭圆形；瘦果扁，常有短柄。（种 40-44）

组 5. 长柄唐松草组 Sect. *Omalophysa* 花序为聚伞圆锥花序；花丝上部棒形；花柱较长；瘦果扁，具长柄，腹缝线强烈凸起。（种 45-47）

组 6. 尖叶唐松草组 Sect. *Physocarpum* 茎生叶少数，或不存在；单歧聚伞花序伞房状，稀圆锥状；花丝分化成上、下两部分；花柱短，稀伸长；瘦果有细长柄。

系 1. 尖叶唐松草系 Ser. *Clavata* 花柱短。（种 48-55）

系 2. 拟盾叶唐松草系 Ser. *Pseudoichangensia* 花柱伸长。（种 56）

组 7. 唐松草组 Sect. *Tripterium* 单歧聚伞花序伞房状，具多数花；花丝棒形；花柱短；瘦果具细长柄和 4 条纵翅。（种 57）

组 8. 芸香唐松草组 Sect. *Rutifolia* 单歧聚伞花序狭长，似总状花序；花丝狭条形或丝形；花柱直，柱头无翅或具翅。

系 1. 小金唐松草系 Ser. *Xiaojinensia* 小叶有毛；花丝狭条形；柱头无翅。（种 58-59）

系 2. 芸香叶唐松草系 Ser. *Rutifolia* 植株无毛；花丝丝形，柱头无翅。（种 60）

系 3. 长柱唐松草系 Ser. *Sinomacrostigmata* 植株无毛；雄蕊约 50 枚；花丝丝形；花柱伸长，比子房长 2 倍，柱头具 1 翅。（种 61）

系 4. 白茎唐松草系 Ser. *Leuconota* 植株无毛；雄蕊 7-16 枚；花丝丝形；花柱比子房短，柱头具 2 条侧生狭翅。（种 62-63）

组 9. 帚枝唐松草组 Sect. *Platystachys* 单歧聚伞花序伞房状或圆锥状；花丝狭条形或丝形；花柱直，柱头无翅；瘦果无柄，稀具短柄，每侧有 2-3 条纵肋，无翅，稀具翅。

系 1. 帚枝唐松草系 Ser. *Virgata* 叶为复叶，小叶不呈圆卵形，基部不呈心形；花序伞房状；萼片长 3-10(-20)mm，白色，稀粉红色；瘦果无翅。（种 64-70）

系 2. 圆叶唐松草系 Ser. *Indivisa* 叶为单叶或复叶，叶片近圆形或圆肾形。花序伞房状；萼片长 5-6.8 mm，黄绿色，瘦果无翅。（种 71-72）

系 3. 偏翅唐松草系 Ser. *Violacea* 叶为复叶，小叶不呈圆卵形；花序圆锥状；萼片长达 16 mm，呈紫色，堇色或粉红色，稀白色；瘦果无翅，稀有翅。（种 73-78）

组 10. 二翅唐松草组 Sect. *Dipterium* 单歧聚伞花序圆锥状；花丝狭条形；瘦果具细长柄，扁平，每侧均具 1 条纵脉，沿腹、背缝线均有 1 条狭翅。（种 79）

组 11. 腺毛唐松草组 Sect. *Thalictrum* 单歧聚伞花序圆锥状；花小；花丝丝形，稀狭条形；柱头全部与花柱贴生，多具 2 条侧生狭或宽翅，稀无翅。

系 1. 川鄂唐松草系 Ser. *Osmundifolia* 柱头无翅。（种 80-85）

系 2. 腋毛唐松草系 Ser. *Thalictrum*　柱头具 2 条侧生翅。（种 86–97）

组 12. 石砾唐松草组 Sect. *Schlagintweitella*　花向顶发育，小；花丝丝形；柱头完全与花柱贴生，具 2 条侧生宽翅。

系 1. 石砾唐松草系 Ser. *Squamifera*　基生叶不存在；下部茎生叶退化为小鳞片，上部茎生叶正常发育；花单生上部茎生叶腋部，向顶发育。（种 98）

系 2. 高山唐松草系 Ser. *Alpina*　叶数枚，全部基生；总状花序生花葶顶端。（种 99）

亚属 2. 单性唐松草亚属 Subgen. *Lecoyerium*　花单性，稀两性；柱头鞭形，细长，伸出萼片外。（种 100–101）

一如前述，在 20 世纪 60 年代通过形态特征分析，我们认为叉枝唐松草 *T. saniculiforme* 具有原始形态特征（雄蕊花丝条形，心皮花柱腹面具不明显的柱头），是唐松草属的原始种。在最近的研究中发现小花唐松草 *T. minutiflorum* 和绢毛唐松草 *T. breviserisericeum* 也具上述原始形态特征。此外，还发现糙叶唐松草 *T. scabrifolium* 和鹤庆唐松草 *T. leve* 的心皮花柱形态与扁果草族 Isopyreae 的心皮花柱构造甚为相似，无疑是原始的类群，这样，唐松草属的原始种就增加到 5 个。（图版 1–5）这 5 个种既具原始形态特征，同时也具进化形态特征，如伸长并钩状弯曲的花柱，分化成上、下两部分的不规则花丝，这种情况就是俄国植物学家 Takhtajan（1991，2009）指出的祖衍征并存现象（heterobathmy），是说明具有原始形态特征的古老植物在其生存并适应环境诸条件的长时间过程中，既保留其原有的原始形态特征，又出现了一些新的进化特征。这些问题在形态特征分析工作中应重点进行研究，并力求得到正确的认识。在认识到唐松草属的原始形态特征，以及其原始种之后，我对 Tamura 分类系统中的原始群就产生了怀疑，在前面介绍此系统曾介绍了他的有关原始群的观点，他认为 Subsect. *Indivisa*（包括特产喜马拉雅山区具单叶的 *T. rotundifolium* DC. 和具复叶的 *T. dalzellii* Hook.、*T. punduanum* Wall.3 种）的植物具单叶、丝形花丝和无柄、不扁压的瘦果，是唐松草属中最原始的。这里，他未提到柱头的特征，实际上，这 3 种的心皮花柱中部以上都有明显的狭长圆形柱头，根据本书上面形态特征的分析，圆叶唐松草的单叶可能是复叶退化的产物。丝形花丝和明显的狭长圆形柱头都是进化的形态特征，具有这些特征的 Subsect. *Indivisa* 不应是唐松草属的原始群。由此可见，在形态特征分析后所认识到的原始和进化形态特征不同，据此建立的分类系统也就会不同。

四、地理分布

A. 唐松草属地理分布简介

此属约有 220 种，分布于亚、欧、非、北美、南美诸洲的温带和寒温带，少数种分布于亚热带。唐松草亚属 Subgen. *Thalictrum* 约有 140 种，超过 100 种分布于亚洲（多数分布于东亚），14 种分布于欧洲，6 种分布于北美洲，更少数种分布于非洲北部。单性唐松草亚属 Subgen. *Lecoyerium* 约有 70 种，多数种分布于北、南美洲，6 种分布于非洲，2 种分布于亚洲的中国西南部，另 1 种分布于欧洲西南部。（Boivin, 1944; Tutin, 1964; Tamura, 1995; Park & Festerling, 1997）

B. 中国唐松草属地理分布

1. 概况

中国唐松草属有约 99 种，20 变种（包括 74 特有种，15 特有变种），广布全国 29 省、区，有隶属 2 亚属，11 组的 74 种分布于西南部的四川、云南、西藏三省区（四川西部、云南和西藏东南部是唐松草属的分布中心）（图 7），自此三省、区向北和向东，种的数目迅速减少。中国面积最大的新疆维吾尔自治区只有 7 种，1 变种；东北三省只有 11 种，4 变种；华东和华南的每省只有 5-6 种，海南省只有 1 种。表 1 列出中国 29 省、区的唐松草植物种和变种的数目，从中可了解中国唐松草属植物地理分布的大致情况。

表 1　中国唐松草属在 29 省、区分布的种和变种数目
Table 1　Numbers of species and varieties of the genus *Thalictrum* in 29 provinces and regions

省、区	四川	云南	西藏	甘肃	陕西	青海	河南	湖南	湖北	重庆	山西	内蒙古	贵州	河北	宁夏	新疆	吉林	黑龙江	安徽	浙江	江西	辽宁	台湾	福建	广西	江苏	广东	山东	海南
种	38	36	33	20	16	15	13	12	12	12	9	9	9	8	8	7	6	6	6	6	6	5	5	5	5	4	4	3	1
变种	4	8	3	3	2	4	2	2	2	2	6	5	4	6	3	1	5	5	2	1	0	4	1	0	0	4	1	3	0
特有种	27	24	18	19	11	6	7	10	10	9	4	2	5	4	4	0	0	0	4	3	5	0	5	4	3	2	2	1	0
特有变种	0	6	2	0	0	0	0	0	0	0	0	2	0	0	0	0	0	0	0	0	0	1	0	0	1	0	0	0	0

2. 广域分布种

在中国，分布最广的唐松草属有以下 7 种和 3 变种：东亚唐松草 *T. minus* var. *hypolecum* 广布 20 省、区；瓣蕊唐松草 *T. petaloideum*，广布 18 省、区（图 8），短梗箭头唐松草 *T. simplex* var. *brevipes* 广布 17 省、区，爪哇唐松草 *T. javanicum* 广布 13 省、区（图 8），盾叶唐松草 *T. ichangens* 和唐松草 *T. aquilegifolium* var. *sibiricum* 均广布 12

图 7 唐松草属的分布中心和五原始种分布图

Figure 7　Map showing the distribution centre and the distribution of five primitive species of the genus *Thalictrum*

—— 分布中心 distribution centre

▲ *T. saniculiforme*　　　*T. brevisericeum*　　　★ *T. scabrifolium*
△ *T. minutiflorum*　　　△ var. *brevisericeum*　　　☆ *T. leve*
　　　　　　　　　　　　　var. *pentagynum*
　　　　　　　　　　　　　var. *angustiantherum*

四、地理分布　　37

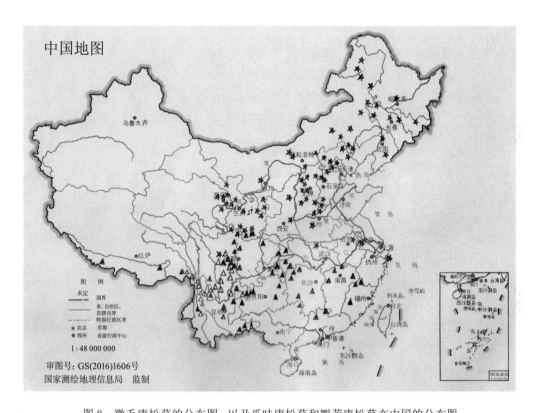

图 8　微毛唐松草的分布图，以及爪哇唐松草和瓣蕊唐松草在中国的分布图
Figure 8　Map showing the distribution of *T. lecoyeri* and those in China of *T. javanicum*. and *T. petaloideum* var. *petaloideum*

△ 微毛唐松草 *T. lecoyeri*
▲ 爪哇唐松草 *T. javanicum*
★ 瓣蕊唐松草 *T. petaloideum* var. *petaloideum*

省、区，贝加尔唐松草 T. baicalense（图9），腺毛唐松草 T. foetidum 和展枝唐松草 T. squarrosum 均广布11省、区，亚欧唐松草 T. minus var. minus 广布9省、区。瓣蕊唐松草的分布区南缘的居群星散分布，其最西的居群位于四川西部新尤，然后向东北在九寨沟有其南缘的第二个居群，再稍向东南在湖北北部枣阳有其南缘的第三个居群，再向东在南京有其第四个居群，最后向东南在杭州湾离开大陆在舟山岛的定海发现其分区南缘的第五个居群（图8）。在唐松草的分布区南缘也看到居群星散分布的情况，但居群较少，只有3个，西侧的第一个居群位于湖北西北部的竹溪，向东越过河南省，在安徽西部的霍山有第二个居群，然后向东南在浙江西北部的天目山发现其第三个居群。推测这种星散分布的情况，可能是在第四纪最后一次冰期来临时，瓣蕊唐松草和唐松草由西伯利亚向南转移到中国南部，在冰期过后返回北部的过程中，一些居群"行动迟缓"，就被星散地遗留在分布区的后（南）面。

此外，箭头唐松草 T. simplex var. simplex 和高山唐松草 T. alpinum var. alpinum（图10）虽在中国只分布于5-6省、区，但前者广布于亚洲和欧洲，后者则广布于亚洲、欧洲和北美洲的高山以及环北极地区。散花唐松草 T. sparsiflorum 在中国分布于河北北部和东北三省，但在国外分布到亚洲东北部和北美洲。

3. 广布的和稍广布的特有种

有7特有种和1特有变种广布7-9省、区：尖叶唐松草 T. acutifolium 广布13省、区，长喙唐松草 T. macrorhynchum 和直梗高山唐松草 T. alpinum var. elatum（图10）均广布9省、区，弯喙唐松草 T. uncinulatum，大叶唐松草 T. faberi，粗壮唐松草 T. robustum 和细唐松草 T. tenue 均广布8省、区，西南唐松草 T. fargesii 和小果唐松草 T. microgynum 均广布7省、区。

另23特有种分布于2-6省、区：华东唐松草 T. fortunei 分布于6省，多枝唐松草 T. ramosum 和丝叶唐松草 T. foeniculaceum（图11）均分布于5省、区，微毛唐松草 T. lecoyeri，钩柱唐松草 T. uncatum，狭序唐松草 T. atriplex，阴地唐松草 T. umbricola，稀蕊唐松草 T. oligandrum，白茎唐松草 T. leuconotum 和偏翅唐松草 T. delavayi 均分布于4省、区，绢毛唐松草 T. brevisericeum，川鄂唐松草 T. osmundifolium，滇川唐松草 T. finetii，星毛唐松草 T. cirrhosum 和鞭柱唐松草 T. smithii 均分布3省、区，峨嵋唐松草 T. omeiense，细柱唐松草 T. megalostigma，网脉唐松草 T. reticulatum，丽江唐松草 T. wangii，大花唐松草 T. grandiflorum，金丝马尾连 T. glandulosissimum，毛发唐松草 T. trichopus 和河南唐松草 T. honanense 均分布于2省。

四、地理分布   39

图 9　细柱唐松草的分布和贝加尔唐松草在中国的分布图
Figure 9　Map showing the distribution of *T. megalostigma* and that in China of *T. baicalense*
△ *T.megalostigma*　▲ *T. baicalense*

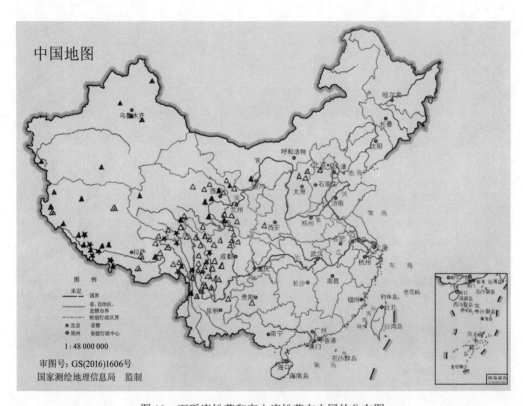

图 10 石砾唐松草和高山唐松草在中国的分布图
Figure 10 Map showing the distribution in China of *T. squamiferum* and *T. alpinum*.

★ *T. squamiferum*　　　*T. alpinum*
　　　　　　　　　　　▲ var. *alpinum*
　　　　　　　　　　　　var. *microphyllum*
　　　　　　　　　　　△ var. *elatum*

4. 狭域分布种

此类种无疑均是特有种，其中有分布于 2-3 个县的特有种，但大多数均分布在一个县的一至少数山头。

8 特有种分布于 2-3 县：糙叶唐松草 *T. scabrifolium* 和鹤庆唐松草 *T. leve*（图 7）均分布于云南鹤庆和洱源一带，云南唐松草 *T. yunnanense* 分布于云南宾川、昆明一带，武夷唐松草 *T. wuyishanicum* 分布于武夷山脉北部的东坡和西坡，矮唐松草 *T. pumilum* 分布于云南东北部东川和 Pe-long-tsin[①] 一带，美花唐松草 *T. callianthum* 分布于西藏东南部米林、林芝一带，散花唐松草 *T. diffusiflorum* 分布于西藏东南部工布江达至波密数县，藏南唐松草 *T. austrotibeticum* 分布于西藏南部吉隆、聂拉木和亚东。

29 特有种特产一县：小花唐松草 *T. minutiflorum* 产贵州德江县一山头，大关唐松草 *T. daguanense* 产云南大关县联合村，新宁唐松草 *T. xinningense* 产湖南新宁县大水岭，巨齿唐松草 *T. grandidentatum* 产四川峨眉山一山头，察隅唐松草 *T. chayuense* 和纺锤唐松草 *T. fusiforme* 均产西藏察隅县一山头，澜沧唐松草 *T. lancangense* 产云南澜沧县一山头，思茅唐松草 *T. simaoense* 产云南思茅一山头，短蕊唐松草 *T. brachyandrum* 产河南商城一山头，螺柱唐松草 *T. spiristylum* 产云南宁蒗县小草坝，钻柱唐松草 *T. tenuisubulatum* 产云南腾冲县狼牙山，宽柱唐松草 *T. latistylum* 产甘肃文县邱家坝，兴山唐松草 *T. xingshanicum* 产湖北兴山县一山头，察瓦龙唐松草 *T. tsawarungense* 产西藏察隅县察瓦龙，毛蕊唐松草 *T. lasiogynum* 产四川平武县王朗自然保护区，岳西唐松草 *T. yuoxiense* 产安徽岳西县一山头，拟盾叶唐松草 *T. pseudoichangense* 产贵州省一山头，小金唐松草 *T. xiaojinense* 产四川小金县一山头，六脉萼唐松草 *T. sexnervisepalum* 产四川崇州鸡冠山，长柱唐松草 *T. sinomacrostigma* 产四川西部一山头，吉隆唐松草 *T. jilongense* 产西藏南部吉隆汝戈喇嘛庙后山，希陶唐松草 *T. tsaii* 产云南西部一山头，攀枝花唐松草 *T. panzhihuaense* 产四川攀枝花苏铁山，粘唐松草 *T. viscosum* 产云南丽江虎跳峡，错那唐松草 *T. cuonaense* 产西藏错那县那布茶场，陕西唐松草 *T. shensiense* 产陕西旬阳庙岭，札达唐松草 *T. zhadaense* 产西藏札达县马阳村，亚东唐松草 *T. yadongense* 产西藏南部亚东县上亚东乡，定结唐松草 *T. dingjieense* 产西藏南部定结县日屋镇附近。

在上述的种中，发表于 1886 年的 *T. scabrifolium* Franch.，发表于 20 世纪初的 *T. pumilum* Ulbr. 和 *T. sinomacrostigma* W. T. Wang（= *T. macrostigma* Finet & Gagnep.,

---

① 法国传教士 E.E. Maire 采集标本的地方，应该也在东川附近。——编辑注

1906, non Edgew., 1851), 以及我们于 20 世纪 70 年代发表的 *T. shensiense*, *T. grandidentatum*, *T. tsawarungense* 和 *T. viscosum*, 到现在只有模式标本, 均未再采到标本。

5. 五原始种

在对唐松草的形态特征进行分析之后发现 5 种唐松草属的原始种, 其中, 叉枝唐松草 *T. saniculiforme* 分布于云南东南至西南部、西藏东南和南部, 也分布喜马拉雅山区; 小花唐松草 *T. minutiflorum* 特产贵州东北部德江县; 绢毛唐松草 *T. brevisericeum* 的模式变种分布四川北部和秦岭西段, 其变种五果绢毛唐松草 var. *pentagynum* 特产四川西部小金县, 另一变种狭药绢毛唐松草 var. *angustiantherum* 特产云南西部维西县; 糙叶唐松草 *T. scabrifolium* 和鹤庆唐松草 *T. leve* 均特产云南西北部洱源至鹤庆县一带。(图 7)

6. 分布中心

在青藏高原东南部边缘的横断山区（包括云南西部、西藏东南部、四川西部、青海东南部和甘肃西南部）以及云南东部分布有 2 亚属、11 组的 57 种, 8 变种。在 57 种中, 有 38 种为中国特有种, 其中, 除了绢毛唐松草 *T. brevisericeum*、多枝唐松草 *T. ramosum*、弯柱唐松草 *T. uncinulatum*、西南唐松草 *T. fargesii*、稀蕊唐松草 *T. oligandrum* 和川鄂唐松草 *T. osmundifolium* 等 6 种之外, 其他 32 种还是本分布中心的特有种（见表 2, 在表中, 对这些分布中心的特有种均加上★符号）, 这 57 种中有唐松草属的 4 个原始种, 叉枝唐松草 *T. saniculiforme*、绢毛唐松草 *T. brevisericeum*、糙叶唐松草 *T. scabrifolium* 和鹤庆唐松草 *T. leve*; 有唐松草亚属 Subgen. *Thalictrum* 不少演化水平低的种, 如大关唐松草 *T. daguanense*, 多枝唐松草 *T. ramosum*, 云南唐松草 *T. yunnanense*、峨嵋唐松草 *T. omeiense* 等种; 不少演化水平中等的种, 如贝加尔唐松草 *T. baicalense*、西南唐松草 *T. fargesii*、丽江唐松草 *T. wangii*、帚枝唐松草 *T. virgatum* 等; 不少花、果实构造复杂（如子房、柱头或瘦果具翅）, 演化水平高的种, 如偏翅唐松草 *T. delavayi*, 二翅唐松草 *T. elegans*, 腺毛唐松草 *T. foetidum*, 高山唐松草 *T. alpinum* 等; 还有单性唐松草亚属 Subgen. *Lecoyerium* 的原始种之一鞭柱唐松草 *T. smithii*。这个物种多样性程度极高、花构造多样性程度极高的地区, 毫无疑问, 当是唐松草属的分布中心。(图 7)

四、地理分布

　　57种这个数目看起来不大，但从下面三个情况一比较就可了解在横断山区加上云南东部这个不太大的地区能分布57种，这个数目是相当大的；(a) 地跨欧洲东北部和亚洲北部的俄罗斯有19种唐松草属植物（Nevski, 1937）；(b) 整个欧洲有15种唐松草属植物（Akeroyd, 1993）；(c) 整个北美洲有28种唐松草属植物（Park & Festerling Jr., 1997）。

表 2　唐松草属在其分布中心的植物名录
（有★符号的种为本分布中心的特有种）
Table 2. List of the species of the genus *Thalictrum* within its distribution centre.
(The species with a star mark are endemic to the distribution centre)

Subgen 1. **Thalictrum**
　Sect. 1. **Leptostigma**
　　1. *T. saniculiforme*
　　2. *T. brevisericeum*
　　　★var. *pentagynum*
　　　★var. *angustiantherum*
　　3. ★*T. scabrifolium*
　　4. ★*T. leve*
　　5. ★*T. daguanense*
　　6. ★*T. ramosum*
　　7. ★*T. yunnanense*
　　8. *T. javanicum*
　　9. ★*T. lecoyeri*
　　10. ★*T. omeiense*
　　11. ★*T. grandidentatum*
　　12. *T. rostellatum*
　　13. ★*T. chayuense*
　　14. ★*T. fusiforme*
　　15. *T. uncinulatum*
　　16. ★*T. simaoense*
　　17. ★*T. uncatum*
　　18. ★*T. atriplex*
　　19. ★*T. spiristylum*
　Sect. 2. **Dolichostylus**
　　20. ★*T. tenuisubulatum*
　Sect. 3. **Platystylus**
　　21. ★*T. latistylum*

　Sect. 4. **Erythrandra**
　　22. ★*T. megalostigma*
　　23. *T. baicalense*
　　24. ★*T. reticulatum*
　　　★var. *hirtellum*
　　25. *T. fargesii*
　　26. ★*T. wangii*
　　27. ★*T. tsawarungense*
　　28. ★*T. pumilum*
　Sect. 5. **Omalophysa**
　　29. ★*T. lasiogynum*
　　30. *T. przewalskii*
　Sect. 6. **Physocarpum**
　　31. *T. oligandrum*
　　32. *T. microgynum*
　　33. *T. acutifolium*
　Sect. 7. **Rutifolia**
　　34. ★*T. xiaojinense*
　　35. ★*T. sexnervisepalum*
　　36. *T. rutifolium*
　　37. ★*T. sinomacrostigma*
　　38. ★*T. leuconotum*
　Sect. 8. **Platystachys**
　　39. *T. virgatum*
　　40. ★*T. tsaii*
　　41. ★*T. grandiflorum*

　　42. ★*T. panzhihuaense*
　　43. ★*T. viscosum*
　　44. ★*T. glandulosissimum*
　　45. ★*T. delavayi*
　　　★var. *acuminatum*
　　　★var. *decorum*
　　　　var. *mucronatum*
　Sect. 9. **Dipterium**
　　46. *T. elegans*
　Sect. 10. **Thalictrum**
　　47. ★*T. trichopus*
　　48. *T. osmundifolium*
　　49. ★*T. finetii*
　　50. ★*T. cirrhosum*
　　51. *T. foliolosum*
　　52. *T. cultratum*
　　53. *T. foetidum*
　　54. *T. minus*
　Sect. 11. **Schlagintweitella**
　　55. *T. squamiferum*
　　56. *T. alpinum*
　　　　var. *microphyllum*
　　　　var. *elatum*
Subgen. 2. **Lecoyerium**
　　57. *T. smithii*

7. 姊妹群

在中国唐松草属中可看到 15 对种在亲缘关系上十分相近，可能是分别由不同祖先演化出的 15 对姊妹群。其中以下 6 对起源于分布中心。

（1）二原始种，糙叶唐松草 *T. scabrifolium* 和鹤庆唐松草 *T. leve*，如前所述，具有相同构造的花，心皮花柱直，只顶端稍向后弯曲，柱头不明显，与扁果草族 Trib. Isopyreae 的心皮甚为相似。二种均为狭域分布种，一同分布在云南洱源与鹤庆交界的山区。（图版 4-5）

（2）帚枝唐松草系 Ser. *Virgata* 的攀枝花唐松草 *T. panzhihuaense* 和粘唐松草 *T. viscosum* 的茎、叶、花梗和子房均密被极短腺毛，二种均为狭域分布，产地距离较近，前者特产四川南部攀枝花金沙江河谷的苏铁山低山区，后者特产云南丽江北部金沙江河谷的虎跳峡山坡上。（图版 68-69）

（3）叉枝唐松草组 Sect. *Leptostigma* 的钩柱唐松草 *T. uncatum*（瘦果扁平，呈新月形）和狭序唐松草 *T. atriplex*（瘦果稍扁，宽椭圆体形，不扁平）均具像总状花序的狭窄复单歧聚伞花序，钩状花柱和不明显的柱头。也具狭窄复单歧聚花序，但具直花柱和明显柱头的芸香叶唐松草组 Sect. *Rutifolia* 可能源出于 *T. uncatum* 群。此二者的分布区大部重叠，分布于云南西北、西藏东南、四川西部，北达青海或甘肃南部。（图版 30-31）

（4）叉枝唐松草组 Sect. *Leptostigma* 的爪哇唐松草 *T. javanicum* 和微毛唐松草 *T. lecoyeri* 二种亲缘关系极为相近。法国植物学家 A. Franchet 在 1889 年发表 *T. lecoyeri*，之后被 Handel-Mazzetti 等不少专家归并于 *T. javanicum*。1991 年我访问法国自然历史博物馆显花植物研究所植物标本馆（P），看到此种的模式标本，其茎节和叶柄基部均密被极短柔毛，瘦果的宿存花柱坚硬，钩状弯曲程度较大，与 *T. javanicum* 有明显区别，正是我在 1979 年发表的变种 *T. javanicum* var. *puberulum*。这时，我才认识到 A. Franchet 描述的这个新种是应予承认的独立种，并推测上述二种是具共同祖先的一对姊妹群。*T. lecoyeri* 分布于云南、四川西部和贵州西部；*T. javanicum* 如前所述是个广布种，广布中国长江流域以南地区（图 8），更向南分布到喜马拉雅山区、印度、斯里兰卡、印度尼西亚爪哇岛等地。（图版 12-13）

（5）瓣蕊唐松草组 Sect. *Erythrondra* 的细柱唐松草 *T. megalostigma* 和贝加尔唐松草 *T. baicalense* 有较近亲缘关系，美国唐松草属专家 B. Boivin（1944）最先给细柱唐松草命名，将其作为 *T. baicalense* 的变种：*T. baicalense* var. *megalostigma*。*T. megalostigma* 的雄蕊花丝狭条形，其演化水平较低，而 *T. baicalense* 的花丝分化成形状不同的上、下两部分，其演化水平较高。*T. megalostigma* 的分布区不大，只分布于四川西部和甘肃南部，*T. baicalense* 则相反，具颇大的分布区，其分布与短尾铁线莲 *Clematis brevicaudata* 的颇为相似，在此中心起源后，自西藏东南部向东北经四川西北部、青海东部，越过秦岭、太行山脉、阴山山脉和小兴安岭，最后到达俄罗斯（西伯利亚）。（王文采，1992）（图 9；图版 35–36）

(6) 石砾唐松草组 Sect. *Schlagintweitella* 的 2 种，石砾唐松草 *T. squamiferum* 和高山唐松草 *T. alpinum* 是由腺毛唐松草系 Ser. *Thalictrum* 在此分布中心演化出的一对姊妹群。*T. squamiferum* 的分布区较小，分布于云南西北、四川西部、青海南部、西藏东部和南部，向南达喜马拉雅山区（图 10），分布海拔高达 5000 m，为流石滩等多石砾山区。*T. alpinum* 在此中心发生了不小的分化，出现了 3 变种。原始变种 var. *elatum* 的总状花序的花梗直，近直立。在此中心的居群的不少植株高达 25–38 cm，有 1 条分枝。其分布自此中心向北经过西藏东部、四川西部、青海东部、甘肃南部和中部，向东越过秦岭和黄土高原到达山西东北部的五台山和河北西部小五台山；在向北分布的过程中，其植株变低矮，高 15–7 cm，同时花葶不具分枝。二进化变种的花葶低矮，花序的花梗向下弧状弯曲。var. *microphyllum*（瘦果有柄）的分布区较小，分布于四川西南、云南西北和西藏的东部和南部，向南达喜马拉雅山区。Var. *alpinum*（瘦果无柄）自此分布中心向西南达喜马拉雅山区，再北经青藏高原达亚洲北部，再向西经亚洲西部高山到达欧洲北部和北美洲高山以及环北极地区，此变种成了唐松草属分布最广的植物（图 10）。另一特产日本的变种 var. *stipitatum* Yabe. 其花序花梗长达 3 cm，向下弯曲，瘦果具柄。Boivin（1944）研究高山唐松草做出一定贡献，但他不承认 var. *elatum* 和 var. *microphyllum*，将此二变种归并于上述特产日本的 var. *stipitatum*，造成一个分类学混乱。

以下 2 对姊妹群都是其中一种分布于分布中心和西藏东南部，另一种为一狭域分布种，特产西藏南部一个山头上。

(7) 芸香叶唐松草组 Sect. *Rutifolia* 的一对柱头具翅的进化种是白茎唐松草 *T. leuconotum*（植株有顶生和侧生花序，花药有短尖头）和吉隆唐松草 *T. jilongense*

（植株无侧生花序，只有顶生花序，花药无短尖头），前者分布于云南西北、西藏东南、四川西部和青海南部，后者特产西藏吉隆一山头。（图版62-63）

（8）单性唐松草亚属 Subgen. *Lecoyerium* 在亚洲分部的二种，鞭柱唐松草 *T. smithii*（小叶较大，长达 15 mm，下面脉上被微硬毛；雄蕊花丝长 1.5-5.5 mm）和定结唐松草 *T. dingjieense*（小叶长 3-9 mm，下面被短腺毛，花丝长 8 mm）的花构造相同，心皮均具细长呈鞭状的柱头。前者分布云南西北、西藏东南和四川西部，后者特产西藏定结的一山头。（图11）在中国科学院植物研究所标本馆（PE）收藏有 *T. smithii* 的采自云南、西藏和四川的 34 号标本，其中 25 号为具花标本，10 号为具果标本，在具花的标本中有 19 号为具两性花的标本，4 号为具雌花的标本，2 号为具雄花的标本。最近发现的 *T. dingjieense* 目前只有 1 号具两性花的标本，根据此二种的花多为两性，再根据二种的分布区位于唐松草属分布中心或邻近地区，以及最近发现的腺毛唐松草组 Sect. *Thalictrum* 特产西藏南部的亚东唐松草 *T. yadongense* 的花部构造（花丝丝形，心皮具细长狭条形柱头）接近 Subgen. *Lecoyerium* 植物的花部构造，推测 Subgen. *Lecoyerium* 可能在唐松草属分布中心自 Sect. *Thalictrum* 演化而出，后来由分布中心分布到北、南美洲等地区，而 *T. smithii* 和 *T. dingjieense* 则是 Subgen. *Lecoyerium* 的原始群遗留在中国西南部的一对原始种。（图11；图版100-101）

Tamura(1995) 在其唐松草属分类系统中将 *T. smithii* 与分布于北美洲的 14 个进化种置于 Subgen. *Lecoyerium* 的进化群 Sect. *Heterogam* 中，并认为 *T. smithii* 是由 Sect. *Heterogamia* 的其他种演化出并在旧世界独立起源。他的这些论断与我上述论断大相径庭。

以下 2 对姊妹群可能起源自青藏高原南部一带。

（9）偏翅唐松系 Ser. *Violacea* 的美丽唐松草 *T. reniforme*（小叶长达 2.5 cm；萼片具 3 条基生脉；花丝狭条形）和堇花唐松草 *T. diffusiflorum*（小叶较小，长达 1.2 cm，萼片具 7 条基生脉，花丝丝形）的瘦果扁平，腹缝线凸起，每侧的 3 条纵脉分枝并网结，与偏翅唐松系的其他种明显不同。美丽唐松草分布于西藏南部和东南部，以及喜马拉雅山区；堇花唐松草特产西藏东南部。（图版76-77）

（10）腺毛唐松草组 Sect. *Thalictrum* 的高原唐松草 *T. cultratum* 和腺毛唐松草 *T. foetidum* 具有十分相似的花序和花部构造（萼片淡黄绿色，速落，花丝丝形，柱

四、地理分布　　　　　　　　　　　　　　47

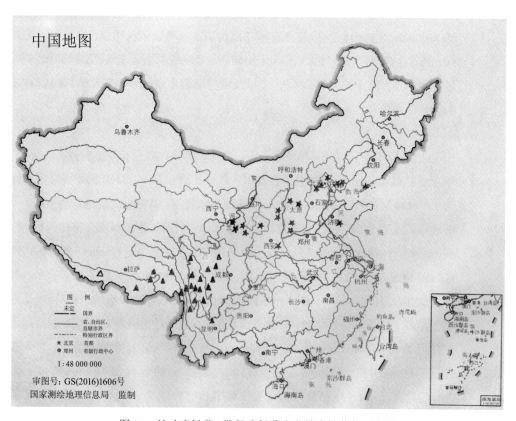

图 11　丝叶唐松草，鞭柱唐松草和定结唐松草的分布图
Figure 11　Map showing the distribution of *T. foeniculaceum*, *T. smithii* and *T. dingjieense*.
★ *T. foeniculaceum*　　　▲ *T. smithii*　　　△ *T. dingjieense*

头完全与花柱贴生，呈狭长圆形，具 2 条狭侧生翅）。前者分布于西藏东南部、云南西北、四川西部、青海和甘肃南部，以及喜马拉雅山区；后者为广布种，分布于亚洲北部和西部，以及欧洲，在中国广布西部和北部。（图版 90–91）

下面一对姊妹群分布于亚洲东北部，可能起源自唐松草属分布中心。

(11) 尖叶唐松草组 Sect. *Physocarpum* 的深山唐松草 *T. tuberiferum* 和花唐松草 *T. filamentasum* 具有相似的植株体态和花部构造，地理分布也相似，均分布于亚洲东北部，前者的分布区较大，分布于中国东北部、俄罗斯远东地区、朝鲜和日本；后者则仅分布于中国吉林、黑龙江，以及俄罗斯远东地区和朝鲜北部，未达日本。（图版 53–54）

以下 3 对姊妹群中，均有一种分布于华东一带，另一种则分布于更向西的地区。

(12) 华东唐松草 *T. fortunei*（瘦果狭长圆形）和多枝唐松草 *T. ramosum*（瘦果狭披针形）二种位居糙叶唐松草系 Ser. *Scabrifolia* 的前列，具相似的体态，茎不高，小叶较小，近圆形，花序有少数花，花丝上部倒披针形，下部丝形，花柱钩状，柱头不明显，二种难于区别，致使两位中国植物区系专家 A. Franchet（1889）和 H. Handel-Mazzetti（1931）先后将采自四川的 *T. ramosum* 标本鉴定为 *T. fortunei*。一直到 1945 年，Boivin 才将分布于西侧的植物命名为 *T. ramosum*。*T. fortunei* 分布于浙江、江苏、安徽南部、河南南部、江西北部、湖北东南部的阳新，向西达到接近 *T. ramosum* 分布区的宜昌；*T. ramosum* 分布广西西北、湖南西部、贵州东北、重庆、四川东部和西部以及甘肃南部（文县、舟曲）。（图 12；图版 7–8）

(13) 大叶唐松草 *T. faberi*（植株全部无毛）和粗壮唐松草 *T. robustum*（茎、叶、花梗均被短柔毛）二种排在 Ser. *Scabrifolia* 的最后位置，其小叶大卵形，长达 8.5–10 cm，边缘具尖齿，花序分枝多，具多数花，花丝的上、下部分化不明显，易与同系的其他种区分。*T. faberi* 分布于浙江、江西、湖南、湖北西北以及河南、安徽和江苏三省的南部，1958 年成立的"秦岭考察队"在秦岭太白山采到 *T. faberi* 的标本，在这里与 *T. robustum* 相遇；*T. robustum* 分布于四川北部、重庆、湖南西北（石门）、湖北西部、甘肃和陕西的南部，以及河南西部。（图版 17–18）

(14) 网脉唐松草系 Ser. *Reticulata* 由 2 亲缘关系甚为相近的种组成，即网脉唐松草 *T. reticulatum*（花序伞房状，具多数分枝和花，瘦果卵球形）和武夷唐松草 *T.*

四、地理分布　　　　　　　　　　　　　　　　　　　　　　　　　　　　49

图 12　华东唐松草，多枝唐松草，长柄唐松草和散花唐松草的分布图
Figure 12　Map showing the distribution of *T. fortunei*, *T. ramosum*, *T. przewalskii* and *T. sparsiflorum*.
△ *T. fortunei*　▲ *T. ramosum*　☆ *T. przewalskii*　★ *T. sparsiflorum*

wuyishanicum（花序具 2-4 花，瘦果纺锤形），2 种的花部构造十分相似，花丝上部倒披针形，下部丝形，心皮无柄，花柱很短，其腹面具密集的柱头组织，好像柱头直接生在子房顶端。二种的分布区有一间断距离，*T. reticulatum* 分布于位在横断山区的云南西北和四川西南部，而 *T. wuyishanicum* 则分布于福建西北部和江西东南部交界武夷山区。（图版 38-39）

下面的一对姊妹群中的一种位于中国西南部，另一种则位于东北部。

（15）长柄唐松草组 Sect. *Omalophysa* 的长柄唐松草 *T. przewalskii*（小叶下面有短柔毛，果梗直）与散花唐松草 *T. sparsiflorum*（植株全部无毛，果梗顶端反曲）具相同的花部构造和具细长柄的瘦果。前者分布于西藏东部、四川西部和北部、重庆、湖北西部、青海东部、甘肃中部和南部、陕西南部、河南西部、山西、河北和内蒙古南部；后者分布于北京西部、河北北部、吉林和黑龙江的东部，以及朝鲜、俄罗斯远东地区和北美洲西部。（Park & Festerling, 1997）二种在河北兴隆县的雾灵山相遇。（图 12；图版 46-47）

五、经济用途

A. 药用

《中国药用植物志》（谢磊，王峥涛等，2016）报道唐松草属多数种含不少类型的生物碱类化合物，主要有双苄基异喹啉型，如华东唐松草碱（thalifortine），芳香花桂林碱（aromoline），狭序唐松草灵（thaliatrine）；阿朴菲型（如前绿心樟碱（preocotine），箭头唐松草米定碱（thalicsimidine）；阿朴菲苄基异喹啉型，如甲氧基铁线蕨叶唐松草碱（methoxyadiantifoline），秋唐松草替定碱（thalmelatidine）；原小檗碱型（如小檗碱（berberine），药根碱（jatrorrhizine）；异帕文碱型如箭头唐松草莫宁（thalimonine）等，但也有少数种（如帚枝唐松草 *T. virgatum*，展枝唐松草 *T. squarrosum*，鞭柱唐松草 *T. smithii*）不含生物碱。此外，从本属植物还分离到许多环菠萝烷三萜皂苷和一些黄酮。由于含上述化学成分，本属中不少植物可供药用，治疗一些疾病。

1961年出版的《中国经济植物志》报道四川中医中药研究所从偏翅唐松草 *T. delavayi* 的根分离出含量高达2.96%的小檗碱，并称此植物的根有杀菌、消炎的功效，可代黄连药用。

1964年，裴鉴、周太炎两位教授在他们编写出版的《中国药用植物志》第七册中介绍了唐松草属10种药用植物：盾叶唐松草 *T. ichangense*，肾叶唐松草 *T. petaloideum*，峨嵋唐松草 *T. omeiense*，多叶唐松草 *T. foliolosum*，偏翅唐松草 *T. delavayi*，贝加尔唐松草 *T. baicalense*，多枝唐松草 *T. ramosum*，毛发唐松草 *T. trichopus*，箭头唐松草 *T. simplex* 和翅果唐松草 *T. contortum*，给出了10种的形态描述、图版和药用用途等。此外，在文献引证中，两位教授将 *T. coreanum* Yabe 归并于 *T. ichangense* Lecoy. ex Oliv.，将根据采自北京十三陵的一号标本描述的新种 *T. supradecompositum* Nakai 和另一学名 *T. petaloideum* L. var. *supradecompositum*（Nakai）Kitagawa 一起归并于 *T. petaloideum* L.，澄清了两个分类学混淆问题。

1965年，肖培根，王文采合作发表了一篇关于中国唐松草属药用植物的文章，对此属26种植物的根和根状茎进行初步化学预试，证明都有生物碱反应，其中高原唐松草 *T. cultratum*，多叶唐松草 *T. foliolosum*，昭通唐松草 *T. chaotugense*，滇川唐松草 *T. finetii*，贝加尔唐松草 *T. baicalense* 及其变种 var. *megalostigma* 的根和根状茎均含有小檗碱，可代替黄连供药用。

1975 年出版的《全国中草药汇编》，1980 年发表的童玉懿等的论文和 1993 年出版的《中药志》都报道了"马尾连"项下的唐松草属十余种到二十余种的药用植物，其中金丝马尾连 *T. glandulosissimum*，昭通唐松草 *T. glandulosissimum* var. *chaotungense*，高原唐松草 *T. cultratum*，多叶唐松草 *T. foliolosum* 的根和根状茎的小檗碱含量最高，其次是贝加尔唐松草 *T. baicalense* 及其变种 var. *megalostigma*，偏翅唐松草 *T. delavayi*，多枝唐松草 *T. ramosum*，滇川唐松草 *T. finetii*，大叶唐松草 *T. faberi*，箭头唐松草 *T. simplex*，峨嵋唐松草 *T. omeiense* 等种。

1988 年出版的《新华本草纲要》收载的唐松草属药用植物增到 30 种。

2016 年出版的《中国药用植物志》收载的唐松草属药用植物增到 43 种，10 变种。其中根和根状茎用于治疗痢疾、黄疸、胃热、肝炎等症的药用植物有多枝唐松草 *T. ramosum*，华东唐松草 *T. fortunei*，峨嵋唐松草 *T. omeiense*，大叶唐松草 *T. faberi*，贝加尔唐松草 *T. baicalense* 及其变种 var. *megalostigma*，瓣蕊唐松草 *T. petaloideum*，网脉唐松草 *T. reticulatum*，金丝马尾连 *T. glandulosissimum*，昭通唐松草 *T. glandulosissimum* var. *chaotungense*，偏翅唐松草 *T. delavayi*，高原唐松草 *T. cultratum*，多叶唐松草 *T. foliolosum*，滇川唐松草 *T. finetii*，短梗箭头唐松草 *T. simplex* var. *brevipes*，黄唐松草 *T. flavum* 等；用于痢疾、肠炎等症的有粗壮唐松草 *T. robustum*，腺毛唐松草 *T. foetidum*，箭头唐松草 *T. simplex*，亚欧唐松草 *T. minus*，用于清热解毒、跌打损伤等症的有爪哇唐松草 *T. javanicum*，微毛唐松草 *T. lecoyeri*，长喙唐松草 *T. macrorhynchum*，盾叶唐松草 *T. ichangense* 等；用于消炎、肾结石等症的有深山唐松草 *T. tuberiferum*；用于诸热症，炭疽病的有芸香叶唐松草 *T. rutifolium*；用于胃病的有帚枝唐松草 *T. virgatum*；全草用于黄疸病的有小果唐松草 *T. microgynum*；以及用下痢、疏肝、清热明目的有鞭柱唐松草 *T. smithii*；花和果实用于肝炎的有长柄唐松草 *T. przewalskii*。

B. 观赏

帚枝唐松草系 Ser. *Virgata* 的大花唐松草 *T. grandiflorum* 的花较大，萼片粉红色，长 13–30 mm（图版 67）；偏翅唐松草系 Ser. *Violacea* 的珠芽唐松草 *T. chelidonii*，美花唐松草 *T. callianthum*，美丽唐松草 *T. reniforme*，堇花唐松草 *T. diffusiflorum* 和偏翅唐松草 *T. delavayi* 等 5 种的花也较大，萼片呈紫色，堇色或粉红色，长 5.5–15 cm（图版 74–78）。这 6 种的花美丽，有观赏价值，其中的偏翅唐松草 *T. delavayi* 有较长的栽培历史，此种在 1886 年由法国植物学家 A. Franchet 根据法国传教士 Abbé Delavay 采自云南大理苍山的一号标本发表，以后经由法国传教士引种到法国，在 20 世纪初著名采集家 E. H. Wilson 和英国爱丁堡植物园的 G. Forrest 也先后引种此植物到英国和欧洲其他地区，并获得英国皇家园艺学会的大奖（Mallett, 1905; Lauener, 1996）。我在 1991 年访问瑞典乌普萨拉大学植物标本馆期间，在瑞典中部一私人花园中见到几株栽培的宽萼偏翅唐松草 T. delavayi var. decorum Franch.，植株高约 2 m，和高翠雀花 Delphinium elatum L.（植株也高约 2 m）种植在一起，其下垂的聚伞圆锥花序的紫色花正盛放中。与我 1962 年初秋在云南丽江低山区见到的此变种野生居群的植株十分相似。（王文采，1993）此变种植株高大，花序也较长，其美丽程度实较模式变种 var. delavayi（高度通常在 1 m 以下）更胜一筹，但在有关文献中罕被提及。美国园艺学家 L. H. Bailey 编著的 *Manual of Cultivated Plants*（1949）中收载了 *T. delavayi* Franch.，此外还收载了 *T. dipterocarpum* Franch.（我已于 1979 年将此学名归并于 *T. delavayi* Franch.），从其给出的描述"植株高达 7ft."和"萼片大"等，我认为他鉴定的"*T. dipterocarpum*"实为 *T. delavayi* var. *decorum*。上述 6 种具美丽花的唐松草属植物，在中国还未见引种、栽培，甚是遗憾。

参考文献

王文采、1979. 金莲花族 . // 中国植物志, 27: 60–88，科学出版社 . [W. T. Wang. 1979. Trib. *Trollieae*. In Fl. Reip. Pop. Sin.27: 60–88. Science Press].

王文采、1992. 东亚植物区系的一些分布式样和迁移路线，植物分类学报，30（1）: 1-24;（2）: 97–117. [W. T. Wang. 1992. On some distribution patterns and some migration routes found in the eastern Asiatic region. Acta Phytotax. Sin. 30(1):1–24; (2)97–117].

王文采、1993. 中国毛茛科植物小志（十五）. 植物分类学报 31（3）: 201–226. [W. T. Wang. 1993. Notulae de *Ranunculaceis* Sinensibus(xV). Acta Phytotax. Sin. 31(3):201–226].

王文采，王蜀秀，1979，唐松草属 . // 中国植物志 27: 502–592，科学出版社 . [W. T. Wang & S. H. Wang. 1979. *Thalictrum*. In Fl. Reip. Pop. Sin. 27: 502–592. Science Press].

中国医学科学院药物研究所 .1993. 马尾连，中药志 1: 264，人民卫生出版社 .

中华人民共和国商业部土产废品局，中国科学院植物研究所，1961. 偏翅唐松草 *Thalictrum* delavayi. 中国经济植物志 2: 1705，科学出版社 .

朱敏，肖培根 .1991. 中国唐松草属植物的化学系统学初探 . 植物分类学报 29（4）: 358–369. [M. Zhu & P. K. Xiao. 1991. Chemosystematic studies on *Thalictrum* L. in China. Acta Phytotax. Sin. 29(4):358–369].

全国中草药汇编编写组 .1975. 马尾连 . 全国中草药汇编 1: 75–77，人民卫生出版社。

肖培根 .1979. 扁果草族 . // 中国植物志 27: 465–502，科学出版社 .[P. K. Hsiao. 1979. Trib. *Isopyreae*.

In Fl. Reip. Pop. Sin. 27: 465–502. Science Press].

肖培根，王文采 .1965. 中国毛茛科药用植物的研究Ⅲ. 唐松草属的药用植物，药学学报 12（11）：745–753。[P. K. Hsiao & W. T. Wang. 1965. A study of the Ranunculaceous medicinal plants in China Ⅲ. The medicinal plants of genus *Thalictrum* Linn. Acta Pharm. Sin. 12(11):745–753].

江苏省植物研究所等，1988. 唐松草属，新华本草纲要 1：133–139，上海科学技术出版社。

郝大程，肖培根 .2017. 药用植物亲缘学导论，化学工业出版社．[D. C. Hao & P. K. Xiao. 2017. An introduction of plant pharmacophylogeny. Chem. Ind. Press].

张芝玉，1982. 星叶草属，独叶草属，鸡爪草属的染色体观察和系统位置的探讨 . 植物分类学报 30（1）：1–24．[C. Y. Chang. 1982. Chromosome observations of three ranunculaceous genera in relation to their systematic positions. Acta Phytotax. Sin. 30(1):1–24].

席以珍 .1973. 唐松草属花粉形态的研究，植物学报 15（2）：155–162．[Y. Z. xi. 1973. A. study of the pollen morphology of *Thalictrum* L.].

谢磊（分类），王峥涛，刘青，李思蒙，李业生，宋瑞凯，赵淼淼，韩丽订，王伟，俞丽，钱景时，杨秀伟，张英涛，乔雪，杨鑫宝（化学），丛潇，马涛（药理），董素哲，王阿娜，何中燕，裴瑾（注评）.2016. 唐松草属 *Thalictrum*. // 艾铁民，韦发南（eds.），中国药用植物志 Med. Fl. China 3: 234–288. 北京大学医学出版社 .

童玉懿，陈碧珠，肖培根，楼之岑 .1980. 马尾连的原植物与生药学研究，药学学报 15（9）：563–570 [Y. Y. Tong. P. C. Chen. P. K. Hsiao & C. C. Lou. 1980. Botanical and Pharmacognostical studies in the Chinese Medicinal drug "Ma Wei Lian". Acta Pharm. Sin. 15(9):563–570].

裴鉴，周太炎，1964，中国药用植物志 7：325 图 –334 图，科学出版社。

Bailey, L. H. 1949. Thalictrum. Manual of cultivated plants, rev. ed., pp.390–391.

Boivin, B. 1944. American *Thalictrum* and their Old World allies. Rhodora 46: 337–377, 391–445, 453–487.

Boivin, B. 1945. Notes on some Chinese and Korean species of *Thalictrum*. J. Arn. Arb. 26: 111–118.

De. Candolle, A. P. 1817. Regni vegetabilis systema naturale 1. Paris.

De. Candolle, A. P. 1824. Prodrumus systematis naturalis regni vegetabilis 1. Paris.

Foster, A. S. & E. M. Gifford Jr. 1959. Vascular anatomy of flower. Comparative morphology of vascular plants, pp.463–467. W. H. Freeman and Company.

Fu, D. Z. 1990. Phylogenetic considerations on the subfamily *Thalictroideae* (Ranunculacede). Cathaya 2: 181–190.

Fu, D. Z. & G. Zhu. 2001. *Thalictrum*. In Z. Y. Wu & P. H. Raven (eds.), Flora of China 6: 282–302. Beijing: Science Press; st. Louis: Missouri Botanical Garden Press.

Gregory, W. C. 1941. Phylogenetic and cytologyical studies in the Ranunculaceae. Trans. Amer. Philos. Soc., n. s., 31: 443–521.

Langlet, O. 1932. Uber Chromosomenverhatnisse und Systematik der Ranunculaceae, Svensk Bot. Tidskr. 26: 381–400.

Lauener, L. A. 1996. The introduction of Chinese plants into Europe. SPB Academic Publishing.

Lecoyer. C. J. 1885. Monographie du genere *Thalictrum*. Bull. Soc. Bot. Belg. 24: 78–324.

Mallett, G. B. 1905. *Thalictrum delavayi*. Gard. Chron. Ⅲ, 38: 450, fig. 169.

Nevski, S. A. 1937. *Thalictrum*. In Flora LIRSS 7: 510–528. Moskwa & Leningrad.

Nowicke, W. & J. J. Skvarla. 1995. Pollen morphology of Ranunculacede. In P. Hiepko(ed.), Die naturlichen Pflanzenfamilien (Zwei Aufl.),17aiv: 129–184. Berlin: Duncker & Humblot.

Okada, H. & M. Tamura. 1979. Karyomorphology and relationship in the Ranunculaceae. J. Jap. Bot. 54: 65–77.

Park, M. M. & D. Festerling Jr. 1997. *Thalictrum*. In Flora of North America 3: 258–271. Oxford University Press.

Prantl, K. 1887. Beitrage sur Morphologie und Systematik der Ranunculaceen. Bot. Jahrb. 9: 225–273.

Takhtajan, A. 1991. Mosaics of the evolutionary trends and heterobathmy of character; the evolution of inflorescences. Evolutionary trends in flowering plants, pp. 112–126 and 227–232. New York: Columbia University Press.

Takhtajan, A. 2009. Weighting of taxonomic character and heterobathmy; Ranunculaceae. Flowering plants, 2nd ed., 1: xxx–xxxi; 86–91. Springer.

Tutin T. G. 1964. *Thalictrum*. in T. G. Tutin et al. Flora Europaea. Cambridge University Press. 1: 240–242.

Tamura, M. 1995. Karyology of Ranunculaceae; Trib. *Isopyreae*; *Thalictrum*. In P. Hiepko(ed.), Die naturlichen Pflanzenfamilien (Zwei. Aufl.)17aiv: 77–88, 451–466, 474–494. Berlin: Duncker & Humblot.

Wang, W, A. M. Lu, Y. Ren, M. E. Endress & Z. D. Chen. 2009. Phylogeny and classification of Ranunculales: Evidence from four molecular loci and morphological data. Persp. Plant Ecol. Evol. Syst. 11: 81–110.

六、分类学处理

唐松草属

Thalictrum L. Sp. Pl. 1: 545. 1753; DC. Syst. 1: 168. 1817; et Prodr. 1: 11. 1824; Benth. & Hook. f. Gen. Pl. 1: 4. 1867; Lecoy. in Bull. Soc Bot. Belg. 24: 78. 1885; Prantl in Bot. Jahrb. 9: 268. 1887; et in Engler & Prantl, Nat. Pflanzenfam. 3(2):66. 1888; Finet & Gagnep. in Bull. Soc. Bot. France 50: 602. 1903; Nevski in Fl. URSS 7: 510. 1937; Boivin in Rhodora 46: 346. 1944; Tamura in Sci. Rep. Osaka Univ. 17: 50. 1968; in Acta Phytotax. Geobot. 43: 55. 1992; et in Hiepko, Nat. Pflanzenfam., Zwei Aufl., 17a IV : 474. 1995; Emura in J. Fac. Sci. Univ. Tokyo. Sect. 3, Bot. 9: 95. 1972; W. T. Wang & S. H. Wang in Fl. Reip Pop. Sin. 27: 502. 1979; Akeroyd in Fl. Europ., 2nd ed., 1: 61. 1993; M. M. Park & D. Festerling Jr. in Fl. N. Amer. 3: 358. 1997.   Lectotype: *T. foetidum* L. (Britton & Brown, 1913)

多年生草本植物。根为须根，有时为块根。根状茎合轴，横走，有时直立，有时不太发育，有时具匍匐茎。茎常有浅纵沟，通常具叶和分枝。叶基生并茎生，稀全部茎生或基生，为1–5回三出复叶或三出羽状复叶，稀为单叶；茎生叶互生，稀对生或轮生；小叶通常3浅裂，稀不分裂，具3–7脉，边缘具齿或全缘；叶柄基部具鞘；托叶存在或不存在。花序通常为少花或多花的伞房状单歧聚伞花序，如多分枝并具多数花呈圆锥状，即形成聚伞圆锥花序，稀为总状花序，稀单朵花生于茎生叶腋部并沿茎向顶发育。花两性、单性、杂性、雌雄同株或雌雄异株。花具三轮，无花瓣：萼片（3–）4–5(–10)，小，早落，稀较大，多少迟落，黄绿色或白色，稀粉红色或紫色，每一萼片具（1–2–）3(–4–7)条不分枝或二歧状分枝的基生脉。雄蕊多数或数枚，直立或下垂；花丝条形、狭条形或丝形，或上部增宽呈狭倒披针形，棒形或倒披针形，下部呈狭条形或丝形；花药长圆形，宽长圆形、狭长圆形或条形，顶端钝或具短尖头。心皮（单雌蕊）2–20(–68)，离生；子房无柄或具柄，具1室和1胚珠；花柱短或长，细圆柱状，直或钩状弯曲，在腹面或近顶端具一不明显或明显柱头，稀扁平，薄片状，长圆形，稀不具柱头；柱头有时全部与花柱贴生，并具1–2条侧生狭或宽膜质翅。花托在果期不或稍增大。瘦果无柄或具柄，不扁或稍扁至强度扁，每侧具1–3条纵肋或纵脉，脉直或弯曲，背缝线通常凸起，腹缝线稍凸起或凹；宿存花柱短或长，直或钩状弯曲。

Perennial herbs. Roots fibrous, sometimes tuberous. Rhizome sympodial, creeping, sometimes erect, sometimes not well-developing, or sometimes with stolones. Stems usually shallowly longitudinally sulcate, leafy and branched. Leaves basal and cauline, rarely all cauline or all basal, 1–5-ternate or 1–5-ternate-pinnate, rarely simple; cauline leaves alternate, rarely opposite or verticillate; leaflets usually 3-lobed, rarely undivided, 3–7-nerved, margin dentate or entire; petioles at base vaginate; stipules present or absent. Inflorescence usually a few- or more-flowered corymbiform monochasium, if much branched and many-flowered and paniculiform, then a thyrse, rarely a raceme or flowers singly axillary to cauline leaves and along stem acropetally developed. Flowers bisexual, unisexual, polygamous, monoecious or dioecious. Flower tricyclic, apetalous: Sepals (3–)4–5(–10) small, caducous, rarely larger and more or less tardily deciduous, yellowish-green or white, rarely pink or purple, each sepal with (1–2–)3(–4–7) simple or dichotomizing basal nerves. Stamens numerous or several, erect or pendulous; filaments linear, narrow-linear, filiform, or above dilated, narrow-oblanceolate, clavate or oblanceolate, below narrow-linear or filiform; anthers oblong, broad-oblong, narrow-oblong or linear, apex obtuse or mucronate. Carpels (simple pistils)2–20(–68), free; ovary sessile or stipitate, 1-locular and 1-ovulate; style short or long, thinly terete, straight or hooked, ventrally or near apex with an inconspicuous or conspicuous stigma, rarely flattened, lamelliform, oblong, rarely lacking a stigma; stigma not winged,sometimes entirely adnate with style and with 1 or 2 lateral narrow or broad membraneous wings. Receptacle not or slightly enlarged in fruit. Achene sessile or stipitate, not or slightly or strongly compressed, on each side longitudinally 1–3-ribbed or 1–3-nerved, with straight or sinuous nerves, dorsal suture usually convex, ventral suture usually slightly convex or concave; persistent style short or long, straight or hooked.

约 220 种，广布亚洲，欧洲，北、南美洲和非洲。中国有 99 种，20 变种（包括 72 特有种，14 特有变种），隶属 2 亚属，12 组，广布全国各省、区，多数种（约 74 种）分布于西南部四川、云南、西藏等省、区，其中有 57 种分布于横断山区和云南一带，这里是唐松草属的一个重要分布中心。

此次中国唐松草属修订所研究的标本主要是中国科学院植物研究所植物标本馆（PE）收藏的唐松草属植物标本，此外，还研究了自哈佛大学植物标本馆（GH）、纽约植物园植物标本馆（NY）和江苏植物研究所植物标本馆（NAS）借用的少量该属植物标本。在本书中每一种的形态描述之后的"标本登录"中列出的上述 GH、NY、NAS 三个标本馆借用的标本，均写出有关标本馆的缩写名，对 PE 收藏的标本则不写出标本馆缩写名。

分亚属、组、系检索表

1. 花两性；柱头内藏，不呈鞭形 ·············· 亚属 1. **唐松草亚属** Subgen. **Thalictrum**
    2. 花柱顶端向后弯曲，通常钩状或拳卷状。
        3. 花柱腹面有不明显柱头 ·············· 组 1. **叉枝唐松草组** Sect. **Leptostigma**
            4. 花组成伞房状复单歧聚伞花序。
                5. 雄蕊花丝条形、狭条形或丝形。
                    6. 花丝条形或狭条形 ·············· 系 1. **叉枝唐松草系** Ser. **Samiculiformia** 种 1–3
                    6. 花丝丝形 ·············· 系 4. **思茅唐松草系** Ser. **Simaoensia** 种 28–29
                5. 花丝分化成形状和大小不同的两部分（在大叶唐松草，花丝常为狭条形）。
                    7. 瘦果无柄 ·············· 系 2. **糙叶唐松草系** Ser. **Scabrifolia** 种 4–18
                    7. 瘦果具柄 ·············· 系 3. **小喙唐松草系** Ser. **Rostellata** 种 19–27
            4. 花组成狭窄、似总状花序的复单歧聚伞花序 ······ 系 5. **钩柱唐松草系** Ser. **Uncata** 种 30–32
        3. 花柱钻状狭条形，比子房长，腹面无柱头 ······ 组 2. **钻柱唐松草组** Sect. **Dolichostylus** 种 33
    2. 花柱直
        8. 花柱扁平，长圆形，腹面无柱头 ·············· 组 3. **宽柱唐松草组** Sect. **Platystylus** 种 34
        8. 花柱不呈长圆形，腹面有明显柱头。
            9. 花丝分化成在形状和大小上不同的上、下二部分，稀狭条形（细柱唐松草，察瓦龙唐松草）。
                10. 瘦果无翅。
                    11. 花柱细圆柱形。
                        12. 瘦果无柄，有粗纵肋 ·············· 组 4. **瓣蕊唐松草组瓣蕊唐松草系**
                            Setc. **Erythrandra** ser. **Petaloidea** 种 37
                        12. 瘦果具柄，有细纵肋。
                            13. 瘦果具短柄。
                                14. 瘦果扁球形。
                                    15. 花丝和柱头均狭条形 ·· 系 1. **细柱唐松草系** Ser. **Megalostigmata** 种 35
                                    15. 花丝上部狭倒披针形，下部丝形，柱头椭圆体形 ················
                                        ·············· 系 2. **贝加尔唐松草系** Ser. **Baicalensia** 种 36
                                14. 瘦果明显扁压 ·············· 系 5. **西南唐松草系** Ser. **Fargesiana** 种 40–44
                            13. 瘦果具细长柄。
                                16. 小叶不呈盾状；花组成多花聚伞圆锥花序；花柱长 0.1–1 mm ·············
                                    ·············· 组 5. **长柄唐松草组** Sect. **Omalophysa** 种 45–47
                                16. 小叶基部盾状；花组成少花单歧聚伞花序；花柱长 3 mm ·············
                                    ·············· 组 6. **尖叶唐松草组拟盾叶唐松草系**
                                    Sect. **Physocarpum** ser. **Pseudoichangensia** 种 56
                    11. 花柱很短，有时近不存在，腹面有密集的柱头组织。
                        17. 瘦果无柄 ·············· 组 4. **瓣蕊唐松草组网脉唐松草系**
                            Sect. **Erythrandra** ser. **Reticulata** 种 38–39
                        17. 瘦果具细长柄 ·············· 组 5. **尖叶唐松草组尖叶唐松草系**

六、分类学处理

                    Sect. **Physocarpum** ser. **Clavata** 种 48–55
  10. 瘦果具 4 纵翅和细长柄；花柱很短 …… 组 7. **唐松草组** Sect. **Tripterium** 种 57
 9. 花丝条形、狭条形或丝形。
   18. 花离顶发育，组成单歧聚伞花序或聚伞圆锥花序。
    19. 花组成狭窄、似总状花序的复单歧聚伞花序 ……………………………………
    …………………………………………… 组 8. **芸香叶唐松草组** Sect. **Rutifolia**
     20. 小叶下面被微硬毛；花丝狭条形
      ………………………………… 系 1. **小金唐松草系** Ser. **Xiaojinensia** 种 58–59
     20. 小叶无毛；花丝丝形。
      21. 柱头无翅 ………………………… 系 2. **芸香叶唐松草系** Ser. **Rutifolia** 种 60
      21. 柱头具翅。
       22. 花具约 50 枚雄蕊；花药比花丝长，花柱比子房长 2 倍；柱头有 1 条翅
        ………………… 系 3. **长柱唐松草系** Ser. **Sinomacrostigmata** 种 61
       22. 花具 7–16 枚雄蕊；花药比花丝短；花柱稍短于子房；柱头有 2 条侧生
        狭翅 ………………… 系 4. **白茎唐松草系** Ser. **Leuconota** 种 62–63
    19. 花组成伞房状复单歧聚伞花序或聚伞圆锥花序。
     23. 瘦果扁平，每侧有 1 条纵脉，沿腹、背缝线各有 1 条狭翅 ………………
     ………………………………… 组 10. **二翅唐松草组** Sect. **Dipterium** 种 79
     23. 瘦果每侧有（2–）3（–4）条纵肋，多无翅。
      24. 花组成少花、伞房状复单歧聚伞花序 …………………………………………
      …………………………………… 组 9. **帚枝唐松草组** Sect. **Platystachys**
       25. 叶为复叶 ………… 系 1. **帚枝唐松草系** Ser. **Virgata** 种 64–70
       25. 叶为单叶 ………… 系 2. **圆叶唐松草系** Ser. **Indivisa** 种 71–72
      24. 花组成多花聚伞圆锥花序。
       26. 萼片较大，长 5–16 mm，紫色或堇色，稀白色；柱头无翅
        ……………………………………… 组 9. **帚枝唐松草组偏翅唐松草系**
        Sect. **Platystachys** ser. **Violacea** 种 73–78
       26. 萼片较小，长 2–4.5 mm，通常淡黄绿色，稀白色或带紫色 …………
        ………………………………… 组 11. **腺毛唐松草组** Sect. **Thalictrum**
        27. 柱头无翅 … 系 1. **川鄂唐松草系** Ser. **Osmundifolia** 种 80–85
        27. 柱头具 2 条侧生翅 ……… 系 2. **腺毛唐松草系** Ser. **Thalictrum**
         种 86–97
   18. 花向顶发育 ……………………………… 组 12. **石砾唐松草组** Sect. **Schlagintweitella**
    28. 基生叶不存在；花单生上部茎生叶腋部 ……………………………………
    ………………………………… 系 1. **石砾唐松草系** Ser. **Squamifera** 种 98
    28. 叶均基生；花组成总状花序 ……… 系 2. **高山唐松草系** Ser. **Alpina** 种 99
1. 花单性，稀两性；柱头伸出，鞭形 ……… 亚属 2. **单性唐松草亚属** Subgen. **Lecoyerium** 种 101–102

Key to subgenera, sections and series

1. Flowers bisexual; stigmas included, not flagelliform ·········· Subgen. 1 .**Thalictrum**
  2. Style apex recurved, usually hooked or circinate.
    3. Style ventrally with an inconspicuous stigma ·········· Sect.1. **Leptostigma**
      4. Flowers in corymbiform compound monochasia.
        5. Stamen filaments linear, narrow-linear or filiform.
          6. Filaments linear or narrow-linear ·········· Ser. 1. **Samiculiformia** spp.1−3
          6. Filaments filiform ·········· Ser. 4. **Simaoensia** spp.28−29
        5. Filament differentiated into upper and lower two parsts different in shape and size (in *T. faberi*, filaments often narrow-linear).
          7. Achenes sessile ·········· Ser. 2. **Scabrifolia** spp.4−18
          7. Achenes stipitate ·········· Ser. 3. **Rostellata** spp.19−27
      4. Flowers in narrow, raceme-like compound monochasia ·········· Ser. 5. **Uncata** spp.30−32
    3. Style narrowly subulate-linear, much longer than ovary, ventrally lacking a stigma ·········· Sect. 2. **Dolichostylus** sp.33
  2. Style straight.
    8. Style flattened, oblong, ventrally lacking a stigma ·········· Sect. 3. **Platystylus** sp.34
    8. Style not oblong, ventrally with a conspicuous stigma.
      9. Filament differentiated into upper and lower two parts different in shape and size, rarely narrow-linear (*T. megalostigma* and *T. tsawarungense*).
        10. Achenes not winged.
          11. Styles thinly terete.
            12. Achenes sessile, with conspicuous, thick longitudinal ribs ·········· Sect. **Erythrandra** Ser. 3.**Petaloidea** sp.37
            12. Achenes stipitate. with thin longitudinal ribs.
              13. Achenes shortly stipitate.
                14. Achenes depressed-globose.
                  15. Filaments narrow-linear; stigma narrow-linear ·········· Ser. 1. **Megalostigmata** sp.35
                  15. Filaments above narrow-oblanceolate, below filiform; stigma ellipsoid ·········· Ser. 2.**Baicalensia** sp.36
                14. Achenes conspicuously compressed ·········· Ser. 3. **Fargesiana** spp.40−44
              13. Achenes with longer, slender stipes.
                16. Leaflets not peltate; flowers in many-flowered thyrses; style 0.1−1 mm long ·········· Sect. 5. **Omalophysa** spp.45−47
                16. Leaflets at base peltate; flowers in few-flowered monochasia; style 3 mm long ·········· Sect. 6. **Physocarpum** Ser. **Pseudoichangensia** sp.56
          11. Style very short, sometimes nearly absent, ventrally with dense stigmatic tissues.

17. Achenes sessile ················ Sect. 4. **Erythrandra** ser. **Reticulata** spp.38–39
17. Achenes slenderly stipitate ········ Sect. 5. **Physocarpum** ser. **Clavata** spp.48–55
10. Achene with 4 longitudinal wings and a slender stipe ······ Sect. 7. **Tripterium** sp.57
9. Filaments linear, narrow-linear of filiform.
 18. Flowers acrofugally developed, in monochasia or thyrses.
  19. Flowers in narrow, raceme-like compound monochasia ········ Sect. 8. **Rutifolia**
   20. Leaflets abaxially hirtellous; filaments narrow-linear ························
    ·· Ser. 1. **Xiaojinensia** spp.58–59
   20. Leaflets glabrous; filaments filiform.
    21. Stigma not winged ··························· Ser. 2. **Rutifolia** sp.60
    21. Stigma winged.
     22. Flower with ca. 50 stamens, with anthers longer than filaments; style twice longer than ovary; stigma with 1 wing ····························
      ·· Ser. 3. **Sinomacrostigmata** sp.61
     22. Flower with 7–16 stamens, with anthers shorter than filaments; style slightly shorter than ovary; stigma with 2 lateral narrow wings ········
      ·· Ser. 4. **Leuconota** spp.62–63
  19. Flowers in corymbiform compound monochasia or thyrses.
   23. Achene flattened, on each side with 1 longitudinal nerve and along ventral and dorsal sutures narrowly winged ············ Sect. 10. **Dipterium** sp.79
   23. Achene on each side with (2–)3(–4) longitudinal ribs, mostly not winged.
    24. Flowers in few-flowered compound monochasia ·····················
     ·· Sect. 9. **Platystachys**
     25. Leaves compound ···················· Ser. 1. **Virgata** spp.64–70
     25. Leaves simple ························ Ser.2. **Indivisa** spp.71–72
    24. Flowers in many-flowered thyrses.
     26. Sepals larger, 5–16 mm long, purple or lilac, rarely white; stigma not winged ····· Sect. 9. **Platystachys** ser. **Violacea** spp.73–78
     26. Sepals small, 2–4.5 mm long, usually yellowish-green, rarely white or tinged with purple ················ Sect. 11. **Thalictrum**
      27. Stigma not winged ········ Ser. 1. **Osmundifolia** spp.80–85
      27. Stigma with 2 lateral wings ·· Ser. 2. **Thalictrum** spp.86–97
 18. Flowers acropetally developed ······················ Sect. 12. **Schlagintweitella**
  28. Basal leaves absent; flowers singly borne in axils of upper cauline leaves ·······
   ·· Ser. 1. **Squamifera** sp.98
  28. Leaves all basal; flowers in racemes ························ Ser. 2. **Alpina** sp.99
1. Flowers unisexual, rarely bisexual; stigmas exserted, flagelliform ···· Subgen. 2. **Lecoyerium** spp.100–101

分种、变种、变型检索表

1. 花两性；柱头内藏，不呈鞭形。
   2. 心皮花柱顶端向后弯曲，常呈钩状或拳卷状。
      3. 柱头小，不明显。
         4. 复单歧聚伞花序伞房状。
            5. 雄蕊花丝条形、狭条形或丝形。
               6. 花丝条形或狭条形。
                  7. 花具 34—47 枚心皮；子房与花柱等长；花丝条形 ·· **1. 叉枝唐松草 T. saniculiforme**
                  7. 花具 6-14 枚心皮；子房比花柱长。
                     8. 花丝条形 ·· **2. 小花唐松草 T. minutiflorum**
                     8. 花丝狭条形，下部稍变狭。
                        9. 小叶宽卵形或倒卵形，长达 3 cm，下面被短柔毛，边缘有小齿 ············
                        ··· **3. 绢毛唐松草 T. brevisericeum**
                           10. 花药长圆形，宽 0.4–0.45 mm。
                              11. 花序分枝和花梗的毛长 0.25-0.5 mm；心皮 6-16，无柄，花柱顶端拳卷状弯曲 ····················· **3a. 模式变种 var. brevisericeum**
                              11. 花序分枝和花梗的毛较短，长 0.05-0.2 mm；心皮 5，有短柄，花柱顶端弧状弯曲 ····················· **3b. 五果绢毛唐松草 var. pentagynum**
                           10. 花药狭长圆形，宽 0.2 mm ········ **3c. 狭药绢毛唐松草 var. angustiantherum**
                        9. 小叶卵形或宽卵形，长达 8.5-10 cm，边缘具尖牙齿。
                           12. 小叶无毛 ·· **17. 大叶唐松草 T. faberi**
                           12. 小叶下面被短柔毛 ·· **18. 粗壮唐松草 T. robustum**
               6. 花丝丝形。
                  13. 植株全部无毛；茎高 7-15 cm；花 2-3 朵组成单歧聚伞花序；花具 9-15 枚雄蕊 ····
                  ··· **28. 思茅唐松草 T. simaoense**
                  13. 茎高 80-85 cm，与叶柄均疏被短柔毛；花多数组成聚伞圆锥花序；花具 40-50 枚雄蕊 ··· **29. 短蕊唐松草 T. brachyandrum**
            5. 花丝分化成在形状和大小上不同的上、下两部分。
               14. 瘦果无柄。
                  15. 植株全部无毛。
                     16. 小叶宽卵形、圆形或楔形，顶端圆形或钝。
                        17. 小叶小，长 3-8 mm ······························· **8. 密叶唐松草 T. myriophyllum**
                        17. 小叶长在 8 mm 以上。
                           18. 小叶长 3-7 cm；聚伞圆锥花序有多数花 ···· **15. 峨眉唐松草 T. omeiense**
                           18. 小叶长 3.4 cm 以下。
                              19. 花柱顶端稍向后弯曲，不呈钩状或拳卷状。
                                 20. 复单歧聚伞花序有 6 花；瘦果新月形，每侧有 2 条纵肋 ··········
                                 ··· **6. 大关唐松草 T. daguanense**

20. 复单歧聚伞花序有多数花；瘦果纺锤形，每侧有 3 条纵肋 ················
·· 5. **鹤庆唐松草 T. leve**
19. 花柱近顶端钩状或拳卷状弯曲。
21. 花丝上部棒状，比花药窄 ································ 7. **新宁唐松草 T. xinningense**
21. 花丝上部倒披针形，比花药宽。
22. 萼片顶端有少数小齿；花柱比子房长约 2 倍 ··· 14. **玉山唐松草 T. sessile**
22. 萼片全缘；花柱比子房短或稍长。
23. 复单歧聚伞花序具多数分枝和多数花；瘦果狭椭圆形 ···············
·· 12. **爪哇唐松草 T. javanicum**
23. 复单歧聚伞花序具少数分枝和少数花。
24. 纤维根顶端变粗；萼片长 3–3.5 mm；花具 2–6 枚心皮；瘦果狭长圆形 ································ 7. **华东唐松草 T. fortunei**
25. 茎无珠芽 ·· 7a. 模式变种 var. **fortunei**
25. 茎具珠芽 ································· 7b. **珠芽华东唐松草** var. **bulbiliferum**
24. 纤维根顶端不变粗；萼片长约 2 mm；花具 8–16 枚心皮；瘦果狭披针形 ·· 8. **多枝唐松草 T. ramosum**
16. 小叶卵形或宽卵形，长达 10–13 cm，顶端渐狭或急尖，边缘有尖牙齿。
26. 小叶不分裂或不明显 3 浅裂，边缘有 2–5 牙齿；花具 10–13 枚心皮 ················
·· 16. **巨齿唐松草 T. grandidentatum**
26. 小叶明显 3 浅裂，边缘有 5–10 牙齿；花具 3–6 枚心皮 ······ 17. **大叶唐松草 T. faberi**
15. 小叶和花梗被短柔毛。
27. 小叶近圆形，基部心形；花柱只顶端稍内弯 ·················· 4. **糙叶唐松草 T. scabrifolium**
27. 小叶圆卵形，卵形或倒卵形，基部不呈心形；花柱近顶端钩状弯曲。
28. 小叶圆卵形或倒卵形，长达 3 cm；顶端不渐狭，边缘有小齿。
29. 小叶和叶鞘的毛长 0.2–0.3 mm；宿存花柱柔弱。
30. 小叶下面被微硬毛；花梗被腺毛 ············· 11. **云南唐松草 T. yunnanense**
31. 花药长圆形；瘦果椭圆形 ············· 11a. 模式变种 var. **yunnanense**
31. 花药狭长圆形；瘦果纺锤形 ············ 11b. **滇南唐松草** var. **austroyunnanense**
30. 小叶下面被短柔毛；花梗无毛 ····· 13b. **柔柱微毛唐松草 T. lecoyeri** var. **debilistylum**
29. 小叶和叶鞘的毛长 0.1 mm；宿存花柱坚硬 ······ 13a. **微毛唐松草 T. lecoyeri** var. **lecoyeri**
28. 小叶卵形，长达 8.5 cm，顶端渐狭，边缘有粗齿 ·················· 18. **粗壮唐松草 T. robustum**
14. 瘦果具柄。
32. 子房比花柱短。
33. 茎无毛；花柱比子房长 2 倍 ································ 19. **小喙唐松草 T. rostellatum**
33. 茎被短无毛。
34. 花柱比子房稍长；瘦果新月形 ································ 21. **察隅唐松草 T. chayuense**
34. 花柱比子房长 2 倍；瘦果纺锤形 ································ 22. **纺锤唐松草 T. fusiforme**
32. 子房比花柱长。

35. 茎、叶和花梗被短柔毛；复单歧聚伞花序有多数密集的花；花梗长 1.5-3.5 mm ………… ……………………………………………………………………… 24. **弯柱唐松草 T. uncinullatum**
35. 植株全部无毛。
　　36. 花组成具 2 花的简单单歧聚伞花序；花具 4-6 枚萼片（长 4.5-4.8 mm）和 50-60 枚心皮 ……………………………………………………………… 26. **南湖唐松草 T. rubescens**
　　36. 较多花组成复单歧聚伞花序；花具 30 枚以下的心皮。
　　　　37. 花柱顶端稍向后弯曲或直。
　　　　　　38. 茎高约 18 cm；小叶长达 1.4 cm；花具 5 枚萼片 ……………………………… ……………………………………………… 20. **菲律宾唐松草 T. philippinense**
　　　　　　38. 茎高达 70 cm；小叶长达 2.2 cm；花具 4 枚萼片 ……………………………… ……………………………………………………… 23. **澜沧唐松草 T. lancangense**
　　　　37. 花柱顶端钩状弯曲。
　　　　　　39. 花具 12-15 枚心皮；心皮柄长约 0.4 mm；瘦果的柄长 0.6-1 mm；花有 4 枚萼片 …………………………………………………… 25. **长喙唐松草 T. macrorhynchum**
　　　　　　39. 花具 15-30 枚心皮，心皮柄长 0.6-0.8 mm；瘦果的柄长 1.5-2 mm ……………… ……………………………………………………… 27. **台湾唐松草 T. urbainii**
　　　　　　　　40. 花具 4 枚萼片（长 3-3.5 mm）……………… 27a. **模式变种 var. urbainii**
　　　　　　　　40. 花具 6 枚萼片（长 8-15 mm）…………… 27b. **大花台湾唐松草 var. majus**
4. 复单歧聚伞花序狭，似总状花序。
　　41. 花具 6-18 枚雄蕊；花丝狭条形。
　　　　42. 萼片具 3 条基生脉；花有 5-11 枚心皮；瘦果新月形 … 30. **钩柱唐松草 T. uncatum**
　　　　　　43. 柱头无翅 ……………………………………………… 30a. **模式变种 var. uncatum**
　　　　　　43. 柱头有 2 侧生狭翅 ……………………… 30b. **狭翅钩柱唐松草 var. angustialatum**
　　　　42. 萼片具 1 条基生脉；花有 3-5 枚心皮；瘦果宽椭圆体形 … 31. **狭序唐松草 T. atriplex**
　　41. 花具 4-5 枚雄蕊；花丝丝形 ……………………………… 32. **螺柱唐松草 T. spiristylum**
3. 花柱狭钻状条形，比子房长许多，腹面无柱头 …………… 33. **钻状唐松草 T. tenuisubulatum**
2. 花柱直（在细柱唐松草和兴山唐松草，花柱稍弧状弯曲）。
　　44. 花柱扁平，长圆形，腹面无柱头 ……………………………… 34. **宽柱唐松草 T. latistylum**
　　44. 花柱通常细圆柱形，稀短，呈三角形或卵形，在腹面具一明显柱头或密集柱头组织。
　　　　45. 花丝分化成在形状和大小上不同的上、下两部分（在细柱唐松草和察瓦龙唐松草，花丝为狭条形）。
　　　　　　46. 瘦果无翅。
　　　　　　　　47. 花柱细圆柱形，明显。
　　　　　　　　　　48. 瘦果无柄，每侧具明显隆起的纵肋 …… 37. **瓣蕊唐松草 T. petaloideum**
　　　　　　　　　　　　49. 小叶倒卵形，菱形或近圆形 ……… 37a. **模式变种 var. petaloideum**
　　　　　　　　　　　　49. 小叶狭椭圆形或狭披针形 …………………………………………… ……………………………… 37b. **狭裂瓣蕊唐松草 var. supradecompositum**
　　　　　　　　　　48. 瘦果具柄，每侧具细肋。
　　　　　　　　　　　　50. 瘦果具短柄。
　　　　　　　　　　　　　　51. 瘦果扁球形。

52. 花丝狭条形，下部稍变狭；花柱细圆柱形，顶端稍向后反曲，在腹面具一狭条形柱头 ………………………………… 35. **细柱唐松草 T. megalostigma**

52. 花丝上部倒披针形，下部丝形；花柱短，直，在腹面顶部具一椭圆形小柱头 ……………………………………… 36. **贝加尔唐松草 T. baicalense**

    53. 瘦果有细肋 ……………………………………… 36a. **模式变型 f. baicalense**

    53. 瘦果光滑，肋不明显 ……………………… 36b. **光果唐松草 f. levicarpum**

51. 瘦果多少扁压，不为扁球形。

    54. 植株全部无毛。

        55. 小叶长 1–3 cm；花有 2–5 心皮 ……… 40. **西南唐松草 T. fargesii**

        55. 小叶长 2–6 cm；花有 5–13 心皮 ‥ 41. **兴山唐松草 T. xingshanense**

    54. 茎和小叶、花梗被短腺毛或无毛；子房被短腺毛或无毛。

        56. 茎和小叶、花梗被短腺毛；花具 4 枚萼片；子房无毛。

            57. 萼片具 1 不分枝的中脉和 2 条 1 或 2 回二歧状分枝的侧生脉；花具 20–25 枚雄蕊，5–6 枚心皮；花丝上部狭倒披针形，下部丝形；瘦果扁平，新月形 ……… 42. **丽江唐松草 T. wangii**

            57. 萼片具 3 条不分枝的基生脉；花具 15 枚雄蕊，2 枚心皮；花丝狭条形；瘦果稍扁压，不扁平 ‥ 43. **察瓦龙唐松草 T. tsawarungense**

        56. 茎和小叶、花梗无毛；花具 5 枚萼片；子房被短腺毛 …………………………………………………… 44. **矮唐松草 T. pumllum**

    50. 瘦果具较长细柄。

        58. 小叶基部盾形；花组成少花伞房状复单歧聚伞花序 …………………………………………………… 56. **拟盾叶唐松草 T. pseudoichangense**

        58. 小叶不为盾形；花组成多花聚伞圆锥花序。

            59. 茎和小叶疏被短柔毛；萼片上部被缘毛，具 3 条基生脉；子房被短柔毛 …………………… 45. **毛蕊唐松草 T. lasiogynum**

            59. 茎、萼片和心皮无毛；萼片具 5 条基生脉。

                60. 小叶下面被短柔毛；果梗直；花具 35–48 枚雄蕊 ……………………………………………………… 46. **长柄唐松草 T. przewalskii**

                60. 植株全部无毛；果梗顶端反曲；花具 10–30 枚雄蕊 ……………………………………………… 47. **散花唐松草 T. sparsiflorum**

47. 花柱很短，不呈细圆柱形，不明显。

    61. 瘦果无柄。

        62. 花组成多花伞房状复单歧聚伞花序 ……………………… 38. **网脉唐松草 T. reticulatum**

            63. 植株全部无毛 ……………………………………… 38a. **模式变种 var. reticulatum**

            63. 小叶下面疏被微硬毛 ……………………… 38b. **毛叶网脉唐松草 var. hirtellum**

        62. 花组成简单单歧聚伞花序或具 4 花的蝎尾状聚伞花序 … 39. **武夷唐松草 T. wuyishornicum**

61. 瘦果具较长细柄。

    64. 小叶基部盾形 ……………………………………… 52. **盾叶唐松草 T. ichangense**

    64. 小叶不为盾形。

65. 茎生叶 2, 为单叶, 对生 ·· 54. **花唐松草 T. filamentosum**
65. 茎生叶为 1-2 回三出复叶, 互生, 稀对生。
    66. 小叶下面被短柔毛 ·· 51. **尖叶唐松草 T. acutifolium**
    66. 植株全体无毛。
        67. 花组成圆锥状聚伞圆锥花序。
            68. 萼片基部楔形, 具 3 条基生脉; 花具 20-24 枚雄蕊; 花柱长 0.2-0.3 mm, 在顶端具一椭圆形小柱头 ·· 50. **岳西唐松草 T. yuoxiense**
            68. 萼片基部近圆形, 具 2 条基生脉; 花具 3-7 枚雄蕊; 花柱钻形, 长 0.8 mm, 腹面具 1 狭条形柱头 ·· 49. **稀蕊唐松草 T. oligandrum**
        67. 花组成伞房状或伞状复单歧聚伞花序。
            69. 复单歧聚伞花序具疏散分枝, 伞房状; 萼片具 3 条基生脉; 瘦果长 3 mm 以上。
                70. 小叶近圆形或倒卵形 ·················· 48. **阴地唐松草 T. umbricola**
                70. 小叶卵形或菱状卵形。
                    71. 块根不存在; 茎生叶互生; 瘦果狭长圆形或近新月形 ·· 51. **尖叶唐松草 T. acutifolium**
                    71. 块根存在; 茎生叶通常对生; 瘦果狭椭圆形 ·· 53. **深山唐松草 T. tuberiferum**
            69. 复单歧聚伞花序具密集分枝, 似伞形花序; 萼片具 1 条基生脉; 瘦果长达 1.8 mm ·· 55. **小果唐松草 T. microgynum**
46. 瘦果具 4 条纵翅, 基部具反曲的细长柄; 花柱很短 ··· 57. **唐松草 T. aquilegifolium** var. **sibiricum**
45. 花丝条形或狭条形 (下部常稍变狭) 或丝形。
    72. 花离顶发育, 组成单歧聚伞花序或伞房圆锥花序。
        73. 花组成狭窄似总状花序的复单歧聚伞花序。
            74. 小叶下面被微硬毛; 花丝狭条形; 柱头无翅。
                75. 萼片 3 条基生脉; 雄蕊全部能育; 子房无毛 ········ 58. **小金唐松草 T. xiaojinense**
                75. 萼片具 6 条基生脉; 部分雄蕊的花药不育; 子房被短柔毛 ·· 59. **六脉萼唐松草 T. sexnervisepalum**
            74. 植株全部无毛; 花丝丝形。
                76. 柱头无翅, 花柱直或顶端稍向后弯曲 ·············· 60. **芸香叶唐松草 T. rutifolium**
                76. 柱头有翅, 花柱直。
                    77. 花有约 50 枚雄蕊, 花药比花丝长; 花柱伸长, 比子房长 2 倍; 柱头具 1 条狭翅 ·· 61. **长柱唐松草 T. sinomacrostigma**
                    77. 花有 7-16 枚雄蕊, 花药比花丝短; 花柱比子房稍长; 柱头具 2 条侧生狭翅。
                        78. 小托叶卵形, 顶端微尖; 复单歧聚伞序顶生, 无侧生复单歧聚伞花序; 花药狭长圆形, 顶端无短尖头 ···················· 62. **吉隆唐松草 T. jilongense**
                        78. 小托叶圆卵形, 顶端圆形; 复单歧聚伞花序顶生并侧生; 花药长圆形, 顶端具短尖头 ·· 63. **白茎唐松草 T. leuconotum**
    73. 花组成伞房状复单歧聚伞花序或圆锥状聚伞圆锥花序。
        79. 瘦果扁平, 近圆形, 每侧具 1 条纵脉, 沿腹、背缝线具狭翅 ·· 77. **二翅唐松草 T. elegans**

六、分类学处理　　　　　　　　　　　　　　　　　　　　67

79. 瘦果每侧具 2–3 条纵肋，无翅。
  80. 花通常组成少花伞房状复单歧聚伞花序。
    81. 叶为单叶，圆肾形；花丝条形 ·················· **71. 圆叶唐松草 T. rotundifolium**
    81. 叶为复叶。
      82. 植株全部无毛。
        83. 基生叶不存在；萼片有 3 条基生脉；花药比花丝短或与花丝等长。
          84. 茎不分枝或在中部以上分枝；茎生叶为 1 回三出复叶；花丝狭条形；心皮有柄 ·················· **64. 帚枝唐松草 T. virgatum**
          84. 茎自基部有 2–4 条长分枝；茎生叶为 2–3 回三出复叶，最上部茎生叶为 5 小叶的羽状复叶；花丝丝形；心皮无柄 ·· **65. 希陶唐松草 T. tsaii**
        83. 基生叶存在；萼片有 5–6 条基生脉；花药比花丝长；叶为 2–3 回三出复叶。
          85. 小叶狭条形或丝形；萼片白色，长 6.5–11 mm；花丝条形 ·················· **66. 丝叶唐松草 T. foeniculaceum**
          85. 小叶卵形；萼片粉红色，长 13–20 mm；花丝狭条形 ·················· **67. 大花唐松草 T. grandiflorum**
      82. 茎、叶和花梗被短腺毛。
        86. 柱头无翅；叶为 3 回三出复叶或 3 回羽状复叶。
          87. 萼片具 3 条不分枝的基生脉；花药狭长圆形，比花丝稍长；心皮无毛 ·················· **43. 察瓦龙唐松草 T. tsawarungense**
          87. 萼片具 1 条不分枝的中脉和 2 条 2 回二歧状分枝的基生脉；花药长圆形，比花丝短；子房被腺毛 ·················· **69. 粘唐松草 T. viscosum**
        86. 柱头具 2 侧生狭翅。
          88. 叶为 1 回羽状复叶，具 5 小叶，或为三出复叶，具 3 小叶；子房密被短腺毛 ·················· **68. 攀枝花唐松草 T. panzhihuaense**
          88. 叶为 3 回羽状复叶；子房被短柔毛 ·················· **70. 金丝马尾连 T. glandulosissimum**
            89. 小叶长和宽 0.7–1.6 cm；叶轴的毛长 0.2 mm；柱头狭三角形 ·················· **70a. 模式变种 var. glandulosissimum**
            89. 小叶长和宽 1.5–2.5 cm；叶轴的毛长 0.4 mm；柱头三角形 ·················· **70b. 昭通唐松草 var. chaotungense**
  80. 花组成多花的聚伞圆锥花序。
    90. 花较大，萼片长 5.5–13 mm；柱头无翅。
      91. 萼片白色，长 10–12.8 mm，每一萼片具 3 条二歧分枝的基生脉；花药条形；瘦果无翅 ·················· **73. 错那唐松草 T. cuonaense**
      91. 萼片紫色或堇色。
        92. 茎、叶和子房被短柔毛；花药条形；瘦果无翅。
          93. 小叶 1.2–2.5×1.5–3.2 cm；萼片具 3 条基生脉；花丝狭条形 ·················· **76. 美丽唐松草 T. reniforme**

93. 小叶较小，0.4–1.2×0.4–1 cm；萼片具 7 条基生脉；花丝丝形 ·························
·· 77. 堇花唐松草 T. diffusiflorum
92. 植株全部无毛。
    94. 茎上部叶腋具珠芽；萼片具 5 条基出脉；花丝狭条形；瘦果无翅 ··················
·· 74. 珠芽唐松草 T. chelidonii
    94. 茎无珠芽。
        95. 花有 10–15 枚心皮；子房和新月形瘦果无翅 ···· 75. 美花唐松草 T. callianthum
        95. 花有 15–22 枚心皮；子房和倒卵形瘦果沿腹、背缝线有狭翅 ···················
·· 78. 偏翅唐松草 T. delavayi
            96. 萼片有 5 条基生脉；花药短尖头长 0.1–0.2 mm。
                97. 萼片顶端稍钝；花药长圆形。
                    98. 花丝丝形 ························· 78a. 模式变种 var. delavayi
                    98. 花丝狭条形 ··················· 78c. 宽萼偏翅唐松草 var. decorum
                97. 萼片顶端渐尖；花药狭长圆形；花丝丝形 ·····························
·· 78b. 渐尖偏翅唐松草 var. acuminatum
            96. 萼片有 3 条基生脉；花药宽长圆形，具长 1–1.8 mm 的短尖头；花丝丝形 ···
·· 78d. 角药偏翅唐松草 var. mucronatum
90. 花较小，萼片长在 5 mm 以下，每一萼片有 3 条基生脉。
    99. 柱头无翅。
        100. 瘦果无翅。
            101. 花丝狭条形；花药狭长圆形；柱头明显。
                102. 植株全部无毛。
                    103. 小叶长 0.8–2.1 cm；花梗丝形，长 14–35 mm；花约有 17 枚雄蕊 ···········
·· 80. 毛发唐松草 T. trichopus
                    103. 小叶较大，长 4.2–6.5 cm；花梗较短，长 4–6 mm；花具 30–35 枚雄蕊 ·········
·· 81. 河南唐松草 T. honanense
                102. 植株只小叶柄疏被短柔毛，其他部分无毛；小叶长 1.5–3 cm；花梗长 8–25 mm ···
·· 82. 川鄂唐松草 T. osmundifolium
            101. 花丝丝形。
                104. 小叶宽 1.2–5.4 cm，基部心形，下面被微硬毛；花药狭长圆形；柱头明显，长
1.8 mm，花柱长 2 mm ··············· 72. 心基唐松草 T. punduanum var. hirtellum
                104. 小叶较小，宽 2 cm 以下，基部不呈心形。
                    105. 植株全部无毛；小叶较小，宽 0.3–1 cm；花药狭长圆形；花柱长 2 mm，柱头
明显，长 1.6 mm ·························· 83. 亚东唐松草 T. yadongense
                    105. 茎、小叶和花梗均被短柔毛；花药条形；花柱长 0.4 cm，柱头不明显 ·········
·· 84. 陕西唐松草 T. shensiense
        100. 瘦果沿背缝线有狭翅；花丝丝形；柱头狭条形 ·············· 85. 滇川唐松草 T. finetii
    99. 柱头全部与花柱贴生，具 2 侧生翅。

106. 柱头狭长圆形或狭三角形，具 2 侧生狭翅。
    107. 小叶下面在隆起的叶脉上密被分枝毛 ·················· 86. **星毛唐松草 T. cirrhosum**
    107. 小叶无分枝毛。
        108. 花药淡黄色，长圆形，密被长约 0.05 mm 的毛 ··· 88. **藏南唐松草 T. austrotibeticum**
        108. 花药无毛。
            109. 柱头狭长圆形；植株全部无毛。
                110. 小叶顶端急尖，边缘具尖齿；聚伞圆锥花序具稀疏少数花；萼片顶端渐狭，具不分枝的中脉和 2 条二歧状分枝的侧生脉；花药狭长圆形 ·· 87. **札达唐松草 T. zhadaense**
                110. 小叶顶端圆，边缘具钝齿；聚伞圆锥花序具多数密集的花；萼片长圆形，顶端圆钝，具 3 条不分枝的基生脉 ·················· 89. **多叶唐松草 T. foliolosum**
            109. 柱头狭三角形。
                111. 小叶下面无毛；瘦果无毛 ·················· 90. **高原唐松草 T. cultratum**
                111. 小叶下面有毛；瘦果背面有短毛 ··· 91a. **腺毛唐松草 T. foetidum var. foetidum**
106. 柱头宽三角形或箭头状正三角形，具 2 侧生宽翅。
    112. 茎生叶向上开展；聚伞圆锥花序狭圆锥状，其分枝与花序轴以 45° 角向上方开展 ·· 95. **箭头唐松草 T. simplex**
        113. 植株全部无毛。
            114. 果梗比瘦果长 2–3 倍。
                115. 小叶倒卵形，顶端以及裂片顶端急尖或钝，基部圆形或宽楔形 ·· 95a. **模式变种 var. simplex**
                115. 小叶楔形，顶端以及裂片顶端锐尖，基部狭楔形 ·· 95b. **锐裂箭头唐松草 var. affine**
            114. 果梗约与瘦果等长；小叶与锐裂箭头唐松草的小叶相似 ·· 95c. **短梗箭头唐松草 var. brevipes**
        113. 茎、叶和花梗被短腺毛 ·················· 95d. **腺毛箭头唐松草 var. glandulosum**
    112. 茎生叶和花序分枝与茎和花序轴均成钝角向上开展；聚伞圆锥花序圆锥状。
        116. 小叶椭圆形，不分裂，全缘；花药条形 ·················· 93. **细唐松草 T. tenue**
        116. 小叶通常 3 浅裂，边缘有齿。
            117. 小叶小，约 4×4 mm ·················· 94. **紫堇叶唐松草 T. isopyroides**
            117. 小叶较大。
                118. 聚伞圆锥花序伞房状，多回二歧状分枝；瘦果长 4–5.2 mm；花梗长 1.5–3 cm；花药短尖头长 0.3–0.5 mm ·········· 97. **展枝唐松草 T. squarrosum**
                118. 聚伞圆锥花序圆锥状，较少二歧状分枝；瘦果长 3–4 mm。
                    119. 小叶草质，脉平，脉网不明显。
                        120. 小叶长 0.6–1(–1.5) cm；瘦果扁平 ······································ 91b. **扁果唐松草 T. foetidum var. glabrescens**

120. 小叶较大，顶生小叶长 4–7 cm；瘦果稍扁压，椭圆体形 ………………………
…………………………………………………………… 96. 黄唐松草 T. flavum
119. 小叶较厚，纸质或近革质，脉在下面隆起，形成明显的脉网 ……………………
……………………………………………………………… 92. 亚欧唐松草 T. minus
121. 花梗长 3–8 mm；花药长圆形或狭长圆形，顶端短尖头长 0.1 mm。
122. 小叶宽 0.4–1.5 cm，下面淡绿色 …………… 92a. 模式变种 var. minus
122. 小叶较大，宽 1.5–4（–5）cm。
123. 小叶边缘有少数尖齿，下面具白霜，脉网明显 …………………
……………………………………… 92b. 东亚唐松草 var. hypoleucum
123. 小叶边缘有少数圆齿，下面淡绿色，脉网不明显 …………………
……………………………………… 92c. 圆齿唐松草 var. rotundidens
121. 花梗长 15–30 mm；花药条形，顶端短尖头长 0.4–0.45 mm；小叶下面淡绿色 …
……………………………………… 92d. 长梗亚欧唐松草 var. stipellatum
72. 花向顶发育；萼片小，淡黄绿色，每一萼片具 3 条基生脉；花丝丝形，花药狭长圆形，稀条形，顶端具短尖头；柱头全部与花柱贴生，箭头状正三角形，具 2 侧生宽翅。
124. 基生叶不存在；下部茎生叶退化成小鳞片，上部茎生叶正常发育，为 3–4 回羽状复叶；花单生叶腋，沿茎向顶发育 ………………………… 98. 石砾唐松草 T. squamiferum
124. 叶全部基生；花组成顶生总状花序 ………………… 99. 高山唐松草 T. alpinum
125. 花葶不分枝，高 6–20 cm；花梗向下弧状弯曲，长 1–10 mm；植株全部无毛。
126. 瘦果无柄 ……………………………………… 99a. 模式变种 var. alpinum
126. 瘦果具柄 ………………………………… 99b. 柄果高山唐松草 var. microphyllum
125. 花葶不分枝或具 1 分枝，高 7–15（–25–38）cm；花梗近直立，直，长 1.8–9 mm；瘦果无柄 ……………………………………………… 99c. 直梗高山唐松草 var. elatum
127. 植株全部无毛 …………………………………… 99ci. 模式变型 f. elatum
127. 小叶下面有毛。
128. 小叶下面被短柔毛 …………………… 99cii. 毛叶高山唐松草 f. puberulum
128. 小叶下面被短柔毛，脉上被微硬毛 …… 99ciii. 毛脉高山唐松草 f. setulosinerve
1. 花两性或单性；柱头伸出，细长呈鞭形。
129. 小叶长达 15 mm，下面脉上被微硬毛；雄蕊花丝长 1.5–5.5 mm …… 100. 鞭柱唐松草 T. smithii
129. 小叶较小，长 3–9 mm，下面被短腺毛；花丝长 8 mm ………… 101. 定结唐松草 T. dingjieense

在中国唐松草属中，五种即圆叶唐松草 T. rotundifolium（叶为圆肾形单叶）、丝叶唐松草 T. foeniculaceurn（小叶狭条形或丝形）、盾叶唐松草 T. ichangense（小叶盾形、花柱极短）、星毛唐松草 T. cirrhosum（小叶背面脉上密被分枝毛）和藏南唐松草 T. austrotibeticum（花药密被小毛）都拥有独特的形态学特征，借助这些独特特征，可即刻使这些种得到正确鉴定。

Key to species, varieties and forms

1. Flowers bisexual; stigma included, not flagelliform.
   2. Carpel style at apex recurved, often hooked or circinate.
      3. Stigma small, inconspicuous.
         4. Compound monochasia corymbiform.
            5. Stamen filaments linear, narrow-linear, or filiform.
               6. Filaments linear or narrow-linear.
                  7. Flower with 34–47 carpels; ovary as long as style; filaments linear ⋯ 1. **T. saniculiforme**
                  7. Flower with 6–14 carpels; ovary longer than style.
                     8. Filaments linear ··· 2. **T. minutiflorum**
                     8. Filaments narrow-linear, below slightly narrowed.
                        9. Leaflets broad-ovate or obovate, up to 3 cm long, abaxially puberulous, margin few-denticulate ··· 3. **T. brevisericeum**
                           10. Anthers oblong, 0.4–0.45 mm broad.
                              11. Monochasium branches and pedicels with hairs 0.25–0.5 mm long; carpels 6–16, sessile; styles at apex circinate ················ 3a. var. **brevisericeum**
                              11. Monochasium branches and pedicels with hairs 0.05–0.2 mm long; carpels 5, shortly stipitate, styles at apex arcuate ················ 3b. var. **pentagynum**
                           10. Anthers narrow-oblong, 0.2 mm broad ················ 3c. var. **angustiantherum**
                        9. Leaflets ovate or broad-ovate, up to 8.5–10 cm long, margin acutely dentate.
                           12. Leaflets glabrous ···17. **T. faberi**
                           12. Leaflets abaxially puberulous ······································ 18. **T. robustum**
               6. Filaments filiform.
                  13. Plants totally glabrous; stems 7–15 cm tall; flowers 2–3 in monochasium; flower with 9–15 stamens ··· 28. **T. simaoense**
                  13. Stems 80–85 cm tall, with petioles puberulous; flowers many in thyrse; flower with 40–50 stamens ·· 29. **T. brachyandrum**
            5. Filament differentiated into upper and lower two parts different in shape and shape.
               14. Achenes sessile.
                  15. Plants totally glabrous.
                     16. Leaflets broad-ovate, orbicular or cuneate, apex rounded or obtuse.
                        17. Leaflets small, 3–8 mm long ·································· 8. **T. myriophyllum**
                        17. Leaflets larger, more than 8 mm long.
                           18. Leaflets 3–7 cm long; thyrse many-flowered ················ 15. **T. omeiense**
                           18. Leaflets less than 3.4 cm
                              19. Style apex slightly recurved, not hooked or circinate.
                                 20. Compound monochasia 6-flowered; achenes lunate, on each side longitudinally 2-ribbed ································ 6. **T. daguanense**

    20. Compound monochasia many-flowered; achenes fusiform, on each side longitudinally 3-ribbed ·· 5. **T. leve**
  19. Style near apex hooked or circinate.
    21. Filaments above clavate, narrower than anthers ················ 7. **T. xinningense**
    21. Filaments above oblanceolate, broader than anthers.
      22. Sepals at apex few-denticulate; style about twice longer than ovary ············ ··· 14. **T. sessile**
      22. Sepals entire; style shorter than ovary or slightly longer than ovary.
        23. Compound monochasia many-branched and many-flowered; achenes narrow-elliptic ··· 12. **T. javanicum**
        23. Compound monochasia few-branched and few-flowered.
          24. Fibrous roots at apex thickened; sepals 3–3.5 mm long; flower with 2–6 carpels; achenes narrow-oblong ····················· 7. **T. fortunei**
            25. Stem without bulbils ····························· 7a. var. **fortunei**
            25. Stem with bulbils ························· 7b. var. **bulbiliferum**
          24. Fibrous roots not thickened; sepals ca. 2 mm long; flower with 8–16 carpels; achenes narrow-lanceolate ····················· 8. **T. ramosum**
16. Leaflets ovate or broad-ovate, up to 10–13 cm long, apex attenuate or acute, margin acutely dentate.
  26. Leaflets undivided or indistinctly 3-lobed, margin 2–5-dentate; flower with 10–13carpels ······· ·· 16. **T. grandidentatum**
  26. Leaflets distinctly 3-lobed, margin 5–10-dentate; flower with 3–6carpels ·········· 17. **T. faberi**
15. Leaflets and pedicels puberulous.
  27. Leaflets suborbicular, base cordate; style at apex slightly recurved, not hooked ················ ··· 4. **T. scabrifolium**
  27. Leaflets rotund-ovate, ovate or obovate, base not cordate; style near apex hooked.
    28. Leaflets rotund-ovate or obovate, up to 3 cm long, apex not attenuate, margin denticulate.
      29. Leaflets and petiole sheaths with hairs 0.2–0.3 mm long; persistent styles delicate.
        30. Leaflets abaxially hirtellous; pedicels with glandular hairs ············· 11. **T. yunnanense**
          31. Anthers oblong; achenes elliptic ······································ 11a. var. **yunnanense**
          31. Anthers narrow-oblong; achenes fusiform ················· 11b. var. **austroyunnanense**
        30. Leatlets abaxially puberulous; pedicels glabrous ······· 13b. **T. lecoyeri** var. **debilistylum**
      29. Leaflets and petiole sheaths with hairs ca. 0.1 mm long; persistent styles rigid ················ ··· 13a. **T. lecoyeri** var. **lecoyeri**
    28. Leaflets ovate, up to 8.5 cm long, margin coarsely dentate ·················· 18. **T. robustum**
14. Achenes stipitate.
  32. Ovary shorter than style.
    33. Stems glabrous; style twice longer than ovary ·································· 19. **T. rostellatum**

33. Stems puberulous.
    34. Style slightly longer than ovary; achenes lunate ·················· 21. **T. chayuense**
    34. Style twice longer than ovary; achenes fusiform ················· 22. **T. fusiforme**
32. Ovary longer than style.
    35. Stems, leaves and pedicels puberulous; compound monochasia densely many-flowered; pedicels 1.5–3.5 mm long ················· 24. **T. uncinullatum**
    35. Plants totally glabrous.
        36. Flowers in simple 2-flowered monochasium; flower with 4–6 sepals 4.5–4.8 mm long and 50–60 carpels ················· 26. **T. rubescens**
        36. Flowers in compound monochasium; flower with carpels less than 30.
            37. Style at apex slightly recurved or straight.
                38. Stem. ca. 18 cm tall; leaflets up to 1.4 cm long; flower with 5 sepals ················· 30. **T. philippinense**
                38. Stem up to 70 cm tall; leaflets up to 2.2 cm long; flower with 4 sepals ················· 32. **T. lancangense**
            37. Style near apex hooked.
                39. Flower with 12–15 carpels; carpel stipe ca. 0.4 mm long; achene stipe 0.6–1 mm long; flower with 4 sepals ················· 25. **T. macrorhynchum**
                39. Flower with 15–30 carpels; carpel stipe 0.6–0.8 mm long; achene stipe 1.5–2 mm long ················· 27. **T. urbainii**
                    40. Flowers with 4 sepals 3–3.5 mm long ················· 27a. var. **urbainii**
                    40. Flower withs 6 sepals 8–15 mm long ················· 27b. var. **majus**
4. Compound monochasia narrow, raceme-like.
   41. Flower with 6–18 stamens; filaments narrow-linear.
      42. Sepal with 3 basal nerves; flower with 5–11 carpels; achenes lunate ·············· 30. **T. uncatum**
         43. Stiama not winged ················· 30a. var. **uncatum**
         43. Stigma with 2 narrow lateral wings ················· 30b. var. **angustialatum**
      42. Sepal with 1 basal nerve; flower with 3–5 carpels; achenes broadly ellipsoidal ················· 31. **T. atriplex**
   41. Flower with 4–5 stamens; filaments filiform ················· 32. **T. spiristylum**
3. Style narrowly subulate-linear, much longer than ovary, ventrally without a stigma ················· 33. **T. tenuisubulatum**
2. Style straight (in *T. megalostigma* and *T. xingshanicum* slightly arcuate).
   44. Style flat, oblong, ventrally with no stigma ················· 34. **T. latistylum**
   44. Style usually thinly terete, rarely short, triangular or ovate, ventrally with a conspicuous stigma or dense stigmatic tissue.
      45. Filaments differentiated into upper and lower two parts different in shape and size (in *T. megalostigma* and *T. tsawarungense*, filaments narrow-linear).

46. Achenes not winged.
　　47. Style thin terete, conspicuous.
　　　　48. Achenes sessile, on each side with conspicuous prominent longitudinal ribs ··············
　　　　　　··· 37. **T. petaloideum**
　　　　　　　49. Leaflets obovate, rhombic or suborbicular ·························· 37a. var. **petaloideum**
　　　　　　　49. Leaflets narrow-elliptic or narrow-lanceolate ············ 37b. var. **supradecompositum**
　　　　48. Achenes stipitate, on each side with thin ribs.
　　　　　　50. Achenes shortly stipitate.
　　　　　　　　51. Achenes depressed-globose.
　　　　　　　　　　52. Filaments narrow-linear, below slightly narrowed; style thinly terete, apex slightly recurved, ventrally with a narrow-linear stigma ···· 35. **T. megalostigma**
　　　　　　　　　　52. Filaments above oblanceolate, below filiform; style short, ventrally on apex with a small elliptic stigma ··· 36. **T. baicalense**
　　　　　　　　　　　　53. Achenes with thin ribs ·· 36a f. **baicalense**
　　　　　　　　　　　　53. Achenes smooth, with inconspicuous ribs ················· 36b. f. **levicarpum**
　　　　　　　　51. Achenes more or less compressed, not depressed-globose.
　　　　　　　　　　54. Plants totally glabrous.
　　　　　　　　　　　　55. Leaflets 1–3 cm long; flower with 2–5 carpels ················· 40. **T. fargesii**
　　　　　　　　　　　　55. Leaflets 2–6 cm long; flower with 5–13 carpels ·········· 41. **T. xingshanicum**
　　　　　　　　　　54. Stems, leaflets and pedicels glandular-puberulous or glabrous; ovary glandular-puberulous or glabrous.
　　　　　　　　　　　　56. Stems, leaflets and pedicels glandular-puberulous; flower with 4 sepals; ovary glabrous.
　　　　　　　　　　　　　　57. Sepal with one simple midrib and two 1–twice dichotomizing lateral basal nerves; flower with 20–25 stamens and 5–6 carpels; filaments above narrow-oblanceolate below filiform; achenes flat, lunate ·············· ··· 42. **T. wangji**
　　　　　　　　　　　　　　57. Sepal with 3 simple basal nerves; flower with 15 stamens and 2 carpels; filaments narrow-linear; achenes slightly compressed, not flat ·············· ··· 43. **T. tsawarungense**
　　　　　　　　　　　　56. Stems, leaflets and pedicels glabrous; flower with 5 sepals; ovary glandular-puberulous ··· 44. **T. pumilum**
　　　　　　50. Achenes with longer thin stipes.
　　　　　　　　58. Leaflets at base peltate; flowers in few-flowered corymbiform compound monochasia ··· 56. **T. pseudoichangense**
　　　　　　　　58. Leaflets not peltate; flowers in many-flowered thyrses.
　　　　　　　　　　59. Stems and leaflets sparsely puberulous; sepal above ciliate, with 3 basal nerves; ovary puberulous ······························· 45. **T. lasiogynum**
　　　　　　　　　　59. Stems, sepals and carpels glabrous; sepal with 5 basal nerves.

     60. Leaflets abaxially puberulous; fruiting pedicels straight; flower with 35–48 stamens ·· 46. **T. przewalskii**
     60. Plants totally glabrous; fruiting pedicels at apex reflexed; flower with 10–30 stamens ··· 47. **T. sparsiflorum**
47. Style very short, not thinly terete, inconspicuous.
  61. Achenes sessile.
    62. Flowers in many-flowered corymbiform compound monochasia ·············· 38. **T. reticulatum**
     63. Plants totally glabrous ·· 38a. var. **reticulatum**
     63. Leaflets abaxially sparsely hirtellous ······································ 38b. var. **hirtellum**
    62. Flowers in simple monochasia or 4-flowered cincinus ······················ 39. **T. wuyishanicum**
  61. Achenes with longer thin stipes.
    64. Leaflets at base peltate ·· 52. **T. ichangense**
    64. Leaflets not peltate.
     65. Cauline leaves 2, simple, opposite ·· 54. **T. filamentosum**
     65. Cauline leaves 1–2-ternate, alternate, rarely opposite.
      66. Leaflets abaxially puberulous ··· 51. **T. acutifolium**
      66. Plants totally glabrous.
       67. Flowers in panicle-like thyrses.
        68. Sepal at base cuneate, with 3 basal nerves; flower with 20–24 stamens; style 0.2–0.3 mm long, apex with a small elliptic stigma ·················· 50. **T. yuoxiense**
        68. Sepal at base nearly rounded, with 2 basal nerves; flower with 3–7 stamens; style subulate, 0.8 mm long, ventrally with a narrow-linear stigma ···· 49. **T. oligandrum**
       67. Flowers in corymbiform or umbelliform compound monochasia.
        69. Compound monochasia with sparse branches, corymbiform; sepal with 3 basal nerves; achenes more than 3 mm long.
         70. Leaflets suborbicular or obovate ································ 48. **T. umbricola**
         70. Leaflets ovate or rhombic-ovate.
          71. Root tubers absent; cauline leaves alternate; achenes narrow-oblong or sublunate ·· 51. **T. acutifolium**
          71. Root tubers present; cauline leaves usually opposite; achenes narrow-elliptic ·· 53. **T. tuberiferum**
        69. Compound monochasia with dense branches, umbel-like; sepal with 1 basal nerve; achenes up to 1.8 mm long ···································· 55. **T. microgynum**
46. Achene with 4 longitudinal wings, base with a reflexed long slender stipe; style very short ·· 57. **T. aquilegifolium** var. **sibiricum**
45. Filaments linear, narrow-linear (below often slightly narrowed) or filiform.
  72. Flowers acrofugally developed, in monochasia or thyrses.
    73. Flowers in narrow raceme-like compound monochasia.

74. Leaflets abaxially hirtellous; filaments narrow-linear; stigma not winged.
    75. Sepal with 3 basal nerves; stamens all fertile; ovary glabrous ············ 58. **T. xiaojinense**
    75. Sepal with 6 basal nerves; some stamens with sterile anthers; ovary puberulous ············ ············ 59. **T. sexnervisepalum**
74. Plants totally glabrous; filaments filiform.
    76. Stigma not winged; style straight or at apex slightly recurved ············ 60. **T. rutifolium**
    76. Stigma winged; style straight.
        77. Flower with ca. 50 stamens; anthers longer than filaments; style elongate, twice longer than ovary; stigma with 1 narrow wing ············ 61. **T. sinomacrostigma**
        77. Flower with 7–16 stamens; anthers shorter than filaments; sytle slightly shorter than ovary; stigma with 2 lateral narrow-wings.
            78. Stipels ovate, apex slightly acute; compound monochasia only terminal, lateral compound monochasia absent; anthers narrow-oblong, apex not mucronate ········ ············ 62. **T. jilongense**
            78. Stipels orbicular-ovate, apex rounded; compound monochasia terminal and lateral; anthers oblong, at apex mucronate ············ 63. **T. leuconotum**
73. Flowers in corymbiform compound monochasia or panicle-like thyrses.
    79. Achenes flat. suborbicular, on each side with a longitudinal nerve, along ventral and dorsal sutures narrowly winged ············ 77. **T. elegans**
    79. Achenes on each side longitudinally 2–3–ribbed, not winged.
        80. Flowers usually in few-flowered, corymbiform compound monochasia.
            81. Leaves simple, orbicular-reniform; filaments linear ············ 71. **T. rotundifolium**
            81. Leaves compound.
                82. Plants totally glabrous.
                    83. Basal leaves absent; sepal with 3 basal herves; anthers shorter than filaments.
                        84. Stem simple or above the middle branched; leaves all ternate; filaments narrow-linear; carpels stipitate ············ 64. **T. virgatum**
                        84. Stem from base long 2–4–branched; leaves 2–3–ternate, the uppermost one 5–foliolately pinnate; filaments filiform; carpels sessile ······ ············ 65. **T. tsaii**
                    83. Basal leaves present; sepal with 5–6 basal nerves; anthers longer than filaments.
                        85. Leaflets narrow-linear or filiform; sepals white, 6.5–11 mm long; filaments linear ············ 66. **T. foeniculaceum**
                        85. Leaflets ovate; sepals pink, 13–20 mm long; filaments narrow-linear ·· ············ 67. **T. grandiflorum**
                82. Stems. leaves and pedicels glandular-puberulous.

86. Stigma not winged; leaves 3-ternate or 3-pinnate.
    87. Sepal with 3 simple basal nerves; anthers narrow-oblong, slightly longer than filaments; carpels glabrous ······ 43. **T. tsawarungense**
    87. Sepal with 1 simple midrib and 2 two times dichotomizing lateral nerves; anthers oblong, shorter than filaments; carpels glandular-puberulous ·· 69. **T. viscosum**
86. Stigma with 2 lateral wings.
    88. Leaves 5-foliolately pinnate or ternate with 3 leaflets; ovary densely glandular-puberulous ··························· 68 **T. panzhihuaense**
    88. Leaves 3-pinnate; ovary puberulous ·········· 70. **T. glandulosissimum**
        89. Leaflets 0.7–1.6 cm long and broad; leaf rachis with hairs 0.2 mm long; stigma narrow-triangular ········ 70 a. var. **glandulosissimum**
        89. Leaflets 1.5–2.5 cm long and broad; leaf rachis with hairs 0.4 mm long; stigma triangular ······················ 70b. var. **chautongense**
80. Flowers in many-flowered thyrses.
  90. Flowers larger, sepals 5.5–13 mm long; stigma not winged.
    91. Sepals white, 10–12.8 mm long, each sepal with 3 dichotomizing basal nerves; anthers linear; achenes not winged ··· 73. **T. cuonaense**
    91. Sepals purple and lilac.
      92. Stems, leaves and ovaries puberulous; anthers linear; achenes not winged.
        93. Leaflets 1.2–2.5 cm long; sepal with 3 basal nerves; filaments narrow-linear ···76. **T. reniforme**
        93. Leaflets 0.4–1.2 cm long; sepal with 7 basal nerves; filaments filiform ·· 77. **T. diffusiflorum**
      92. Plants totally glabrous.
        94. Stem in upper leaf axils with bulbils; sepal with 5 basal nerves; filaments narrow-linear; achenes not winged ··· 74. **T. chelidonii**
        94. Stem without bulbils.
          95. Flower with 10–15 carpels; ovary and achene not winged ········ 75. **T. callianthum**
          95. Flower with 15–22 carpels; ovary and achene along ventral and dorsal sutures narrowly winged ·· 78. **T. delavayi**
            96. Sepal with 5 basal nerves; anther apex with mucrones 0.1–0.2 mm long.
              97. Sepals at apex slightly obtuse; anthers oblong.
                98. Filaments filiform ······························· 78a. var. **delavayi**
                98. Filaments narrow-linear ··················· 78c. var. **decorum**
              97. Sepals at apex acuminate; anthers narrow-oblong ······ 78b. var. **acuminatum**
            96. Sepal with 3 basal nerves; anthers broad-oblong, apex with mucrones 1–1.8 mm

　　　　　　long; filaments filiform ·· 78d. var. **mucronatum**
90. Flowers smaller, sepals less than 5 mm long, each sepal with 3 basal nerves.
　　99. Stigma not winged.
　　　　100. Achenes not winged.
　　　　　　101. Filaments narrow-linear; anthers narrow-oblong; stigmas conspicuous.
　　　　　　　　102. Plants totally glabrous.
　　　　　　　　　　103. Leaflets 0.8–2.1 cm long; pedicels 14–35 mm long; flower with ca. 17 stamens ·· ·· 80. **T. trichopus**
　　　　　　　　　　103. Leaflets 4.2–6.5 cm long; pedicels 4–6 mm long; flower with 30–35 stamens ···· ·· 81. **T. honanense**
　　　　　　　　102. Plants only on petiolules sparsely puberulous, elsewhere glabrous; leaflets 1.5–3 cm long; pedicels 8–25 mm long ······························· 82. **T. osmundifolium**
　　　　　　101. Filaments filiform.
　　　　　　　　104. Leaflets 1.2–5.4 cm broad, at base cordate, abaxially hirtellous; anthers narrow-oblong; stigma 1.8 mm long ······················ 72. **T. pundanum** var. **hirtellum**
　　　　　　　　104. Leaflets smaller, less than 2 cm broad, base not cordate.
　　　　　　　　　　105. Plants totally glabrous; leaflets smaller. 0.3–1 cm broad; anthers narrow-oblong; style 2 mm long; stigma 1.6 mm long ····················· 83. **T. yadongense**
　　　　　　　　　　105. Stems, leaflets and pedicels puberulous; leaflets 0.7–2 cm broad; anthers linear; style 0.4 mm long; stigma inconspicuous ······················· 84. **T. shensiense**
　　　　100. Achenes along dorsal suture narrowly winged; stigma linear ·················· 85. **T. finetii**
　　99. Stigma entirely adnate with style, with 2 lateral wings.
　　　　106. Stigma narrow-oblong or narrow-triangular, with 2 narrow lateral wings.
　　　　　　107. Leaflets abaxially on prominent nerves with branched hairs ··················· 86. **T. cirrhosum**
　　　　　　107. Leaflets without branched hairs.
　　　　　　　　108. Anthers yellowish, densely with hairs ca. 0.05 mm long ············ 88. **T. austrotibeticum**
　　　　　　　　108. Anthers glabrous.
　　　　　　　　　　109. Stigma narrow-oblong; plants totally glabrous.
　　　　　　　　　　　　110. Leaflets at apex acute, margin with acute teeth; thyrses sparse-flowered; sepal at apex attenuate, with simple midrib and 2 dichotomizing lateral nerves, anthers narrow-oblong ································· 87. **T. zhadaense**
　　　　　　　　　　　　110. Leaflets at apex rounded, margin with obtuse teeth; thyrses densely many-flowered; sepal at apex rounded-obtuse; with 3 simple basal nerves ················ ·· 89. **T. foliolosum**
　　　　　　　　　　109. Stigma narrow-triangular.
　　　　　　　　　　　　111. Leaflets abaxially glabrous; achenes glabrous ····················· 90. **T. cultratum**
　　　　　　　　　　　　111. Leaflets abaxially hairy; achenes dorsally puberulous ································· ·· 91a. **T. foetidum** var. **foetidum**

106. Stigma broad-triangular or sagittate-deltoid, with 2 broad lateral wings.
  112. Cauline leaves nearly erect; thyrses narrow-paniculiform, with branches by angle of 45° from thyrse axis upwards spreading ·· 95. **T. simplex**
   113. Plants totally glabrous.
    114. Fruiting pedicels 2–thrice longer than achenes.
     115. Leaflets obovate, apex and lobe apex acute or obtuse, base rounded or broadly cuneate ·· 95a. var. **simplex**
     115. Leaflets cuneate, apex and lobe apex pungent, base narrowly cuneate ·················
        ·· 95b. var. **affine**
    114. Fruiting pedicels ca. as long as achenes; leaflets similar to those of var. *affine* ···············
        ·· 95c. var. **brevipes**
   113. Stems, leaves and pedicels glandular-puberulous ····················· 95d. var. **glandulosum**
  112. Cauline leaves and thyrse branches all by obtuse angle from stem or thyrse axis upwards spreading; thyrses paniculiform.
   116. Leaflets undivided and entire; anthers linear ······································· 93. **T. tenue**
   116. Leaflets usually 3–lobed and margin with teeth.
    117. Leaflets small, ca. 4×4 mm ·· 94. **T. isopyroides**
    117. Leaflets larger.
     118. Thyrses corymbiform, more times dichotomously branched; achenes 4–5.2 mm long; pedicels 1.5–3 cm long ·· 97. **T. squarrosum**
     118. Thyrses paniculiform, with fewer dichotomous branches; achenes 3–4 mm long.
      119. Leaflets herbaceous, with flat nerves and inconspicuous nervous nets.
       120. Leaflets 0.6–1(–1.5)cm long; achenes flat ·······································
          ·· 91b. **T. foetidum** var. **glabrescens**
       120. Leaflets larger, terminal leaflet 4–7 cm long; achenes slightly compressed, ellipsoidal ·· 96. **T. flavum**
      119. Leaflets thicker, papery or subcoriaceous, nerves abaxially prominent and forming conspicuous nervous nets ······················· 92. **T. minus**
       121. Pedicels 3–8 mm long; anthers oblong or narrow-oblong, apex with mucrones 0.1 mm long.
        122. Leaflets 0.4–1.5 cm broad, abaxially greenish ········ 92a. var. **minus**
        122. Leaflets larger, 1.5–4(–5) cm broad.
         123. Leaflets margin with few acute teeth, abaxially glaucous and with conspicuous nervous nets ············ 92b. var. **hypoleucum**
         123. Leaflets at margin with few rotund teeth, abaxially greenish and with inconspicuous nervous nets ··········· 92c. var. **rotundidens**
       121. Pedicels 15–30 mm long; anthers linear, apex with mucrones 0.4–0.45 mm long; leaflets abaxially greenish ······················· 92d. var. **stipellatum**

72. Flowers acropetally developed; sepals small, yellowish-greenish, each sepal with 3 simple basal nerves; filaments filiform, anthers narrow-oblong, rarely linear, apex mucronate; stigma entirely adnate with style, sagittate-deltoid, with 2 broad lateral wings.
  124. Basal leaves absent; lower cauline leaves reduced to small scales, upper cauline leaves normally developed, 3–4–pinnate; flowers singly axillary, along stem acropetally developed ·················· ··· 98. **T. squamiferum**
  124. Leaves all basal; flowers in terminal raceme ·· 99. **T. alpinum**
    125. Scapes not branched, 6–20 cm tall; pedicels downwards arcuate, 1–10 mm long; plants totally glabrous.
      126. Achenes sessile ··· 99a. var. **alpinum**
      126. Achenes stipitate ··· 99b. var. **microphyllum**
    125. Scapes simle or 1–branched, 7–15 (–25–38) cm tall; pedicels erect, straight, 1.8–9 mm long; achenes sessile ··· 99c. var. **elatum**
      127. Plants totally glabrous ··· 99ci. f. **elatum**
      127. Leaflets abaxially hairy.
        128. Leaflets abaxially puberulous ·· 99cii. f. **puberulum**
        128. Leaflets abaxially puberulous, on nerves hirtellous ············ 99ciii. f. **setulosinerve**
1. Flowers bisexual or unisexual; stigma exserted, elongate and flagelliform.
  129. Leaflets up to 15 mm long, abaxially on nerves hirtellous; filaments 1.5–5.5 mm long ················ ·· 100. **T. smithii**
  129. Leaflets smaller, 3–9 mm long, abaxially glandular-puberulous; filaments 8 mm long ················ ·· 101. **T. dingjieense**

亚属 1. 唐松草亚属

Subgen. **Thalictrum**. Tamura in Sci. Rep. Osaka Univ. 17: 50. 1968; in Acta Phytotax. Geobot. 43: 55. 1992; et in Hiepko, Nat. Pflanzenfam., 17a Ⅳ : 477. 1995; Emura in J. Fac. Sci. Univ. Tokyo, Sect. 3, Bot. 9: 97. 1972; W. T. Wang & S. H. Wang in Fl. Reip Pop. Sin. 27: 517. 1979. Type: *T. foetidum* L.

Sect. *Thalictrum* DC. Syst. 1: 172. 1818; Ledeb. Fl Ross. 1: 6.1842.— Subgen. *Thalictrum* (DC.) Reihb. Consp. Reg. Veg. 192. 1828; Boivin in Rhodora 46: 349. 1944.

Sect. *Euthalictrum* DC. Prodr. 1: 12. 1824.—Subgen. *Euthalictrum* (DC.) Peterm. Deutsch. Fl. 4. 1846.

Sect. *Microgynes* Lecoy. in Bull. Soc. Bot. Belg. 24: 112. 1885.

花两性. 柱头内藏, 不呈鞭形。

Flowers bisexual. Stigma included, not flagelliform.

中国本亚属有 96 种, 隶属 12 组。

组 1. 叉枝唐松草组

Sect. **Leptostigma B.** Boivin in Rhodora 46: 360. 1944; W. T. Wang & S. H. Wang in Fl. Reip Pop. Sin. 27: 517. 1979. Type: *T. saniculiforme* DC.

花组成单歧聚伞花序或聚伞圆锥花序。心皮具 1 钩状花柱, 稀具直花柱; 花柱腹面上部具 1 小而不明显的柱头。萼片具 3 条基生脉。瘦果在每侧具 2-3 条纵肋。

Flowers in monochasia or thyrses. Carpel with a hooked, rarely straight style; style adaxially above with a small inconspicuous stigma. Sepal with 3 basal nerves. Achene on each side with 2–3 longitudinal ribs.

本组在中国有 32 种, 隶属 6 系, 多数种分布于西南部, 少数分布于长江中、下游诸省及华北、台湾地区。

系 1. 叉枝唐松草系

Ser. **Saniculiformia** W. T. Wang & S. H. Wang in Fl Reip. Pop. Sin. 27: 517, 616. 1979.—Subsect. *Saniculiformia* (W. T. Wang & S. H. Wang) Tamura in Acta Phytotax. Geobot. 43: 55. 1992. Type: *T. saniculiforme* DC.

花组成伞房状复单歧聚伞花序。花丝条形或狭条形。花柱钩状弯曲。瘦果无柄。

Flowers in corymbiform compound monochasia. Filaments linear or narrow-linear. Achenes sessile.

3 种, 分布于中国西南部, 其中 1 种分布达喜马拉雅山区。

1. **叉枝唐松草**(《中国植物志》) 图版 1

Thalictrum saniculiforme DC. Prodr. 1: 12. 1824; Hook. f. & Thoms. Fl. Ind. 14. 1855; et in Hook. f. Fl. Brit Ind. 1: 18. 1872; Lecoy. in Bull. Soc. Bot. Belg. 24: 215, t. 5: 10. 1885;

Finet & Gagnep. in Bull. Soc. Bot. France 50: 611. 1903; W. T. Wang in Acta Phytotax. Sin. 6 (4): 371. 1957; W. T. Wang & S. H. Wang in Fl. Reip. Pop. Sin. 27: 517, pl. 121: 1, pl. 122: 7–11. 1979; Grierson, Fl. Bhutan 1(2): 297. 1984; W. T. Wang in Fl. Xizang. 2: 65. 1985; in Vasc. Pl. Hengduan Mount. 1: 498. 1993; in Fl. Yunnan. 11: 157. 2000; et in High. Pl. China 3: 463, fig. 730. 2000; Rau in Sharma et al. Fl. India 1: 141. 1993; D. Z. Fu & G. Zhu in Fl China 6: 297. 2001. Described from Nepal, no type specimen cited.

T. radiatum Royle, Ill. Bot. Himal. 52. 1838. Described from Mussooree, Himalaya.

多年生草本植物，全部无毛。茎高 18–32 cm，自下部或中部近两歧状分枝。基生叶和下部茎生叶为 2–3 回三出复叶，具长柄；叶片长 4.5–12 cm；小叶纸质，宽倒卵形或菱状宽倒卵形，1–2.5×0.8–2.3 cm，顶端圆形，基部宽楔形或圆形，偶尔浅心形，3 浅裂，裂片顶端钝或圆形，全缘或有 1–2 圆齿，脉下面稍隆起，脉网明显；小叶柄长 0.8–1.6 cm；叶柄长 4.5–13 cm，基部有狭鞘。单歧聚伞花序 2.5–4 cm 宽，2–3 回二歧状分枝，有 2–4 花；苞片小，椭圆形；花梗长 1.5–2.3 cm。花：萼片 4，脱落，白色，椭圆形，4–5×2.2 mm。雄蕊约 30；花丝条形，1.2–2×0.2–0.3 mm；花药长圆形或狭长圆形，1–1.2×0.2–0.3 mm，顶端微尖。心皮 25–68；子房披针形，1.2×0.5 mm；花柱长约 1.2 mm，顶端钩状或稍拳卷；柱头不明显。瘦果狭椭圆形，3–4.8×0.8–1 mm，每侧有 3 条细纵肋；宿存花柱长约 2.2 mm，上部钩状弯曲。7 月开花。

Perennial herbs, totally glabrous. Stems 18–32 cm tall, from below or from the middle 2–thrice dichotomously branched. Basal and lower cauline leaves 2–3–ternate, long petiolate; blades 4.5–12 cm long; leaflets papery, broad-obovate or broadly rhombic-obovate, 1–2.5×0.8–2.3 cm, at apex rounded, at base broad-cuneate or rounded, occasionally subcordate, 3–lobed, lobes at apex obtuse or rounded, entire or 1–2–rotund-dentate, nerves abaxially slightly prominent, nervous nets conspicuous; petiolules 0.8–1.6 cm long; petioles 4.5–13 cm long, at base narrowly vaginate. Monochasia 2.5–4 cm broad, 2–thrice dichotomously branched, 2–4–flowered; bracts small, elliptic; pedicels 1.5–2.3 cm long. Flower: Sepals 4 deciduous, white, elliptic, 4–5×2.2 mm. Stamens ca. 30; filaments linear, 1.2–2×0.2–0.3 mm; anthers oblong or narrow-oblong, 1–1.2×0.2–0.3 mm, apex slightly acute. Carpels 25–68; ovary lanceolate, 1.2×0.5 mm; style ca. 1.2 mm long, above hooked or slightly circinate; stigma inconspicuous. Achene narrow-elliptic, 3–4.8×0.8–1 mm, on each side with 3 thin longitudinal ribs; persistent style ca. 2.2 mm long, hooked.

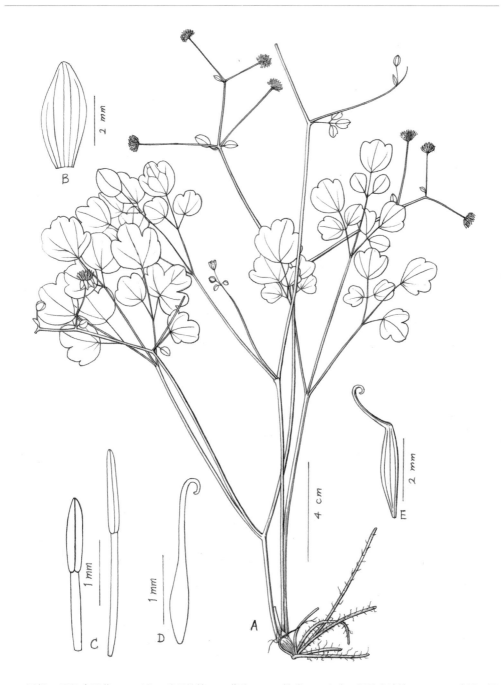

图版 1. 叉枝唐松草　A. 开花、结果植株，B. 萼片，C. 二雄蕊，D. 心皮（据倪志诚等 770），E. 瘦果（据西藏中草药队 1421）。

Plate 1. **Thalictrum saniculiforme**　A. Flowering and fruiting plant, B. sepal, C. two stamens, D. carpel (from Z. C. Ni et al. 770), E. achene (from Xizang Med. Pl. Exped. 1421)

分布于中国云南和西藏东南部和南部；以及不丹、尼泊尔、印度北部。生山地草坡或林中，海拔 1900–2500 m。

标本登录：云南：凤庆，俞德浚 16740；麻栗坡，税玉民等 32997。西藏：察隅，倪志诚等 770；定结，PE- 西藏队 3072，3205；聂拉木，西藏中草药队 1421；聂拉木，樟木，PE- 西藏队 4341。

2. 小花唐松草（《广西植物》） 图版 2

Thalictrum minutiflorum W. T. Wang in Guihaia 37(4):407, fig. 1. 2017. Holotype: 贵州：德江县，桶井乡，高开，alt. 600 m，2003-03-26，安明光 3048（PE）。

多年生草本植物，全部无毛。茎高 11–25 cm，不分枝或有 1–2 条分枝，有 1–3 叶。下部茎生叶为 3 回三出复叶，上部茎生叶为 2 回三出复叶；叶片三角形，1.5–7×2–5 cm；小叶薄纸质，卵形，宽卵形或宽菱形，0.3–1×0.3–1 cm，基部圆形、截形或浅心形，3 浅裂或 3 微裂（裂片全缘或有 1–2 小齿），基出脉 3 或 5 条，上面平，下面稍隆起；小叶柄丝形，长 1–6 mm；叶柄长 1–3.5 cm；托叶膜质，宽 0.2–2 mm。复单歧聚伞花序顶生长 2.5–5 cm，伞房状，有 4–8 花；苞片叶状，或简单，椭圆形或披针形，长 0.5–2 cm；花梗长 4–10 mm。花有香味：萼片 4，白色，椭圆形，2×1 mm，在顶端钝，具 3 条不分枝的基生脉。雄蕊约 18；花丝条形或狭条形，1–2×0.15–0.3 mm，比花药窄；花药淡黄色，长圆形，0.8×0.35–0.5 mm，顶端钝。心皮 8–14；子房近纺锤形，1.2–1.5×0.35 mm；花柱长约 0.8 mm，顶端钩状弯曲，腹面近顶端有不明显柱头。瘦果扁平，纺锤形，2–3.2×0.6–1 mm，每侧有 3 条细纵肋；宿存花柱长 0.4–0.6 mm，顶端钩状弯曲。2–3 月开花，3 月结果。

Perennial herbs, totally glabrous. Stems 11–25 cm tall, simple or 1–2-branched, 1–3-leaved. Lower cauline leaves 3-ternate, upper cauline leaves 2-ternate; blades triangular, 1.5–7×2–5 cm; leaflets thinly papery, ovate, broad-ovate or broad-rhombic, 0.3–1×0.3–1 cm, at base rounded, truncate or subcordate, 3-lobed or 3-lobulate (lobes entire or 1–2-denticulate), basal nerves 3 or 5, adaxially flat, abaxially slightly prominent; petiolules filiform, 1–6 mm long; petioles 1–3.5 cm long; stipules membranous, 0.2–2 mm broad. Compound monochasia terminal, 2.5–5 cm long, corymbiform, 4–8-flowered; bracts foliaceous, or simple, elliptic or lanceolate, 0.5–2 cm long; pedicels 4–10 mm long. Flower fragrant: Sepals 4, white, elliptic, 2×1 mm, at apex obtuse, each sepal with 3 simple basal nerves. Stamens ca. 18; filaments linear or narrow-linear, 1–2×0.15–0.3 mm, narrower than anthers; anthers yellowish, oblong, 0.8×0.35–0.5 mm, apex obtuse. Carpels 8–14; ovary subfusiform, 1.2–1.5×0.35 mm; style

六、分类学处理　　　　　　　　　　　　　　　　　　　　85

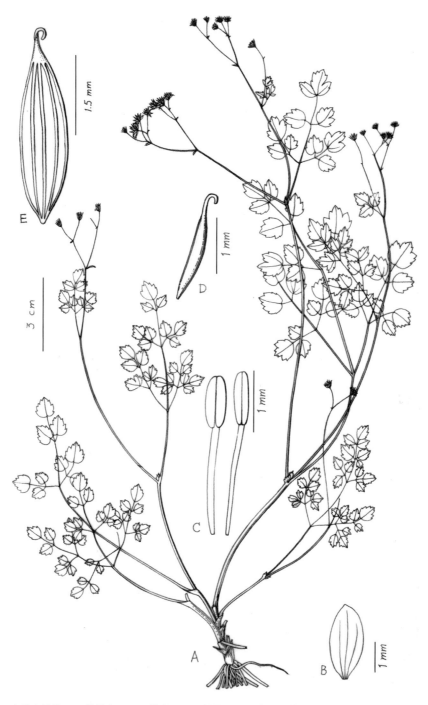

图版 2. **小花唐松草**　A. 植株全形，B. 萼片，C. 二雄蕊，D. 心皮，E. 瘦果。（据 holotype）
Plate 2. **Thalictrum minutiflorum**　A. habit, B. sepal, C. 2 stamens, D. carpel, E. achene. (from holotype)

ca. 0.8 mm long, at apex hooked, adaxially near apex with a inconspicuous stigma. Achene flattened, fusiform, 2–3.2×0.6–1 mm, on each side with 3 thin longitudinal ribs; persistent style 0.4–0.6 mm long, apex hooked.

特产贵州德江县。生山地疏林下，海拔 600 m。

3. **绢毛唐松草**（《秦岭植物志》） 图版 3

Thalictrum brevisericeum W. T. Wang & S. H. Wang in Fl. Tsinling. 1(2):239,603, fig. 203. 1974; et in Fl. Reip. Pop. Sin. 27: 528, pl. 123: 1–4 1979, p.p.; W. T. Wang in Iconogr. Corm. Sin. Suppl. 1: 421, fig. 8605. 1982; et in High. Pl. China 3: 464. 2000; D. Z. Fu & G. Zhu in Fl. China 6: 287. 2001, p.p.; L. Q. Li et al. Pl. Baishuijiang St. Nat. Reserve Gansu Prov. 70. 2014. Holotype：甘肃：天水，麦积山，1950–09–13，吴中伦 20346（PE）. Paratypes：甘肃：平凉，崆峒山，alt. 1700 m，1942–08–10，王作宾 13393。陕西：太白山，alt. 900–1100 m，1937–08–20，刘慎谔，钟补求 60（PE）；同地，alt. 1300 m，1933–07–27，王作宾 1389（PE）。

3a. var. **brevisericeum**

多年生草本植物。茎高 30–77 cm，被短柔毛，分枝。中部茎生叶为 3 回三出复叶；叶片长 12–18 cm；小叶纸质，宽倒卵形、倒卵形或宽卵形，1–3×0.8–3 cm，基部圆形、宽楔形或浅心形，3 浅裂，边缘有小牙齿，下面在隆起的脉网上被短柔毛；小叶柄密被短柔毛；叶柄长 3.5–6.5 cm，疏被短柔毛；托叶边缘具小流苏。聚伞圆锥花序长 12–25 cm；轴和花梗均被绢毛（毛长 0.25–0.5 mm）；苞片钻状条形，长 1.5–2 mm；花梗长 3–4 mm。花：萼片 5，近长圆形，约 2.8×1 mm，顶端圆钝，每萼片具 3 条不分枝的基生脉。雄蕊约 20，无毛；花丝狭条形，长 3.5–4.2 mm，上部宽 0.2–0.4 mm，下部变狭，宽 0.1 mm；花药长圆形或椭圆形，1×0.4 mm，顶端钝。心皮 6–16；子房狭倒卵形，1.5×0.5 mm；花柱长约 1.2 mm，近顶端钩状弯曲，有不明显柱头。瘦果纺锤形，约 2.6×0.8 mm，每侧有 2–3 条纵肋；宿存花柱长约 1 mm，钩状弯曲。6 月开花。

Perennial herbs. Stems 30–77 cm tall, puberulous, branched. Middle cauline leaves 3-ternate; blades 12–18 cm long; leaflets papery, broad-obovate, obovate or broad-ovate, 1–3×0.8–3 cm, at base rounded, broad-cuneate or subcordate, 3-lobed, at margin denticulate, abaxially on prominent nervous nets puberulous; petiolules densely puberulous; petioles 3.5–6.5 cm long, sparsely puberulous; stipules on margin fimbrillate. Thyrses 12–25 cm long; rachis and pedicels sericeous (hairs 0.25–0.5 mm long); bracts subulate-linear, 1.5–2 mm long;

六、分类学处理　　87

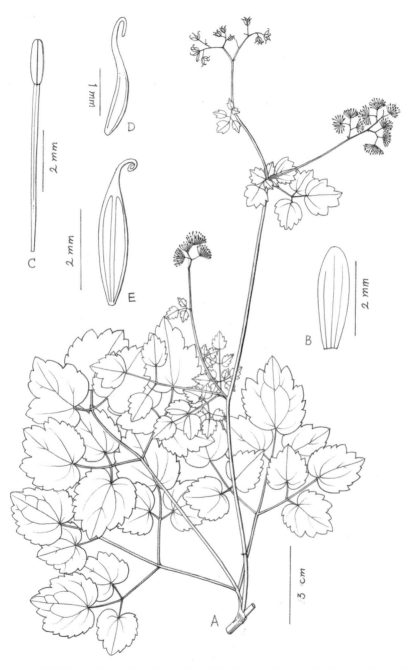

图版 3. **绢毛唐松草**　A.茎生叶及其腋生的聚伞圆锥花序，B.萼片，C.雄蕊，D.心皮，E.瘦果。（据 isotype）

Plate 3. **Thalictum brevisericeum** var. **brevisericeum**　A. cauline leaf and its axillary thyrse, B. sepal, C. stamen, D. carpel, E. achene. (from isotype)

pedicels 3–4 mm long. Flower: Sepals 5, suboblong, ca. 2.8×1 mm, at apex rounded-obtuse, each sepal with 3 simple basal nerves. Stamens ca. 20, glabrous; filaments narrow-linear, 3.5–4.2 mm long, above 0.2–0.4 mm broad, below narrowed, 0.1 mm broad; anthers oblong or elliptic, 1×0.4 mm, apex obtuse. Carpels 6–16; ovary narrow-obovate, 1.5×0.5 mm; style ca. 1.2 mm long, near apex hooked and with a inconspicuous stigma. Achene fusiform, ca. 2.6×0.8 mm, on each side with 2–3 longitudinal ribs; persistent style ca. 1 mm long, hooked.

分布于四川西北部、甘肃东南部和陕西秦岭。生山地林边或灌丛中，海拔950–3000 m。

标本登录：四川：南坪，alt. 1450 m，李全喜 2072。甘肃：舟曲，alt. 2400 m，王作宾 15175；天水，夏纬瑛 5852，刘继孟 10041，李全喜 726；文县，唐昌林 1627；平凉，黄河队 56–1870。陕西：宁陕，傅坤俊 11638；太白山，alt. 2600–3000 m，应俊生等 136（GH）。

3b. 五果绢毛唐松草（变种） 图版 3a: A–E

var. **pentagynum** W. T. Wang. var. nov. Holotype：四川（Sichuan）：黑水县，林中，直立草本高 1 m（Heishui xian, in forest, erect herbs 1 m tall），2009-07-26，Gu Lei & Li Zhong-rong 1331（GH）.

A. var. *brevisericeo* differt monochasiorum compositorum ramis et pedicellis pilis 0.05–0.2 mm long is tectis, floris sepalis 3 late ovatis, 5–nervibus et 1 angste elliptico, 3–nervi, carpellis 5 per florem, stylis apice arcuatis nec uncatis nec circinatis.

本变种与绢毛唐松草的区别在于：本变种的复单歧聚伞花序的分枝和花梗被长 0.05–0.2 mm 的毛，花的 3 枚萼片宽卵形，具 5 脉；1 枚萼片狭椭圆形，具 3 条脉；每花具 5 枚心皮，花柱弧状弯曲，既不钩状，也不拳卷状弯曲。

特产四川黑水县。生山地林中。

3c. 狭药绢毛唐松草（变种） 图版 3a: F–J

var. **angustiantherum** W. T. Wang. var. nov. Type：云南 (Yunnan)：维西县，康普 (Weixi Xian, Kangpa), alt. 2300m, 1935-07, 王启无 (C. W. Wang) 64180 (holotype, PE); 同地 (same locality), 康普——叶枝 (on the way from Kangpu to Yezhi), alt. 1900m, 1940-06-27, 冯国楣 (K. M. Feng) 4983 (PE).

T. brevisericeum auct. non W. T. Wang & S. H. Wang: W. T. Wang in Vasc. Pl. Hengduan Mount. 1:498. 1993; et in Fl. Yunnan. 11:161.2001.

A var. *brevisericeo* differt antheris anguste oblongis 1 mm longis 0.2 mm latis apice obtusis vel minute mucronatis.

六、分类学处理 89

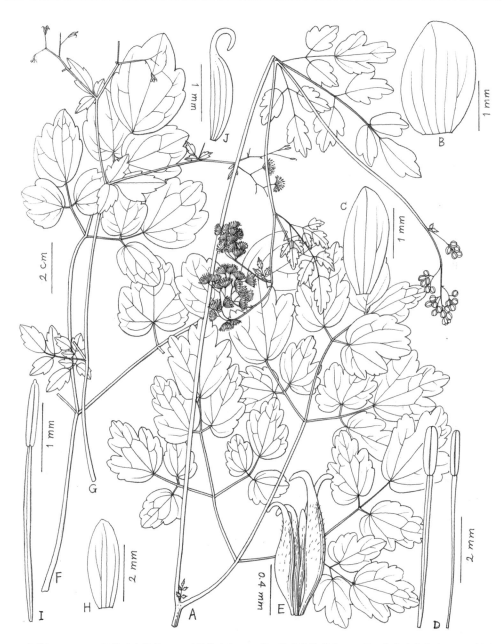

图版 3a. A–E. **五果绢毛唐松草** A. 开花茎上部，B. 具 5 基生脉的萼片，C. 具 3 基生脉的萼片，D. 二雄蕊，E. 具 5 心皮的雌蕊群。（据 holotype）F–J. **狭药绢毛唐松草** F. 茎生叶，G. 聚伞圆锥花序，H. 萼片，I. 雄蕊，J. 心皮。（据 holotype）

Plate 3a. A–E. **Thalictrum brevisericeum** var. **pentagynum** A. apper part of flowering stem, B. sepal with 5 basal nerves, C. sepal with 3 basal nerves, D. two stamens, E. gynoecium consisting of 5 carpels. (from holotype) F–J. **T. brevisericeum** var. **angustiantherum** F. cauline leaf, G. thyres, H. sepal, I. stamen, J. carpel. (from holotype)

本变种与绢毛唐松草的区别在于：本变种的雄蕊花药呈狭长圆形，长 1 mm，宽 0.2 mm，顶端钝或很小短尖头。

特产云南维西县。生山地林边或灌丛中，海拔 1900–2300 m。

系 2. 糙叶唐松草系

Ser. **Scabrifolia** W. T. Wang & S. H. Wang in Fl. Reip. Pop. Sin. 27: 519,616. 1979. Type: *T. scabrifolium* Franch.

Subsect. *Actaeifolia* Tamura in Sci. Rep. Osaka Univ. 17: 50. 1968; et in Hiepko, Nat. Pfanzenfam., 17aiv: 482. 1995. ——Sect. *Actaeifolia* (Tamura) Emura in J. Fac. Sci. Univ. Tokyo, Sect. 3, Bot. 9: 126. 1972. Type: *T. actaeifolium* Sieb. & Zucc.

复单歧聚伞花序伞房状或圆锥状。花丝分化成倒披针形或棒状的上部和狭条形或丝形的下部。瘦果无柄。

Compound monochasia corymbiform or paniculiform. Filament differentiated into oblanceolate or clavate upper part and narrow-linear or filiform lower part. Achenes sessile.

本系在中国有 15 种，广布于秦岭以南多数省区。

4. 糙叶唐松草（《中国植物志》） 图版 4

Thalictrum scabrifolium Franch. in Bull. Soc. Bot, Franch. 33: 371. 1886; Finet & Gagnep. in Bull. Soc. Bot. France 50: 610. 1903; W. T. Wang in Fl. Yunnan. 11: 158. 2000; D. Z. Fu & G.Zhu in Fl. China 6: 297. 2001. Type: 云南：近大理，大坪子之北，1882-08-22，Delavay Thalictrum 16 (syntype, P, not seen)；大坪子之北，1883-09-04，Delavay 222 (syntype, P; isosyntype, PE).

多年生草本植物。茎高约 70 cm，上部分枝，与叶柄和小叶柄均被短柔毛（毛长约 0.1 mm）。基生叶和下部茎生叶为 2 回三出复叶；小叶薄纸质，近圆形或卵圆形，1.2–4×2–6 cm，顶端圆形，基部浅心形或心形，不明显 3 浅裂，边缘具多数圆齿或浅圆齿，上面被短柔毛（毛长 0.1–0.2 mm），下面在隆起的脉上被微硬毛（毛长 0.2–0.5 mm）；小叶柄长 0.5–3 cm；叶柄长 8–13 cm，基部具狭鞘。复单歧聚伞花序伞房状，宽 10–15 cm，有多数花，分枝和花梗被短柔毛（毛长 0.05–0.1 mm）；苞片叶状，长 4–9 cm；花梗细，长 2.5–6 mm。花：萼片 4，圆卵形，约 2×1.8 mm，顶端圆钝，背面上部被短柔毛，每一萼片具 3 条基生脉（中脉 2 回二歧状分枝，一侧脉不分枝，另一侧脉 1 回二歧状分枝）。雄蕊 18–28；花丝长 2.5–3.6 mm，上部倒披针形，宽 0.3–0.4 mm，下部丝形，粗 0.1 mm；花药长圆形，约 0.5×0.25–0.3 mm，顶端钝。心皮 10–15；子房斜狭倒卵形，0.6–1.2×0.3–0.6

六、分类学处理 91

图版 4. 糙叶唐松草　A. 开花茎，B. 萼片，C. 二雄蕊，D. 二心皮，E. 二瘦果（据 isosyntype）。
Plate 4. **Thalictrum scabrifolium**　A. flowering stem, B. sepal, C. two stamens, D. two carpels, E. two achenes (from isosyntype).

mm；花柱长 0.3-0.8 mm，顶端稍向后弯曲。瘦果扁，纺锤形，2-3×0.5-0.8 mm，疏被短柔毛（毛长约 0.1 mm），每侧具 2 条纵肋；宿存花柱长 0.4-1 mm，顶端钩状弯曲。8-9 月开花。

Perennial herbs. Stem ca. 70 cm tall, above branched, with petioles and petiolules puberulous (hairs ca. 0.1 mm long). Basal and lower cauline leaves 2-ternate; leaflets thinly papery, suborbicular or orbicula-ovate, 1.2-4×2-6 cm, at apex rounded, at base subcordate or cordate, inconspicuously 3-lobed, at margin many-rotund-dentate or crenate, adaxially puberulous (hairs 0.1-0.2 mm long), abaxially on prominent nerves hirtellous (hairs 0.2-0.5 mm long); petiolules 0.5-3 cm long; petioles 8-13 cm long, at base narrowly vaginate. Compound monochasia corymbiform, 10-15 cm broad, many-flowered; branches and pedicels puberulous (hairs 0.05-0.1 mm long); bracts foliaceous, 4-9 cm long; pedicels slender, 2.5-6 mm long. Flower: Sepals 4, orbicular-ovate, ca. 2×1.8 mm, at apex rounded, abaxially above puberulous, each sepal with 3 basal nerves (midrib twice dichotomizing, a lateral nerve simple, and the another lateral nerve once dichotomizing). Stamens 18-28; filaments 2.5-3.6 mm long, above oblanceolate, 0.3-0.4 mm broad, below filiform, 0.1 mm across; anthers oblong, ca. 0.5×0.25-0.3 mm, apex obtuse. Carpels 10-15; ovary obliquely narrow-obovate, 0.6-1.2×0.3-0.6 mm; style 0.3-0.8 mm long, apex slightly recurved. Achene compressed, fusiform, 2-3×0.5-0.8 mm; sparsely puberulous, on each side with 2 longitudinal ribs; persistent style 0.4-1 mm long, apex hooked.

特产于云南洱源至鹤庆一带山区。

本种和下面第 5 种，鹤庆唐松草的心皮花柱大部直，只顶端稍向内弯曲，腹面柱头不明显，这种形态构造与扁果草族 Trib. Isopyreae 中的心皮花柱相似。

5. 鹤庆唐松草（《植物分类学报》）　图版 5

Thalictrum leve (Franch.) W. T. Wang in Acta Phytotax. Sin. 31(3): 211. 1993; et in Fl. Yunnan. 11: 158. 2000; D. Z. Fu & G. Zhu in Fl. China 6: 293. 2001.—*T. scabrifolium* Franch. var. *leve* Franch. Pl. Delav. 17. 1889, sphalm. levis; Finet & Gagnep. in Bull. Soc. Bot. Franch 50: 610. 1903; W. T. Wang & S. H. Wang in Fl. Reip. Pop. Sin. 27: 521. 1979; C. Y. Wu et al. Ind. Fl. Yunnan. 1: 131. 1984. Type：云南：鹤庆，大坪子之北，Choui-ma-ouan, 1887-08-06, Delavay s. n. (holotype, P; isotype, PE).

多年生草本植物，全部无毛。茎高 45-60 cm，分枝。基生叶和下部茎生叶为 2-3 回三出复叶；叶片长 10-18 cm；小叶纸质，肾形，近圆形或宽卵形，1-2.8×1.2-4.6 cm，在

六、分类学处理　　　　　　　　　　　　　　　　　93

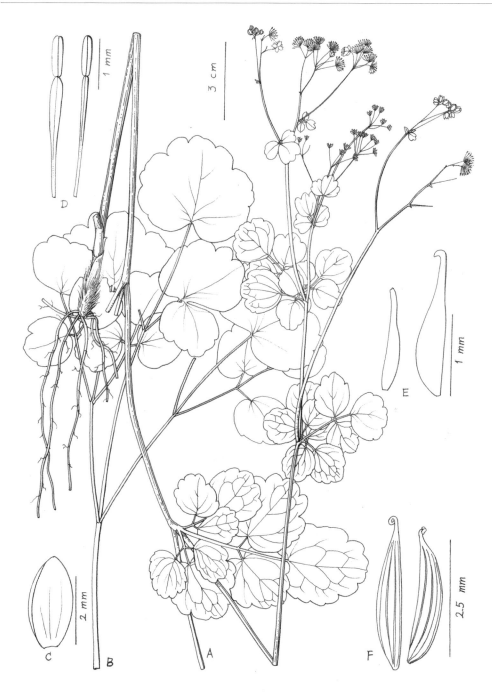

图版 5. **鹤庆唐松草**　A. 开花茎，B. 基生叶，C. 萼片，D. 二雄蕊，E. 二心皮，F. 二瘦果。（据 isotype）
Plate 5. **Thalictrum leve**　A. flowering stem, B. basal leaf, C. sepal, D. two stamens, E. two carpels, F. two achenes. (from isotype)

顶端圆形,在基部浅心形或圆形,通常不明显三浅裂(裂片顶端圆形,全缘或有1-2圆齿),脉下面隆起,脉网明显;小叶柄长0.9-4.6 cm;叶柄长6-10 cm,基部具狭鞘。复单歧聚伞花序顶生和腋生,伞房状,宽4-10 cm,具多数花;花梗长2-7 mm。花:萼片4,早落,绿白色,椭圆形,约2.5×1.5 mm,顶端微尖,每一萼片具3条不分枝或二歧状分枝的基生脉。雄蕊25-30;花丝长1.5-3.5 mm,上部倒披针形或狭倒披针形,宽0.3-0.4 mm,下部狭条形或丝形,宽0.1-0.15 mm;花药长圆形,0.7-0.9×0.3-0.5 mm,顶端钝。心皮3-5;子房半椭圆形,0.6-1.2×0.3-0.6 mm;花柱长0.6-0.7 mm,顶端稍向后弯曲或直。瘦果纺锤形,1.8-2.2×0.4-0.7 mm,每侧有2-3条纵肋;宿存花柱长0.3-0.8 mm,顶端钩状或直。7-8月开花。

Perennial herbs, totally glabrous. Stems 45–60 cm tall, branched Basal and lower cauline leaves 2–3-ternate; blades 10–18 cm long; leaflets papery, reniform, suborbicular or broad-ovate, 1–2.8×1.2–4.6 cm, at apex rounded, at base subcordate or rounded, usually inconspicuously 3-lobed (lobes at apex rounded, entire or 1–2-rotund-dentate), nerves abaxially prominent, nervous nets conspicuous; petiolules 0.9–4.6 cm long; petioles 6–10 cm long, at base narrowly vaginate. Compound monochasia terminal and axillary, corymbiform, 4–10 cm broad, many-flowered; pedicels 2–7 mm long. Flower: Sepals 4, caducous, greenish-white, elliptic, ca. 2.5×1.5 mm, apex slightly acute, each sepal with 3 simple or dichotomizing basal nerves. Stamens 25–30; filaments 1.5–3.5 mm long, above oblanceolate or narrow-oblanceolate, 0.3–0.4 mm broad, below narrow-linear or filiform, 0.1–0.15 mm broad; anthers oblong, 0.7–0.9×0.3–0.5 mm, apex obtuse. Carpels 3–5; ovary semi-elliptic, 0.6–1.2×0.3–0.6 mm; style 0.6–0.7 mm long, apex slightly recurved or straight. Achene fusiform, 1.8–2.2×0.4–0.7 mm, on each side with 2–3 longitudinal ribs; persistent style 0.3–0.8 mm long, apex hooked or straight.

特产云南洱源至鹤庆一带山地。生山地松栎林间草地或林边,海拔1900-2900 m。

标本登录:云南:鹤庆,黄坪区,均华乡,白土坡至大坪子,滇西北金沙江队63-6483;洱源,南大坪至焦石洞,黑山门附近,滇西北金沙江队63-6283。

6. 大关唐松草(《广西植物》)　　图版6

Thalictrum daguanense W. T. Wang in Guihaia 37(6): 676, fig. 1. 2017. Holotype:云南:大关县,联合村,罩子沟,alt. 1800 m,1973-06-17,孙必兴644(PE)。

多年生草本植物,全部无毛。茎高约30 cm,具3茎生叶。茎生叶为2回三出复叶,

六、分类学处理 95

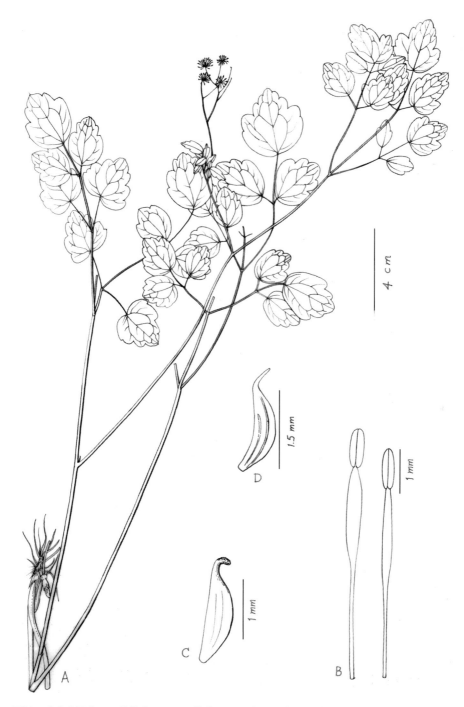

图版 6. **大关唐松草** A. 植株全形, B. 二雄蕊, C. 心皮, D. 瘦果。（据 holotype）
Plate 6. **Thalictrum daguanense** A. habit, B. two stamens, C. carpel, D. achene. (from holotype)

具长柄；叶片 7-15×6-11 cm；小叶纸质，宽卵形、卵形或椭圆形，1.2-2.8×0.5-2.6 cm，顶端微尖，基部圆形，近截形或浅心形，3 浅裂（中央裂片具 2-3 钝齿，侧裂片具 1 小齿或全缘），具五出脉，脉上面平，下面隆起，与细脉形成明显脉网；小叶柄细，长 0.1-2.1 cm；叶柄长 5.2-13.5 cm。复单歧聚伞花序顶生，伞房状，约 2.5×2 cm，有 6 花；苞片常早落，钻形，长约 4 mm；花梗长 0.8-1.1 cm。花：萼片早落，未见。雄蕊约 28；花丝长 4-5 mm；上部条形，宽 0.2-0.35 mm，下部丝形，粗约 0.1 mm；花药长圆形，0.8-1×0.3-0.5 mm，顶端钝。心皮 7-11；子房斜狭倒卵形，约 1.5×0.5 mm；花柱长约 0.5 mm，顶端稍向后弯曲，腹面具不明显柱头。瘦果新月形，约 1.5×0.5 mm，在每侧具 2 条纵肋；宿存花柱长约 0.3 mm。5-6 月开花。

Perennial herbs, totally glabrous. Stem ca. 30 cm tall, 3-leaved. Cauline leaves 2-ternate, long petiolate; blades 7-15×6-11 cm; leaflets papery, broad-ovate, ovate or elliptic, 1.2-2.8×0.5-2.6 cm, at apex slightly acute, at base rounded, subtruncate or subcordate, 3-lobed (central lobe obtusely 2-3-dentate and lateral lobes 1-denticulate or entire), 5-nerved, nerves adaxially flat, abaxially prominent and with nervules forming conspicuous nervous nets; petiolules slender, 0.1-2.1 cm long; petioles 5.2-13.5 cm long. Compound monochasia termianal, corymbiform, ca. 2.5×2 cm, 6-flowered; bracts often caducous, subulate, ca. 4 mm long; pedicels 0.8-1.1 cm long. Flower: Sepals caducous, not seen. Stamens ca. 28; filaments 4-5 mm long, above linear, 0.2-0.35 mm broad, below filiform, 0.1 mm across; anthers oblong, 0.8-1×0.3-0.5 mm, apex obtuse. Carpels 7-11; ovary obliquely narrowhy obovate, ca 1.5×0.5 mm; style ca. 0.5 mm long, at apex slightly recurved, adaxially with an inconspicuous stigma. Achene lunate, ca. 1.5×0.5 mm, on each side with 2 longitudinal ribs; persistent style ca. 0.3 mm long.

特产云南大关县。生山谷林边或灌丛中，海拔 1800 m。

7. **华东唐松草**（《药学学报》） 图版 7

Thalictrum fortunei S. Moore in J. Bot. 16: 130. 1878; Lecoy. in Bull. Soc. Bot. Belg. 24: 162. 1885; Forbes & Hemsl. in J. Linn. Soc. Bot. 23: 8. 1886; Finet & Gagnep. in Bull. Soc. Bot. France 50: 612. 1903, p.p.; C. Pei et al. Man. Pl. South. Jiangsu 281, fig. 443. 1959; Iconogr. Corm. Sin. 1: 672, fig. 1344. 1972; S. H. Fu, Fl. Hupeh. 1: 356. 1976; W. T. Wang & S. H. Wang in Fl. Reip. Pop. sin. 27: 523, pl.121: 4 pl. 122: 1-3.1979; S.R. Lin & X. Z. Zhao in Fl. Fujian. 2: 20, fig. 16. 1985; X. W. Wang in Fl. Anhui 2: 323, fig. 625. 1987; Z. H. Li in

六、分类学处理　　　　　　　　　　　　　　　　　　　97

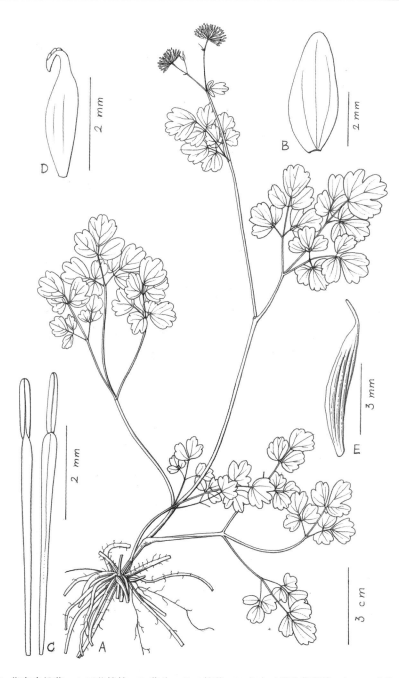

图版7. **华东唐松草**　A.开花植株，B.萼片，C.二雄蕊，D.心皮（据凌苹苹等68），E. 瘦果（据刘晓龙484）。

Plate 7. **Thalictrum fortunei** var. **fortunei**　A. flowering plant, B. sepal, C. two stamens, D. carpel (from P. P. Ling & S. X. Sun 68), E. achene (from X. L. Liu 484).

Fl. Zhejiang 2: 376, fig. 2-366. 1992; W. T. Wang in High. Pl. China 3: 464, fig. 732. 2000; D. Z. Fu & G. Zhu in Fl. China 6: 290. 2001; W. T. Wang in Fl. Jiangxi 2: 174, fig. 179. 2004; T. Fang & J. H. Chen, Field Guid. Wild Pl. China: Gutian Shan, Zhejiang Prov. 322. 2013; M. B. Deng & K. Ye in Fl. Jiangsu 2: 85, fig. 2-137: 1-5. 2013; W. T. Wang in High. Pl. China in Colour 3: 384. 2016; L. Xie in Med. Fl. China 3: 237, with fig. & photo. 2016. Syntypes：浙江：宁波, C. W. Everard 216 (K, not seen; photo, PE); 无准确地点, R. Fortune 28（not seen）.

7a. var. **fortunei**

多年生草本植物，全部无毛。茎高 20-66 cm，自下部或中部分枝。基生叶有长柄，为 2-3 回三出复叶；叶片宽 5-10 cm；小叶草质，宽倒卵形、近圆形、倒卵形或宽卵形，0.6-2.5×0.5-3 cm，基部圆形或浅心形，不明显 3 浅裂，裂片全缘或有 1-2 浅圆齿，脉在下面稍隆起，脉网明显；小叶柄长 0.5-10 mm；叶柄长约 6 cm，基部有短鞘；托叶膜质，宽约 2 mm。复单歧聚花序伞房状，宽 3-8 cm，有 3-15 花；花梗丝形，长 0.6-1.6 cm。花：萼片 4，白色或淡堇色，狭卵形或倒卵形，3-4.5×2.2 mm，顶端圆钝，每一萼片有 3 条不分枝的基生脉。雄蕊 35-47；花丝长 2-4 mm，上部倒披针形或近条形，宽 0.35-0.6 mm，下部狭条形，宽 0.2 mm；花药椭圆形，0.5-1.2×0.2-0.5 mm，顶端钝。心皮 2-6；子房长圆形，2-2.5×0.5 mm；花柱长约 1 mm，直或顶端弯曲，腹面顶端有不明显柱头。瘦果狭长圆形，4-5×1 mm，每侧有 2-3 条细纵肋；宿存花柱长 1-1.2 mm，顶端钩状弯曲或直。3-5 月开花。

Perennial herbs, totally glabrous. Stems 20-66 cm tall, from below or from the middle branched. Basal leaves long petiolate, 2-3-ternate; blades 5-10 cm long; leaflets herbaceous, broad-obovate, suborbicular, obovate or broad-ovate, 0.6-2.5×0.3-3 cm, at base rounded or subcordate, inconspicuously 3-lobed, lobes entire or 2-3-crenate, nerves abaxially slightly prominent and nervous nets conspicuous; petiolules 0.5-10 mm long; petioles ca. 6 cm long, at base shortly vaginate; stipules membranous, ca. 2 mm broad, Compound monochasia corymbiform, 3-8 cm broad, 3-15-flowered; pedicels filiform, 0.6-1.6 cm long. Flower: Sepals 4, white or violetish, narrow-ovate or obovate, 3-4.5×2.2 mm, apex rounded-obtuse, each sepal with 3 simple basal nerves. Stamens 35-47; filaments 2-4 mm long, above oblanceolate or nearly linear, 0.35-0.6 mm broad, below narrow-linear, 0.2 mm broad; anthers elliptic, 0.5-1.2×0.2-0.5 mm, apex obtuse. Carpels 2-6; ovary oblong, 2-2.5×0.5 mm; style ca. 1 mm long, straight or near apex curved, adaxially near apex with an inconspicuous stigma.

Achene narrow-oblong, 4–5×1 mm, on each side with 2–3 longitudinal thin ribs; pesistent style 1–1.2 mm long, at apex hooked or straight.

分布于湖北、江西北部、安徽南部、江苏南部、浙江和福建。生丘陵或山地林下或较阴湿处，海拔 100–1600 m。

根和全草有清湿热、消肿止痛之效，用于痢疾、腹泻、目赤等症。(《中国药用植物志》，2016)

标本登录：湖北：宜昌，郑重 2；阳新，马明华 507。

江西：庐山，钟观光 4289，关克俭 77-082，聂敏祥 92-051，94-022，谭策明 95-215。

安徽：金寨，? 287；霍山，刘焱 A50124；岳西，刘晓龙 984；铜陵，张振华 s. n.；南陵，李林华 s. n.；黄山，alt. 1600 m，刘慎谔，钟补求 2642；绩溪，刘焱 A10062；歙县，刘焱 A110025；宁国，刘焱 A70074。

江苏：句容，宝华山，左景烈 139，关克俭 150，陈彬 44135；苏州，凌苹苹 68。

浙江：龙泉，王景祥 1449；昌化，杭州植物园队 29043；遂昌，植物资源队 59-25686；镇海，钟观光 942；宁波，钟观光 s. n.；天目山，钟观光 116，贺贤育 21730；安吉，邓懋彬 90179。

7b. 珠芽华东唐松草（《植物分类学报》，变种）

var. **bulbiliferum** B. Chen & X. J. Tian in Acta Phytotax. Sin. 43(3):281. 2005; M. B. Deng & K. Ye in Fl. Jiangsu 2: 85, fig. 2-137: 6–7. 2013. Type：江苏：句容，宝华山，2004-05-23，陈彬 523001（holotype, NAS, not seen; isotype, PE）；同地，2004-04-03，陈彬 0403007（paratype, not seen）。

本变种的茎在叶腋具珠芽而与华东唐松草不同。

This variety differs from var. *fortunei* in its stems in leaf axils with bulbils.

特产江苏句容县宝华山。

8. 多枝唐松草（《中国药用植物志》，1964） 图版 8

Thalictrum ramosum B. Boivin in J. Arn. Arb. 26: 115, pl. 1: 12–15. 1945; C. Pei & T. Y. Chou, Icon. Chin. Med. Pl. 7: fig. 328. 1964; Iconogr. Corm. Sin. 1: 672, fig .1343. 1972; W. T. Wang & S. H. Wang in Fl. Reip. Pop. Sin. 27: 523, pl. 122: 4–6. 1979; W. T. Wang in Fl. Guangxi 1: 276, pl. 118: 1–3. 1991; in Vasc. P1. Hengduan Mount. 1: 498. 1993; in Keys Vasc. P1. Wuling Mount. 164. 1995; et in High. P1. China 3: 463, fig. 731. 2000; K. M. Liu, F1. Hunan 2: 649, fig. 2–506. 2000; D. Z. Fu & G. Zhu in Fl. China 3: 275. 2001; Z. Y. Li & M.

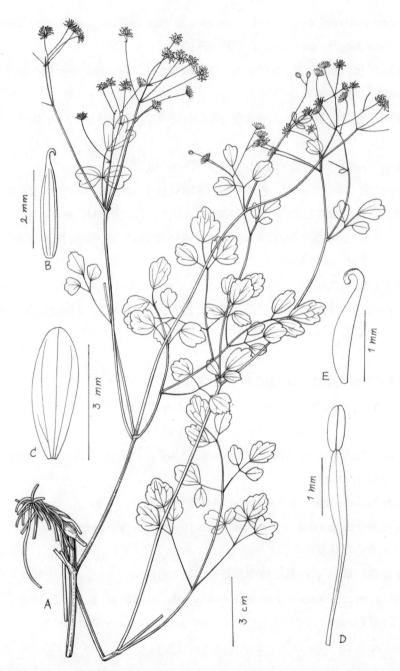

图版8. **多枝唐松草** A. 结果植株,B. 瘦果(据isotype),C. 萼片,D. 雄蕊,E. 心皮。(据四川植物考察队8599)

Plate 8. **Thalictrum ramosum** A. fruiting plant, B. achene (from isotype), C. sepal, D. stamen, E. carpel. (from Sichuan Pl. Exped. 8599)

Ogisu, Pl. Mount Emei 255. 2007; C. X. Yang et al. Keys Vasc. Pl. Chongqing 216. 2009; L. Q. Li et al. Pl. Baishuijiang st. Nat. Reserve Gansu Prov. 71. 2014; L. Q. Li in High. Pl. Daba Mount 118. 2014; L. Xie in Med. Fl. China 3: 236, with fig. & photo. 2016. Type：四川：灌县，alt. 850–950 m, 1930–04–14，汪发瓒 20378（holotype, GH; isotype, PE）；峨眉山，alt. 900 m, 1932–04–15，俞德浚 274（paratypes, GH, PE）；成都，1937–03–15，钱崇澍 5880（paratype, GH, not seen）.

T. fortunei auct. non S. Moore: Franch. Pl. Delav. 17. 1889; Hand.-Mazz. Symb. Sin. 7: 311. 1931.

多年生草本植物，全部无毛。茎高 12–45 cm，自基部之上分枝。基生叶数个，与下部茎生叶为 2–3 回三出复叶；叶片长 7–15 cm；小叶草质，宽卵形、近圆形或宽倒卵形，0.7–2×0.5–1.5 cm，顶端圆钝，有短尖头，基部圆形或浅心形，不明显 3 浅裂，裂片多全缘或有 1–2 小齿，脉上面平，下面稍隆起；小叶柄长 0.6–1.5 cm；叶柄长 7–9 cm，基部有短鞘。复单歧聚伞花序伞房状，宽 8–12 cm，有 5–20 花；花梗丝形，长 5–10 mm。花：萼片 4，早落，淡堇色或白色，长椭圆形，约 3.2×1.6 mm，顶端圆钝，每一萼片具 3 条不分枝的基生脉。雄蕊约 30；花丝长 3–5 mm，上部狭倒披针形，宽 0.3–0.4 mm，下部变为丝形；花药淡黄色，长圆形，约 0.8×0.4 mm，顶端钝。心皮（6–）8–16；子房狭倒卵形，长约 1 mm；花柱长约 0.8 mm，弯曲，腹面有不明显柱头。瘦果披针形，长 3–4.5 mm，每侧有 3 条细纵肋；宿存花柱长 0.5–0.8 mm，钩状弯曲。4 月开花，5–6 月结果。

Perennial herbs, totally glabrous. Stems 12–45 cm tall, above base branched. Basal leaves several, with lower cauline leaves 2–3-ternate; blades 7–15 cm long; leaflets herbaceous, broad-ovate, suborbicular or broad-obovate, 0.7–2×0.5–1.5 cm, at apex rounded-obtuse, mucronate, at base rounded or subcordate, inconspicuously 3-lobed, lobes mostly entire or 1–2-denticulate, nerves adaxially flat, abaxially slightly prominent; petiolules 0.6–1.5 cm long; petioles 7–9 cm long, at base shortly vaginate. Compound monochasia corymbiform, 8–12 cm broad, 5–20-flowered; pedicels filiform, 5–10 mm long. Flower: Sepals 4, caducous, violetish or white, long elliptic, ca. 3.2×1.6 mm, apex rounded-obtuse, each sepal with 3 simple basal nerves. Stamens ca. 30; filaments 3–5 mm long, above narrow-oblanceolate, 0.3–0.4 mm broad, below filiform; anthers yellowish, oblong, ca. 0.8×0.4 mm, apex obtuse. Carpels (6–)8–16; ovary narrow-obovate, ca. 1 mm long; style ca. 0.8 mm long, curved, adaxially with an inconspicuous stigma. Achene lanceolate, 3–4.5 mm long, on each side with 3 thin longitudinal ribs; persistent style 0.5–0.8 mm long, hooked.

分布于甘肃南部、四川、重庆、湖南西部、贵州东北部和广西西北部。生丘陵或低山灌丛中、竹林中或沟边，海拔 300–1800 m。

全草和根有清热解毒之效，用于痢疾、黄疸、目赤等症。(《中国药用植物志》，2016)

标本登录：甘肃：舟曲，铁坝，alt. 1450 m，白龙江队 1311；文县，刘家坪，白水江队 2286。

四川：石棉，alt. 1800 m，赵佐成 2736，谢朝俊 39528；屏山，alt. 500 m，植物资源队 59–784；峨眉山，alt. 400–1100 m，方文培 13974，14084，15960，徐洪贵 180，2418；茂县，四川植物队 8599；南江，巴山队 5550，5625；嘉定，乌龙寺，俞德浚 203。

重庆：巴县，俞德浚 2855；万州，植物所三峡队 08–741。

湖南：隆回，邵阳药检所队 s. n.；洪江，? s. n.；黔阳，安江农校队 5；雪峰山，李泽棠 1738；新宁，alt. 300 m，罗林波 502。

贵州：江口，张志松等 40006。

广西：凤山，黄维岳 30452。

9. 新宁唐松草 (《广西植物》)　图版 9

Thalictrum xinningense W. T. Wang in Guihaia 37(4):408, fig.2. 2017. Holotype：湖南：新宁县，万丰林场，大水岭，alt. 900 m，1995-06-10，罗林波 777 (PE).

多年生草本植物，全部无毛。茎高 60–100 cm，近基部处直径 6.5 mm。上部茎生叶为 3–4 回三出羽状复叶，具短柄；叶片宽卵形，11–32×12–25 cm；小叶具柄，纸质，卵形、宽卵形、倒卵形或椭圆形，1.5–3.4×1.2–4 cm，顶端圆形或截状圆形，有小短尖头，基部浅心形、圆形或宽楔形，3 浅裂（裂片顶端钝，全缘，或中央裂片每侧边缘具 1 钝齿），具三出脉或五出脉，基生脉上面平，下面隆起，并与细脉形成明显脉网；小叶柄长 0.1–3 mm；叶柄长 1–4 cm。聚伞圆锥花序顶生和腋生，包括花序梗长 18–22 cm，伞房状，4–5 回分枝，有多数花；花序梗长 1–12 cm；基部苞片叶状，上部苞片简单，长椭圆形或钻形，长 1–3 mm；花梗长 1.2–6 mm。花：萼片 4，狭椭圆形，3–3.5×0.8–1.5 mm，每一萼片具 3 基生脉（中脉 1 回二歧状分枝，2 侧生脉不分枝）。雄蕊约 30；花丝长 1.8–3 mm，上部狭条状棒形，宽 0.15–0.25 mm，下部丝形，粗 0.1 mm；花药黄色，长圆形，稀宽长圆形，(0.8–)1×0.3–0.45 mm，顶端钝。心皮 3–6；子房长椭圆形，长约 1 mm；花柱长约 1 mm，顶端钩状弯曲，腹面具不明显柱头。瘦果扁，狭倒卵形，约 2×1 mm，每侧具 2 条纵肋；宿存花柱长约 1 mm，顶端钩状弯曲。5–6 月开花。

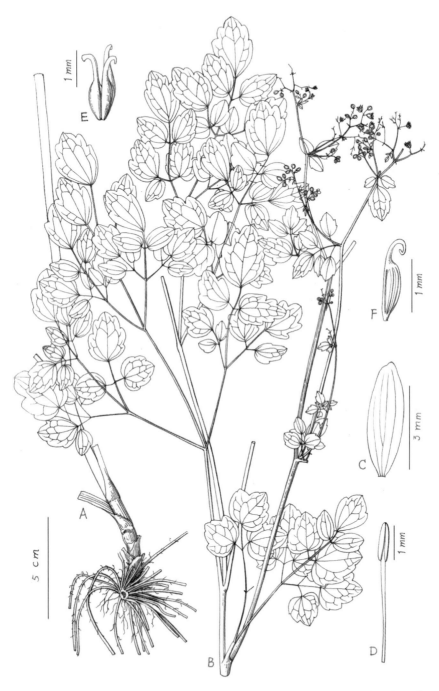

图版9. **新宁唐松草** A.开花植株下部，B.开花茎上部，C.萼片，D.雄蕊，E.雌蕊群，F.瘦果。（据hototype）

Plate 9. **Thalictrum xinningense** A. lower part of flowering plant, B. upper part of flowering stem, C. sepal, D. stamen, E. gynoecium, F. achene. (from holotype)

Perennial herbs, totally glabrous. Stems 60–100 cm tall, near base 6.5 mm in diam. Upper cauline leaves 3–4-ternate-pinnate, shortly petiolate; blades broad-ovate, 11–32 × 12–25 cm; leaflets petiolulate, papery, ovate, broad-ovate, obovate or elliptic, 1.5–3.4 × 1.2–4 cm, at apex rounded or truncate-rounded, at base subcordate, rounded or broadly cuneate, 3-lobed (lobes at apex obtuse, entire or central lobe obtusely 1-dentate at margin of each side), 3–5-nerved, basal nerves adaxially flat, abaxially prominent, with nervules forming conspicuous nervous nets; petiolules 0.1–3 mm long; petioles 1–4 cm long. Thyrses terminal and axillary, including peduncles 18–22 cm long, corymbiform, 4–5 times branched, many-flowered; peduncles 1–12 cm long; basal bracts foliaceous, upper bracts simple, long elliptic or subulate, 1–3 mm long; pedicels 1.2–6 mm long. Flower: Sepals 4, first reddish, later turning white, narrow-elliptic, 3–3.5 × 0.8–1.5 mm, each sepal with 3 basal nerves (midrib once dichotomizing, and 2 lateral nerves simple). Stamens ca. 30; filaments 1.8–3 mm long, above narrowly linear-clavate, 0.15–0.25 mm broad, below filiform, 0.1 mm across; anthers yellow, oblong, rarely broad-oblong, (0.8–) 1 × 0.3–0.45 mm, apex obtuse. Carpels 3–6; ovary long elliptic, ca. 1 mm long; style ca. 1 mm long, apex hooked, adaxially with an inconspicuous stigma. Achene compressed, narrow-obovate, ca. 2 × 1 mm, on each side with 2 longitudinal ribs; persistent style ca. 1 mm long, apex hooked.

特产湖南新宁县万丰林场。生山谷溪边，海拔 900 m。

10. 密叶唐松草（《中国高等植物图鉴》） 图版 10

Thalictrum myriophyllum Ohwi in Acta Phytotax. Geobot. 2: 156. 1933; Iconogr. Corm. Sin. 1: 673, fig. 1345. 1972; Liu & Hsieh in Fl. Taiwan 2: 510. 1976; W. T. Wang & S. H. Wang in Fl. Reip. Pop. Sin. 27: 526, pl. 125: 1–3. 1979; T. Y. Yang & T. C. Huang in Taiwania 34: 61, fig. 3. 1989; et in Fl. Taiwan, 2nd ed., 2: 567, pl. 271. 1996; D. Z. Fu & G. Zhu in Fl. China 6: 294. 2001.Type: 台湾：宜兰，南湖大山 . J. Ohwi 4188 (holotype, KYO, not seen); 同地，J. Ohwi 3997 (paratype, KYO; photo, PE).

多年生草本植物。茎高 34–70 cm，疏被短柔毛，变无毛，分枝，稀不分枝。茎生叶和下部茎生叶为 4–5 回三出复叶，具长柄；叶片长 5–10 cm；小叶纸质，卵形、倒卵形或近圆形，3–8×2.5–5 mm，基部楔形或圆形，3 浅裂或不分裂（裂片全缘），上面无毛，下面有白霜，在脉上疏被短柔毛或变无毛，脉稍隆起；叶柄长 2.4–5.8 cm。上部茎生叶较小，为 2–3 回三出复叶。复单歧聚伞花序顶生并腋生，伞房状，有稀疏的花；花梗长达 1 cm。花：

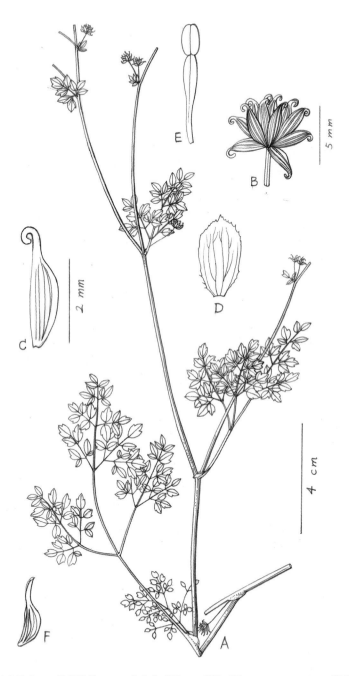

图版10. **密叶唐松草** A.结果植株，B.一朵花的瘦果，C.瘦果（据S. Suzuki 2386），D.萼片，E.雄蕊，F.心皮（抄自 *Flora of Taiwan*, 2nd ed.）

Plate 10. **Thalictrum myriophullum** A. fruiting plant, B. achenes of a flower, C. achene (from S. Suzuki 2386), D. sepal, E. stamen, F. carpel (redrawn from *Flora of Taiwan*, 2nd ed.)

萼片 4，早落，淡绿黄色，宽倒卵形，2–3×1.3–1.5 mm，顶端有小齿，近无毛，每一萼片有 3 条 2–4 回二歧状分枝的基生脉。雄蕊约 25，无毛；花丝长 2–2.5 mm，上部狭倒披针形，下部狭条形；花药长圆形，长 0.5–0.7 mm，顶端钝。心皮约 20，无毛，无柄，长 1.5–1.7 mm，花柱钩状弯曲。瘦果无柄，纺锤形，长 2–3 mm，每侧有 3 条纵肋；宿存花柱拳卷。7 月开花。

Perennial herbs. Stems 34–70 cm tall, sparsely puberulous, glabrescent, branched, rarely simple. Basal and lower cauline leaves 4–5-ternate, long petiolate; blades 5–10 cm long; leaflets papery, ovate, obovate or suborbicular, 3–8×2.5–5 mm, at base cuneate or rounded, 3-lobed or undivided (lobes entire), adaxially glabrous, abaxially glaucous and on nerves sparsely puberulous or glabrescent, nerves slightly prominent; petioles. 2.4–5.8 cm long. Compound monochasia terminal and axillary, corymbiform, with loose flowers; pedicels up to 1 cm long. Flower: Sepals 4, caducous, greenish-yellowish, broad-obovate, 2–3×1.3–1.5 mm, apex denticulate, subglabrous, each sepal with 3 2–4 times dichotomizing basal nerves. Stamens ca. 25, glabrous; filaments 2–2.5 mm long, above narrow-oblanceolate, below narrow-linear; anthers oblong, 0.5–0.7 mm long, apex obtuse. Carpels ca. 20, glabrous, sessile, 1.5–1.7 mm long, style hooked. Achene sessile, fusiform, 2–3 mm long, on each side with 3 longitudinal ribs; persistent style at apex circinate.

特产台湾中央山脉南湖大山、雪山、玉山等山区。生高山林中或林边草地，海拔 3000–3500 m。

标本登录：台湾：南投，自天池至南湖大山，alt. 3184 m，S. L. Kelley, Y. C. Kao & C. T. Huang 220–98(GH)；Mt. Kiraisyu-nampo, 1929–08–24, S. Suzuki 2386（NAS）。

11. 云南唐松草（《植物分类学报》） 图版 11

Thalictrum yunnanense W. T. Wang in Acta Phytotax. Sin. 32(5):471. 1994; et in Fl. Yunnan. 11: 158, pl. 48: 46. 2000; D. Z. Fu & G. Zhu in Fl. China 6: 302. 2001. Type：云南：昆明，西山，华亭寺，1946–08–16，刘慎谔 16668（holotype, PE），19727（paratype, PE）；昆明，太华山，钟观光 2140（paratype, PE）；宾川，鸡足山，钟观光 2466（paratype, PE）。

T. scabrifolium auct. non Franch.: W. T. Wang & S. H. Wang in Fl. Reip. Pop. Sin. 27: 519, pl. 123: 5–8. 1979.

11a. var. **yunnanense**

多年生草本植物。茎高 25–40 cm，疏被短柔毛，变无毛，上部分枝。基生叶和下部

六、分类学处理　　　　　　　　　　　　　　　　　　　　107

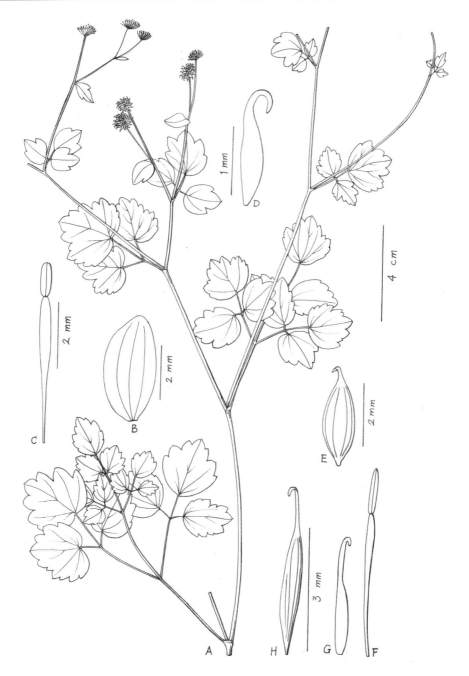

图版11. 云南唐松草 A–E. **模式变种** A. 开花茎，B. 萼片，C. 雄蕊，D. 心皮，E. 瘦果。（据holotype）
F–H. **滇南唐松草** F. 雄蕊，G. 心皮，H. 瘦果。（据holotype）

Plate 11. **Thalictrum yunnanense** A–E. var. **yunnanense** A. flowering stem, B. sepal, C. stamen, D. carpel, E. achene. (from holotype)　F–H. var. **austro-yunnanense** F. stamen, G. carpel, H. achene. (from holotype)

茎生叶为 2-3 回三出复叶；叶片长约 9 cm；小叶纸质或草质，近圆形，宽卵形、宽菱形或卵形，1-2.2×0.9-2.8 cm，基部浅心形或宽楔形，3 浅裂，裂片有 2-3 小齿，稀全缘，上面疏被短柔毛（毛长不达 0.1 mm），下面在脉网上疏被短硬毛（毛长 0.2-0.4 mm）；小叶柄长 0.3-1.8 cm，与叶柄均有短毛；叶柄长 5.5-7 cm。复单歧聚伞花序顶生和腋生，伞房状，宽约 10 cm；花梗细，长 0.7-2 cm，有短毛。花：萼片 4，早落，白绿色，宽椭圆形，约 3.4×2 mm，顶端钝，每一萼片有 3 条不分枝的基生脉。雄蕊约 25，无毛；花丝长 4-4.5 mm，上部条状倒披针形，宽 0.3-0.4 mm，下部丝形，粗 0.1 mm；花药长圆形，0.8-1×0.3-0.4 mm，顶端钝。心皮 11-40，无毛；子房长椭圆形，约 1.2×0.4 mm；花柱长 1 mm，上部钩状弯曲。瘦果扁，椭圆形或狭卵形，2.2×0.7-1 mm，无毛，每侧有 2 条纵肋；宿存花柱长约 1 mm，顶端向后弯曲。8-9 月开花。

Perennial herbs. Stems 25-40 cm tall, sparsely puberulous, glabrescent, above branched. Basal and lower cauline leaves 2-3-ternate; blades ca. 9 cm long; leaflets papery or herbaceous, suborbicular, broad-ovate, broad-rhombic or ovate, 1-2.2×0.9-2.8 cm, at base subcordate or broad-cuneate, 3-lobed (lobes 2-3-denticulate, rarely entire), adaxially sparsely puberulous (hairs less than 0.1 mm long), abaxially on nervous nets hirtellous (hairs 0.2-0.4 mm long); petiolules 0.3-1.8 cm long, with petioles sparsely puberulous; petioles 5.5-7 cm long. Compound monochasia terminal and axillary, corymbiform, ca. 10 cm broad; pedicels slender, 0.7-2 cm long, with short hairs. Flower: Sepals 4, caducous, white-greenish, broad-elliptic, ca. 3.4×2 mm, apex obtuse, each sepal with 3 simple basal nerves. Stamens ca. 25, glabrous; filaments 4-4.5 mm long, above linear-oblanceolate, 0.3-0.4 mm broad, below filiform, 0.1 mm across; anthers oblong, 0.8-1×0.3-0.4 mm, apex obtuse. Carpels 11-40, glabrous; ovary long elliptic, ca. 1.2×0.4 mm; style ca. 1 mm long, above hooked. Achene compressed, elliptic or narrow-ovate, 2.2×0.7-1 mm, glabrous, on each side with 2 longitudinal ribs; persistent style ca. 1 mm long, apex recurved.

特产云南宾川至昆明一带山地。生山地林边，海拔 2000 m。

11b. 滇南唐松草（《植物分类学报》，变种）

var. austroyunnanense Y. Y. Qian in Acta Phytotax. Sin. 35(3): 262. 1997; W. T. Wang in Fl. Yunnan. 11: 160. 2000; D. Z. Fu & G. Zhu in Fl. China 6: 302. 2001. Type：云南：澜沧县，alt. 2000 m, 1993-07-08, 钱义咏 2927（holotype, PE; isotype, SMAO），3042（PE, SMAO）.

本变种与云南唐松草的区别在于本变种的雄蕊花药呈狭长圆形，瘦果呈纺锤形，具较长宿存花柱（长 1.5 mm）。

This variety differs from var. *yunnanense* in its narrow-oblong anthers and fusiform achenes with longer persistent style 1.5 mm in length.

特产云南澜沧县。生山地林中，海拔 2000 m。

12. 爪哇唐松草（《中国高等植物图鉴》） 图版 12

Thalictrum javanicum Bl. Bijdr. 2. 1825; Hook. f. & Thoms. Fl. Ind. 16. 1855; et in Hook. f. Fl. Brit. Ind. 1: 13. 1872; Lecoy in Bull. Soc. Bot. Belg. 24: 167, pl. 6: 15. 1885; Collett, Fl. Siml. 8, fig. 3. 1902; Finet & Gagnep. in Bull. Soc. Bot. France 50: 611. 1903; Finet in J. de Bot. 21: 20. 1908; Hand.-Mazz. Symb. Sin. 7: 310. 1931, p.p. excl. syn. *T. lecoyeri* Franch.; Boivin in J. Arn. Arb. 26: 115. 1945; Backer & Bakh. f. Fl. Java 1: 143. 1963; Iconogr. Corm. Sin. 1: 673, fig. 1346. 1972; S. H. Fu, Fl. Hupeh. 1: 356, fig. 497. 1976; W. T. Wang & S. H. Wang in Fl. Reip. Pop. Sin. 27: 521, pl. 24. 1979, p.p. excl. syn. *T. lecoyeri* Franch. et *T. sessile* Hayata; Grierson, Fl. Bhutan 1(2):298. 1984; W. T. Wang in Fl. Xizang. 2: 66. 1985; S. R. Lin & X. Z. Zhao in Fl. Fujian. 2: 20, fig. 15. 1985; Y. K. Li in Fl. Guizhou. 3: 69, pl. 28: 1–3. 1990; W. T. Wang in Fl. Guangxi 1: 276. 1991; Z. H. Lin in Fl. Zhejiang 2: 377, fig. 2–368. 1992; Rau in Sharma et al. Fl. India 1: 136 1993; W. T. Wang in Vasc. Pl. Hengduan Mount. 2: 498. 1993; et Keys Vasc. Pl. Wuling Mount. 164. 1995; in Fl. Yunnan. 11: 160. 2000; et in High. Pl. China 3: 464, fig. 733. 2000; K. M. Liu in Fl. Hunan 2: 652, fig. 2–507. 2000; D. Z. Fu & G. Zhu in Fl. China 6: 292. 2001; p.p. excl. syn.*T. lecoyeri* Franch.; Y. M. Shui & W. H. Chen, Seed Pl. Honghe Reg. S.–E. Yunnan 78. 2003; R. J. Wang in Fl. Guangdong 5: 20. 2003, p.p. excl. syn. *T. lecoyeri* Franch.; W. T. Wang in Fl. Jiangxi 2: 174, fig. 180. 2004; Z. Y. Li & M. Ogisu, Pl. Mount. Emei 255. 2007; Z. C. Luo & Y. B. Luo, Pl. Xinning 142. 2008; C. x. Yang et al. Keys Vasc. Pl. Chongqing 216. 2009; Q. L. Gan, Fl. Zhuxi Suppl. 224, fig. 292. 2011; L. Q. Li in High. Pl. Daba Mount. 118. 2014; L. Q. Li et al. Pl. Baishuijiang St. Nat. Reserve Gansu Prov. 71.2014; W. T. Wang in High. Pl. China in Colour 3: 384. 2016; L. Xie in Med. Fl. China 3: 234, with fig. & photo. 2016; X. T. Ma et al. Wild Flow. Xizang 27, photo 53. 2016. Described from Java, Indonesia.

T. argyi Lévl. in Bull. Herb. Boiss., ser. 2, 6: 504. 1906; et in Mem. Real. Acad. Ciene. Art. Barcelona 12(22): 19. 1916. Holotype：江苏，Gong-chan，d'Argy s. n. (E, not seen; photo, PE).

多年生草本植物，全部无毛。茎高（30–）50–100 cm，中部以上分枝。茎生叶为 3–4

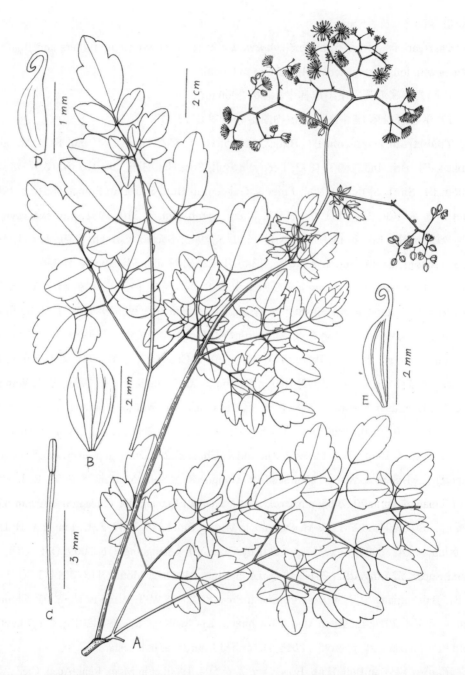

图版12. **爪哇唐松草** A. 开花茎上部，B. 萼片，C. 雄蕊，D. 心皮（据张无休等70），E. 瘦果（据中甸队62-1139）。

Plate 12. **Thalictrum javanicum** A. upper part of flowring stem, B. sepal, C. stamen, D. carpel (from W. X. Zhang et al. 70), E. achene (from Zhongdian Exped. 62-1139).

回三出复叶；叶片长 6–25 cm；小叶纸质，倒卵形、椭圆形、菱形或近圆形，1–2.5×0.7–1.8 cm，基部宽楔形、圆形或浅心形，3 浅裂（裂片有 1–3 小齿或全缘），脉上面平，下面隆起，脉网明显；小叶柄长 0.5–1.4 cm；叶柄长达 5.5 cm；托叶褐色，宽 2–3 mm，边缘流苏状。复单歧聚伞花序在形状和大小等方面有甚大变异，有时伞房状，宽 3–25 cm，有少数至多数花，有时呈圆锥状，长 10–40 cm，有少数至多数花；花梗长 3–7(–10–28)mm。花：萼片 4，早落，倒卵形，2.5–3×1.5–1.8 mm，顶端圆形，每一萼片具 3 条不分枝或二歧状分枝的基生脉。雄蕊约 30；花丝长 2.5–4 mm，上部狭倒披针形，比花药稍宽或与花药等宽，下部变狭呈狭条形；花药长圆形，长 0.6–1 mm，顶端钝。心皮 8–15；子房斜椭圆形，1×0.6 mm；花柱长约 1 mm，钩状弯曲；柱头长约 0.5 mm，不明显。瘦果半椭圆形，约 2.5×0.8 mm，每侧有 2–3 条纵肋；宿存花柱长 0.6–1 mm，拳卷状或钩状弯曲。4–7 月开花。

Perennial herbs, totally glabrous, Stems (30–)50–100 cm tall, above the middle branched. Cauline leaves 3–4–ternate; blades 6–25 cm long; leaflets papery, obovate, elliptic, rhombic or suborbicular, 1–2.5×0.7–1.8 cm, at base broad-cuneate, rounded or subcordate, 3–lobed, lobes 1–3–denticulate or entire, nerves adaxially flat, abaxially prominent, nervous nets conspicuous; petiolules 0.5–1.4 cm long; petioles up to 5.5 cm long; stipules brown, 2–3 mm broad, margin fimbrillate. Compound monochasia variable both in shape and size, sometimes corymbiform, 3–25 cm broad, few- to many-flowered, sometimes paniculiform, 10–40 cm long, also few- to many-flowered; pedicels 3–7(–10–28)mm long. Flower: Sepals 4, caducous, obovate, 2.5–3×1.5–1.8 mm, apex rounded, each sepal with 3 simple or dichotomizing basal nerves. Stamens ca. 30; filaments 2.5–4 mm long, above narrow-oblanceolate, slightly broader than or as broad as anthers, below narrowed and narrow-linear; anthers oblong, 0.6–1 mm long, apex obtuse. Carpels 8–15; ovary obliquely elliptic, 1×0.6 mm; style ca. 1 mm long, hooked; stigma ca. 0.5 mm long. inconspicuous. Achene semi-elliptic, ca. 2.5×0.8 mm, on each side with 2–3 longitudinal ribs; persistent style 0.6–1 mm long, circinate or hooked.

分布于中国西藏南部、云南、广西、广东、福建、浙江、江苏、江西、湖南、湖北、贵州、重庆、四川、甘肃南部；以及不丹、尼泊尔、印度、斯里兰卡、印度尼西亚等国。生山地林中、沟边或陡崖边阴湿处，海拔 800–3400 m。

根有清热解毒之效，用于跌打损伤等症。全草用于关节痛。根含小檗碱等生物碱。（《中国药用植物志》，2016）

标本登录：西藏：察隅，察瓦龙，王启无 65288；米林，西藏中草药队 4080；聂拉木，张永田, 郎楷永 4523。

云南: 镇康, 俞德浚 17217; 腾冲, Goligong Shan Biodiv. Survey 29876; 泸水, 滇西北队 10415; 剑川, 秦仁昌 23067; 福贡, 蔡希陶 68455; 维西, 蔡希陶 59734, 植物所横断山队 81-1785; 贡山, 姜恕 9248; 德钦, 谢磊, 毛建丰 150; 中甸, 中甸队 62-1139, 杨亲二, 袁琼 1822; 丽江, 韩裕丰 81-1374; 永胜, 韩裕丰 81-929; 宾川, 谢磊, 毛建丰 18; 石林, 税玉民等 64187; 永善, 蔡希陶 51079; 大关, 孙必兴 758; 东川, 蓝顺彬 100; 巧家, 孙必兴 899; 文山, 吉占和 417; 屏边, 蔡希陶 62603。

广东: 阳山, 韶关队 1399。

福建: 崇安, 武夷山, 简焯坡等 51146。

江西: 铅山, 武夷山, 简焯坡等 400982。

湖南: 新宁, 罗仲春 941, 罗毅波 2495; 永顺, 北京队 1007; 桑植, 刘林翰 9086, 北京队 3841, 4511; 慈利, 湘西队 84-1010; 石门, 壶瓶山队 87-1458。

湖北: 宣恩, 李洪钧 5038, 王映明 4873; 鹤峰, 王映明 5916。

贵州: 盘县, 北京队 90; 雷山, 简焯坡等 51146; 梵净山, 张无休 70。

重庆: 南川, 金佛山, 曲桂龄 1440, 李国凤 61995, 62585, 熊济华, 周子林 92918, 刘正宇 1790; 武隆, 易思荣 367; 酉阳, 西师生物系 2049; 黔江, 赵佐成 88-1938; 丰都, 溥发鼎, 曹亚玲 658; 云阳, 三峡队 2736; 开县, 巴山队 2275, 2445; 巫溪, 陈耀东等 103, 2275, 2457。

四川: 乡城。D. E. Boufford, B. Bartholomew et al. 28973; 木里, 韩裕丰等 83-104; 西昌, 俞德浚 1157; 冕宁, 武素功 1768; 越西, 植物资源队 59-4516; 甘洛, 植物资源队 59-4076; 德昌, 西师生物系 12159; 石棉, 谢朝俊 41994; 美姑, 管中天 9092; 雷波, 俞德浚 3834, 管中天 8997; 马边, 俞德浚 5514; 峨边, 四川植被队 13221; 屏山, 植物资源队 59-1056; 洪雅, 张超 1016; 峨眉山, 方文培 15662, 汪发瓒 23459, 杨光辉 55446; 康定, 方文培 3627; 二郎山, 姜恕 1915, 关克俭, 王文采等 1833; 贡嘎山, 郎楷永, 李良千, 费勇 1353; 宝兴, 曲桂龄 3156, 张秀实 6377; 绰斯甲, 第八森林队 2317; 布拖, 植物资源队 59-5848; 崇州, 冯正波 1263; 都江堰, 姚淦 9008。

甘肃: 文县, 傅竞秋 2604。

13. 微毛唐松草(《植物分类学报》)　图版 13

Thalictrum lecoyeri Franch. Pl. Delav. 16, pl. 5. 1889; W. T. Wang in Acte. Phytotax. Sin. 31(3): 212. 1993; in Fl. Yunnan. 11: 161. 2000; et in High. Pl. China 3: 464. 2000. Holotype:

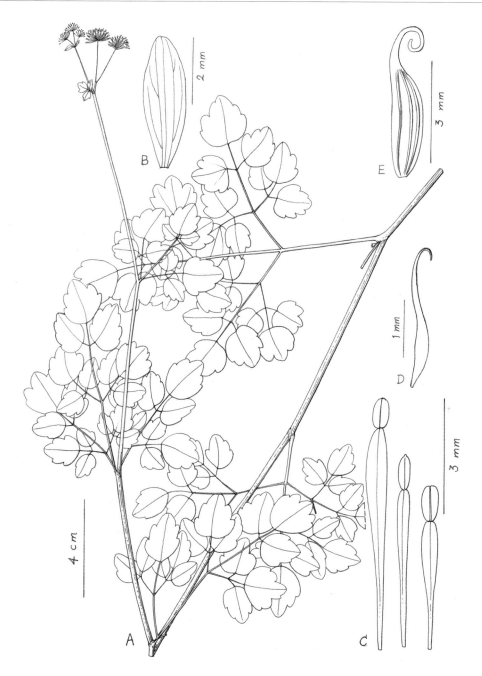

图版13. **微毛唐松草** A. 开花茎上部，B. 萼片，C. 三雄蕊，D. 心皮（据俞德浚9987），E. 瘦果（据植物所横断山队3436）。

Plate 13. **Thalictrum lecoyeri** var. **lecoyeri** A. upper part of flowering stem, B. sepal, C. three stamens, D. carpel (from T. T. Tü 9987), E. achene (from Hengduan Shan Expex. Inst. Bot. 3436).

云南：洱源之北 Hia-lo-pin, alt. 2500 m, 1886-07-04, Delavay 2124 (P; photo, PE).

T. javanicum Bl. var. *puberulum* W. T. Wang in Fl. Reip. Pop. Sin. 27: 521, 617, pl. 119: 1, pl. 120: 1, pl. 127: 1-4. 1979; Y. K. Li in Fl. Guizhou. 3: 69. 1986; W. T. Wang in Vasc. Pl. Hengduan Mount. 1: 498. 1993; D. Z. Fu & G. Zhu in Fl. China 6: 292. 2001; L. Xie in Med. Fl. China 3: 235, with photo. 2016. Type: 云南：洱源,alt. 2600 m, 1963-07-25, 金沙江队 6188（holotype, PE）；中甸，哈巴山队 62-1138（paratype,PE）。四川：木里，俞德浚 7156（paratype,PE）；西昌，俞德浚 1085（paratype, PE）；康定，榆林宫，胡文光，何铸 10547（paratype, PE）.

T. javanicum auct. non Bl.: Finet & Gagnep. in Bull. Soc. Bot. France 50: 611. 1903, p.p. quoad Delaway 2124; Hand.-Mazz. Symb. Sin. 7: 310. 1931, p.p. quoad syn. *T. lecoyeri* Franch.; W. T. Wang & S. H. Wang in Fl. Reip Pop. Sin. 27: 521. 1979, p.p. quoad syn. *T. lecoyeri* Franch.; D. Z. Fu & G. Zhu in Fl. China 6: 292. 2001, p.p. quoad syn. *T. lecoyeri* Franch.

13a. var. **lecoyeri**

多年生草本植物。茎高 30-100 cm，在近节处与叶鞘、小叶柄均被短柔毛（毛长约 0.1 mm），通常分枝。基生叶和下部茎生叶均为 2-3 回三出羽状复叶；叶片 6.5-11×10-16 cm；小叶纸质，圆倒卵形、菱形、近圆形或卵形，0.6-2.2×0.8-2.5 cm，基部圆形、浅心形或宽楔形，3 浅裂，稀 3 中裂，边缘有少数钝齿，两面无毛或下面脉上疏被短毛，具 3 或 5 出脉，脉下面隆起；小叶柄长 1.5-10 mm；叶柄长达 9 cm，基部有狭鞘。复单歧聚伞花序顶生，伞房状，宽 2-12 cm，有 3-12 花，有时 4-5 个形成顶生聚伞圆锥花序，长达 30 cm；花梗长 0.3-3 cm，无毛。花无毛：萼片 4，早落，淡绿色，宽长圆形或长圆状倒卵形，2.5-4×1.5-2 mm，顶端圆形，每一萼片具 3 条不分枝或二歧状分枝的基生脉。雄蕊 24-50；花丝长 2.5-5.8 mm，上部倒披针形，宽 0.3-0.5 mm，下部狭条形，宽 0.15 mm；花药宽长圆形，0.6-1×0.3-0.5 mm，顶端钝。心皮（9-）16-30；子房新月形，约 1.2×0.5 mm；花柱长约 1.5 mm；柱头长 0.4-0.5 mm，不明显。瘦果扁，斜长圆形，约 3×1 mm，每侧有 3 条纵肋；宿存花柱长 1.8-2 mm，坚硬，拳卷状弯曲。7-8 月开花。

Perennial herbs. Stems 30-100 cm tall, near nodes with leaf sheath and petiolules puberulous (hairs ca. 0.1 mm long), usually branched. Basal and lower cauline leaves 2-3-ternate-pinnate; blades 6.5-11×10-16 cm; leaflets papery, orbicular-obovate, rhombic, suborbicular or ovate, 0.6-2.2×0.8-2.5 mm, at base rounded, subcordate or broad-cuneate,

3-lobed, rarely 3-fid, at margin obtusely few-dentate, on both surfaces glabrous or abaxially on nerves sparsely puberulous, 3-5-nerved, nerves abaxially prominent; petiolules 1.5-10 mm long; petioles up to 9 cm long, at base narrowly vaginate. Compound monochasia terminal, corymbiform, 2-12 cm broad, 3-12-flowered, sometimes 4-5 arranged to a terminal thyrse up to 30 cm long; pedicels 0.3-3 cm long. Flower glabrous: Sepal 4, caducous, greenish, broad-oblong or oblong-obovate, 2.5-4×1.5-2 mm, apex rounded, each sepal with 3 simple or dichotomizing basal nerves. Stamens 24-50; filaments 2.5-5.8 mm long, above oblanceolate, 0.3-0.5 mm broad, below narrow-linear, 0.15 mm broad; anthers broad-oblong, 0.6-1×0.3-0.5 mm, apex obtuse. Carpels (9-)16-30; ovary lunate, ca. 1.2×0.5 mm; style ca. 1.5 mm long; stigma 0.4-0.5 mm long, inconspicuous. Achene compressed, obliquely oblong, ca. 3×1 mm, on each side with 3 longitudinal ribs; persistent style 1.8-2 mm long, rigid, circinate.

分布于西藏东南部、云南、四川西部、贵州西部。生山地林中、灌丛中、草坡或沟边；海拔分布：在云南 2500-3200 m，在四川西部，2000-2300 m，在贵州西部 1500 m。

标本登录：西藏：察隅，倪志诚等 297a，837a。

云南：腾冲，Gaoligong Shan Biodiv. Survey 29876，29908（GH）；凤庆，俞德浚 16604；大理，谢磊，毛建丰 0.3-0.5；兰坪，植物所横断山队 81-856；维西，王启无 63927，植物所横断山队 81-1350；德钦，俞德浚 9574，9987，冯国楣 6294，植物所横断山队 81-3436；中甸，俞德浚 12588，谢磊，毛建丰 89，D. E. Boufford et al. 35394；丽江，俞德浚 15310，秦仁昌 20728；宁蒗，谢磊，毛建丰 55；鹤庆，金沙江队 4705；昆明，冯国楣 10469。

四川：盐源，青藏队 83-12131；雷波，管中天 8488；昭觉，四川植物队 12831；美姑，四川植物队 13119；洪溪，植物资源队 59-1057；贡嘎山，郎楷永，李良千，费勇 1084；木里，俞德浚 14409，D. E. Boufford et al. 42714；稻城，四川植物队 2062；九龙，D. E. Boufford et al. 33173；石棉，谢朝俊 41219；康定，刘振书 808；雅江，郎楷永，李良千，费勇 2975；丹八，D. E. Boufford et al. 37922。

贵州：威宁，毕节队 125。

13b. **柔柱微毛唐松草**（变种）

var. **debilistylum** W. T. Wang, var. nov. Type：台湾（Taiwan）：南投县，仁爱乡（Nantou Xian, Renai Xiang），alt. 2370 m，林边（at forest margin），1998-09-03，杨宗愈（T. Y. Yang）11184（holotype, PE）；同地（same locality），alt. 2850 m，2001-09-07，T. Y. Yang，

Y. Kadota, G. P. Hsieh, Y. C. Hsu et al. 13935(PE); 高雄, 桃源乡, 关山 (Gaoxiong Xian, Taoyuan Xiang, Guanshan), alt. 3500 m, on forest floor, rocky and mossy shaded, moist, 1998-08-20, 钟国芳等 (K. F. Chung et al.) 1061 (PE).

T. javanicum Bl. var. *puberulum* auct. non W. T. Wang: T. Y. Yang et T. C. Huang in Taiwania 34: 59. 1989; et in Fl. Taiwan, 2nd ed., 2: 565, pl. 270. 1996.

A var. *lecoyeri* differt caulibus et foliorum vaginis pilis longioribus usque ad 0.2-0.3 mm longis, flore staminibus pauciorbus 13-15 praedito, acheniorum stylis persistentibus debilibus. In var. *lecoyeri*, caules et foliorum vaginae pilis ca. 0.1 mm longis tectae, flos staminibus pluribus 24-50 praeditus, et acheniorum styli persistentes rigidi sunt.

本变种与模式变种的区别在于本变种的茎和叶鞘的毛较长, 长达 0.2-0.3 mm, 花的雄蕊较少, 为 13-15 枚, 瘦果的宿存花柱柔软。

特产台湾中央山脉。生山地林边、林中或草坡上, 海拔 2000-3500 m。

14. 玉山唐松草 (**Taiwania**)　图版 14

Thalictrum sessile Hayata, Icon. Pl. Formos. 3: 6, fig. 2. 1913; Liu & Hsieh. in Fl. Taiwan 2: 512. 1976; T. Y. Yang & T. C. Huang in Taiwania 34: 63, fig. 5. 1989; et in Fl. Taiwan, 2nd 2, 2: 570, pl. 273. 1996. Syntypes: 台湾: 嘉义, 玉山, 1909-12, U. Mori s. n. (T1, not seen; photo, PE); Tohozan 1909-10, S. Sasaki s. n.(T1, not seen)

T. javanicum auct. non Bl.:W. T. Wang & S. H. Wang in Fl. Reip. Pop. Sin. 27: 521. 1979; D. Z. Fu & G. Zhu in Fl. China 6: 292. 2001.

多年生草本植物, 全部无毛。茎高 50-150 cm, 通常分枝。茎生叶为 3-4 回三出复叶; 小叶纸质, 倒卵形、卵形或椭圆形, 0.7-3.4×0.5-3.8 cm, 基部宽楔形或圆钝, 3 浅裂, 裂片顶端圆形, 全缘或具 1-2 小齿, 脉下面隆起; 叶柄长达 10 cm, 基部具鞘。复单歧聚伞花序顶生并腋生, 伞房状, 约有 7 花; 花梗长约 10 mm。花: 萼片 4, 早落, 淡黄绿色, 倒卵形, 长 2-2.7 mm, 顶端微凹, 有少数牙齿, 每一萼片具 3 条 3 回二歧状分枝的基生脉。雄蕊多数, 长约 5 mm; 花丝长约 4 mm, 上部狭倒披针形, 与花药等宽, 下部狭条形; 花药长圆形, 长约 1 mm。心皮 8-10 (-15), 无柄, 长 2-2.7 mm, 花柱比子房长, 顶端钩状弯曲。瘦果纺锤形, 长 2.5-5.8 mm, 每侧有 3 条纵肋; 宿存花柱长约 1.5 mm, 顶端钩状或拳卷状弯曲。

Perennial herbs, totally glabrous. Stems 50-150 cm tall, usually branched. Cauline leaves 3-4-ternate; leaflets papery, obovate, ovate or elliptic, 0.7-3.4×0.5-3.8 cm, at base broadly

六、分类学处理　　　　　　　　　　　　　　　　117

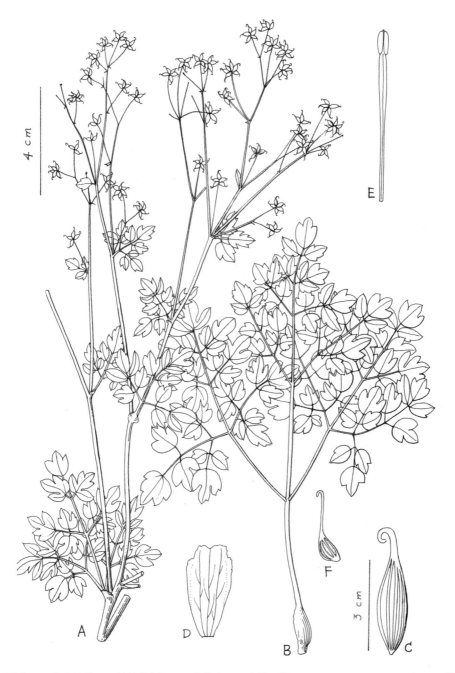

图版14. **玉山唐松草**　A. 结果茎上部，B. 茎生叶，C. 瘦果（据Endo, Sakai & Iketani 18791），D. 萼片，E. 雄蕊，F. 心皮。（抄自*Flora of Taiwan*, 2nd ed.）

Plate 14. **Thalictrum sessile**　A. upper part of fruiting stem, B. cauline leaf. C. achene (from Endo, Sakai & Iketani 18791), D. sepal, E. stamen, F. carpel (redrawn from *Flora of Taiwan*, 2nd ed.)

cuneate or rounded-obtuse, 3-lobed, lobes at apex rounded, entire or 1-2-denticulate, nerves abaxially prominent; petioles up to 10 cm long, at base vaginate. Compound monochasia terminal and axillary, corymbiform, ca. 7-flowered; pedicels ca. 10 mm long. Flower: Sepals 4, caducous, yellowish-green, obovate, 2-2.7 mm long, apex retuse and few-dentate, each sepal with 3 thrice dichotomizing basal nerves. Stamens numerous, ca. 5 mm long; filaments ca. 4 mm long, above narrow-oblanceolate, as broad as anthers, below narrow-linear; anthers oblong, ca. 1 mm long. Carpels 8-10(-15), sessile, 2-2.7 mm long; style longer than ovary, apex hooked. Achene sessile, fusiform, 2.5-5.8 cm long, on each side with 3 longitudinal ribs; persistent style ca. 1.5 mm long, apex hooked or circinate.

产于台湾中央山脉的玉山、阿里山、雪山等山区。生山谷沟边、林中等阴湿处，分布海拔在 2500 m 以上。

标本登录：台湾：台中，南湖大山，主峰，alt. 3300-3740 m，1984-09-11，Y. Tateishi, J. Murata et al. 18791（GH）.

15. 峨眉唐松草（《中国药用植物志》，1964） 图版 15

Thalictrum omeiense W. T. Wang & S. H. Wang in Fl. Reip. Pop. Sin. 27: 524, 617, pl. 126: 1-5. 1979; C. Pei & T. Y. Chou, Icon. Chin. Med. Pl. 7: fig. 329. 1964; D. Z. Fu & G. Zhu in Fl. China 6: 294. 2001; Z. Y. Li & M. Ogisu, Pl. Mount Emei 255. 2007; L. Xie in Med. Fl. China 3: 239, with fig. 2016. Type：四川：峨眉山，alt. 2400 m，1957-08-09，杨光辉 5627（holotype, PE）；同地，alt. 2000 m，1931-07-11，汪发缵 23313（paratype, PE）；洪雅，瓦屋山，1939-08-07，姚仲吾 4058（paratype, PE）。

多年生草本植物，全部无毛。茎高 50-80 cm，分枝。基生叶 1，和下部茎生叶为 3 回三出复叶，具长柄；叶片长 10-25 cm；小叶纸质，倒卵形、菱状倒卵形或卵形，2-6.8×1.6-5 cm，顶端圆形，基部宽楔形，3 浅裂，裂片有 1-2 粗齿，脉上面平，下面隆起，脉网明显；小叶柄长 0.3-3 cm；叶柄长 10-12 cm，基部具狭鞘；托叶与鞘同长，宽 1-2.5 mm。复单歧聚伞花序顶生，伞房状，宽 5-15 cm，有多数花；苞片叶状；花梗长 4-5 mm。花：萼片 4-5，早落，白色，倒卵形，约 3.5×2 mm，顶端圆形，每一萼片具 3 条基生脉（中脉 2 回二歧状分枝，1 侧生脉不分枝，另 1 侧生脉 1 回二歧状分枝）。雄蕊 25-32；花丝长 2-2.8 mm，上部条状倒披针形，宽 0.2-0.3 mm，下部丝形，粗约 0.1 mm；花药长圆形，0.8-1×0.3-0.5 mm，顶端钝。心皮 10-16；子房新月形，0.5-0.7×0.3-0.4 mm；花柱长 0.5-0.6 mm，顶端钩状弯曲，腹面有不明显柱头。瘦果狭倒卵形，1.5-

六、分类学处理 119

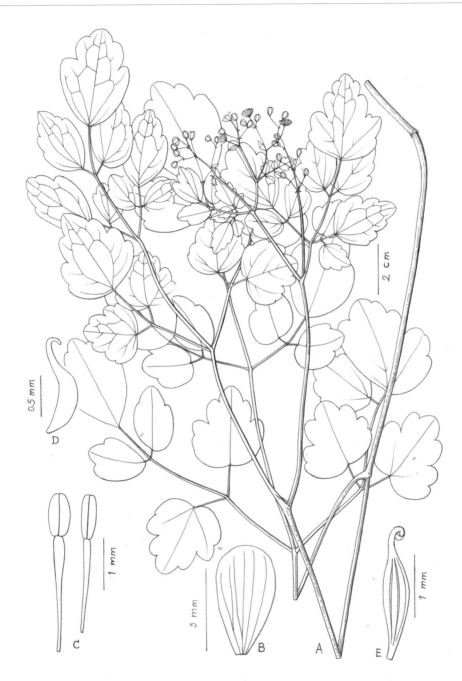

图版15. **峨眉唐松草** A. 开花茎，B. 萼片，C. 二雄蕊，D. 心皮（据王蜀秀579），E. 瘦果（据Ching & Shun 36）。

Plate 15. **Thalictrum omeiense** A. flowering stem, B. sepal, C. two stamens, D. carpel (from S. H. Wang 579), E. achene (from Ching & Shun 36).

1.8×0.4–0.6 mm, 每侧有 2–3 条纵肋；宿存花柱长 0.6–0.8 mm, 顶端拳卷状弯曲。7 月开花，8 月结果。

Perennial herbs, totally glabrous. Stems 50–80 cm tall, branched. Basal leaf 1, with lower cauline leaves 3-ternate, long petiolate; blades 10–25 cm long; leaflets papery, obovate, rhombic-obovate or ovate, 2–6.8×1.6–5 cm, at apex rounded, at base broad-cuneate, 3-lobed, lobes with 1–2 coarse teeth, nerves adaxially flat, abaxially prominent, nervous nets conspicuous; petiolules 0.3–3 cm long; petioles 10–12 cm long, at base narrowly vaginate; stipules as long as leaf sheath, 1–2.5 mm broad. Compound monochasia terminal, corymbiform, 5–15 cm broad, many-flowered; bracts foliaceous; pedicels 4–5 mm long. Flower: Sepals 4–5, caducous, white, obovate, ca. 3.5×2 mm, apex rounded, each sepal with 3 basal nerves (midrib twice dichotomizing, 1 lateral nerve simple, and the other lateral nerve once dichotomizing). Stamens 25–32; filaments 2–2.8 mm long, above linear-oblanceolate, 0.2–0.3 mm broad, below filiform, ca. 0.1 mm across; anthers oblong, 0.8–1×0.3–0.5 mm, apex obtuse. Carpels 10–16; ovary lunate, 0.5–0.7×0.3–0.4 mm; style 0.5–0.6 mm long, at apex hooked, adaxially with an inconspicuous stigma. Achene narrow-obovate, 1.5–1.8×0.4–0.6 mm, on each side with 2–3 longitudinal ribs; persistent style 0.6–0.8 mm long, apex circinate.

分布于四川西部和云南东北部，生山地溪边或石崖边阴湿处，海拔 720–2000 m。

全草药用，用于疟疾寒热、湿热发黄、腹痛泻痢等症。(《中国药用植物志》，1964)

标本登录：四川：峨眉山，Ching & Shun 36，中苏植物考察队 1823，王蜀秀 519。

云南：彝良，滇东北队 72-823。

16. 巨齿唐松草（《中国植物志》） 图版 16

Thalictrum grandidentatum W. T. Wang & S. H. Wang in Fl. Reip. Pop. Sin. 27: 530, 617, pl. 129: 7–9. 1979; D. Z. Fu & G. Zhu in Fl. China 6: 291. 2001; Z. Y. Li & M. Ogisu, Pl. Mount. Emei 255. 2007. Holotype：四川：峨眉山，1930-07-11，方文培 7265（PE）.

多年生草本植物，全部无毛。茎高约 80 cm，上部分枝。中部茎生叶为 2 回三出复叶，有短柄；叶片长约 24 cm；小叶纸质，菱形、菱状卵形或卵形，5.5–13×3–6 cm，顶端急尖或短渐尖，基部宽楔形或楔形，不分裂或不明显 3 浅裂，边缘每侧有 2–5 个不等大的牙齿，牙齿正三角形或宽卵形，顶端钝或圆形，有短尖头，脉上面平，下面隆起，脉网明显；小叶柄长 1.9–4.5 cm；叶柄长 3–4.8 cm。复单歧聚伞花序多少伞房状，宽约 12 cm，有多数花；基部苞片为三出复叶，小叶菱形，长 1.5–4 cm，上部苞片简单，披针形或条形，长 4–11 mm；花梗长 6–10 mm。花：萼片 4，早落，倒卵形，约 3×2 mm，顶端钝，每一

六、分类学处理　　　　　　　　　　　　　　　　　　　121

图版16. **巨齿唐松草**　A. 开花茎上部，B. 中部茎生叶，C. 萼片，D. 2雄蕊，E. 心皮，F. 瘦果（据holotype）。

Plate 16. **Thalictrum grandidentatum**　A. upper part of flowering stem, B. middle cauline leaf, C. sepal, D. two stamens, E. carpel, F. achene (from holotype).

萼片具3条基生脉（中脉不分枝，2侧生脉1回二歧状分枝）。雄蕊20-35；花丝长2.5-4.5 mm，上部狭倒披针形，宽0.3-0.5 mm，下部丝形；花药狭长圆形，0.8-1×0.25 mm，顶端钝。心皮10-13；子房斜倒卵形，约1×0.5 mm；花柱长1 mm，钩状弯曲，腹面上部有不明显柱头。瘦果椭圆形，2.2×0.8-1 mm，每侧有3条细纵肋；宿存花柱长约1 mm，顶端向后弯曲。7月开花。

Perennial herbs, totally glabrous. Stem ca. 80 cm tall, above branched. Middle cauline leaves 2-ternate, shortly petiolate; blades ca. 24 cm long; leaflets papery, rhombic, rhombic-ovate or ovate, 5.5-13×3-6 cm, at apex acute or shortly acuminate, at base broadly cuneate or cuneate, undivided or indistinctly 3-lobed, margin at each side unequally 2-5-dentate (teeth deltoid or broadly ovate, at apex obtuse or rounded, mucronate), nerves adaxially flat, abaxially prominent, nervous nets conspicuous; petiolules 1.9-4.5 cm long; petioles 3-4.8 cm long. Compound monochasium more or less corymbiform, ca. 12 cm broad, many-flowered; basal bract ternate, leaflets rhombic, 1.5-4 cm long, upper bracts simple, lanceolate or linear, 4-11 mm long; pedicels 6-10 mm long. Flower: Sepals 4, caducous obovate, ca. 3×2 mm, apex obtuse, each sepal with 3 basal nerves (midrib simple, and 2 lateral nerves once dichotomizing). Stamens 20-35; filaments 2.5-4.5 mm long, above narrow-oblanceolate, 0.3-0.5 mm broad, below filiform; anthers narrow-oblong, 0.8-1×0.25 mm, apex obtuse. Carpels 10-13; ovary obliquely obovate, ca. 1×0.5 mm; style 1 mm long, hooked, adaxially with an inconspicuous stigma, Achene elliptic, 2.2×0.8-1 mm, on each side with 3 thin longitudinal ribs; persistent style ca. 1 mm long, apex recurved.

特产四川峨眉山。

17. **大叶唐松草**（《中国药用植物志》，1964） 图版17

Thalictrum faberi Ulbr. In Notizbl. Bot. Gart. Berl. 9: 222. 1925; Iconogr. Corm. Sin. 1: 674, fig. 1348. 1972; W. T. Wang & S. H. Wang in Fl. Reip. Pop. Sin. 27: 530, pl. 128. 1979; W. T. Wang in Bull. Bot. Lab. W.-E. Forest. Inst. 8: 23. 1980; B. Z. Ding et al. Fl. Henan. 1: 468, fig. 500. 1981; S. R. Lin & X. Z. Zhao in Fl. Fujian. 2: 19. 1985; X. W. Wang in Fl. Anhui 2: 322, fig. 624. 1987; W. T. Wang in Bull. Bot. Res. Harbin 9(2):5. 1989; Z. H. Lin in Fl. Zhejiang 2: 375, fig. 2-365. 1992; K. M. Liu, Fl Hunan 2: 649, fig. 2-504. 2000; W. T. Wang in High. Pl. China 3: 465, fig. 736. 2000; D. Z. Fu & G. Zhu in Fl. China 6: 289. 2001; W. T. Wang in Fl. Jiangxi. 2: 175, fig. 181. 2004; Z. C. Luo & Y. B. Luo, Pl. Xinning 102. 2008; Q. L.

六、分类学处理 123

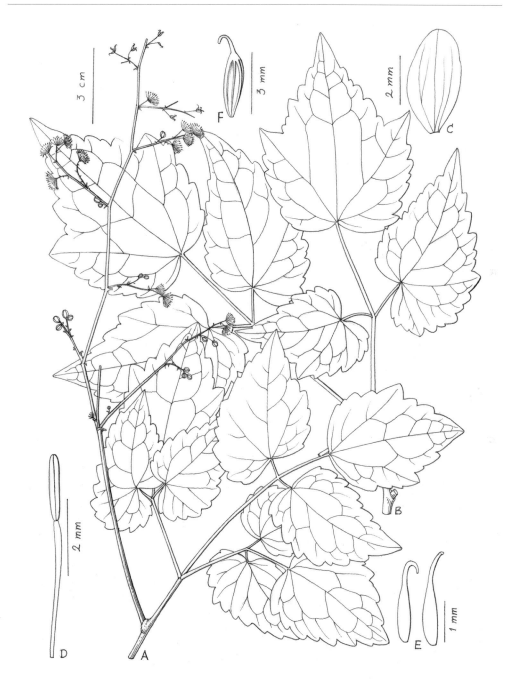

图版17. **大叶唐松草** A. 开花茎上部，B. 中部茎生叶，C. 萼片，D. 雄蕊，E. 二心皮（据植物资源队 29035），F. 瘦果（据王文采 s.n.）。

Plate 17. **Thalictrum faberi** A. upper part of flowering stem, B. middle cauline leaf, C. sepal, D. stamen, E. two carpels (from Pl. Resour. Exped. 29035), F. achene (from W. T. Wang s.n.).

Gan, Fl Zhuxi Suppl. 243, fig. 290. 2011; M. B. Deng & K. Ye in Fl. Jiangsu 2: 83, fig. 2-132. 2013; T. Fang & J. H. Chen, Field Guid. Wild Pl. China: Gutian shan 322. 2013; W. T. Wang in High. Pl. China in Colour 3: 385. 2016; L. Xie in Med. Fl. China 3: 243, with fig. & photo. 2016. Type: 浙江：宁波山地，1888 年，E. Faber s. n.（holotype, not seen），江西：上犹，alt. 1300 m, 1921-07-27, 胡先骕 1344(paratype, not seen).

T. macrophyllum Migo in J. Shanghai Sci. 14(2):136. 1944. Holotype: 江西：庐山，黄龙寺，1941-09-26, H. Migo s. n. Paratype: 浙江：西天目山，1935-08-28. H. Migo s. n. (not seen)

多年生草本植物，全部无毛。茎高（35-）45-110 cm，上部分枝。下部茎生叶为 2-3 回三出复叶；叶片长达 30 cm；小叶大，纸质，卵形或宽卵形，有时近菱形，4-10×3.5-9 cm，基部圆形、浅心形或截形，3 浅裂，边缘每侧有 5-10 个不等大的尖牙齿，脉上面平，下面隆起，脉网明显；小叶柄长 1.5-4 cm；叶柄长 3-6 cm，基部具鞘；托叶狭，全缘。聚伞圆锥花序长 20-40 cm，有多数花；花梗细，长 3-7 mm。花：萼片 4，早落，白色，宽椭圆形，约 3.2×2 mm，顶端圆形，每一萼片具 3 条基生脉（中脉和 1 侧生脉 1 回二歧状分枝，另一侧生脉不分枝）。雄蕊约 45；花丝长 3-4.2 mm，上部条形，宽约 0.2 mm，下部丝形；花药狭长圆形，1-1.2×0.2-0.3 mm，顶端钝。心皮 3-6；子房狭卵形，1.5-1.8×0.5-0.7 mm；花柱长 0.8-1 mm，钩状弯曲或近直。瘦果狭椭圆形或狭卵形，3-3.3×1-1.5 mm，每侧有 3 条细纵肋；宿存花柱长约 1.4 mm，顶端稍钩状弯曲。7-8 月开花。

Perennial herbs, totally glabrous. Stems (35-)45-110 cm, above branched. Lower cauline leaves 2-3-ternate; blades up to 30 cm long; leaflets large, papery, ovate or broad-ovate, sometimes nearly rhombic, 4-10×3.5-9 cm, at base rounded, subcordate or truncate, 3-lobed, margin at each side with 5-10 acute teeth unequal in size, nerves adaxially flat, abaxially prominent, nervous nets conspicuous; petiolules 1.5-4 cm long; petioles 3-6 cm long, at base vaginate; stipules narrow, entire. Thyrses 20-40 cm long, many-flowered; pedicels slender, 3-7 mm long. Flower: Sepals 4, caducous, white, broad-elliptic, ca. 3.2×2 mm, apex rounded, each sepal with 3 basal nerves (midrib and 1 lateral nerve once dichotomizing, and the another lateral nerve simple). Stamens ca. 45; filaments 3-4.2 mm long, above linear, ca. 0.2 mm broad, below filiform; anthers narrow-oblong, 1-1.2×0.2-0.3 mm, apex obtuse. Carpels 3-6; ovary narrow-ovate, 1.5-1.8×0.5-0.7 mm; style 0.8-1 mm long, hooked or nearly straight. Achene narrow-elliptic or narrow-ovate, 3-3.3×1-1.5 mm, on each side with 3 thin longitudinal ribs;

persistent style ca. 1.4 mm long, apex slightly hooked.

分布于湖南、江西、福建、浙江、江苏南部、安徽、河南南部、陕西南部（太白山）。生山地林边或沟边，海拔 100−1700 m。

根和根状茎有清湿解毒、清热之效，用于下痢腹痛、目赤肿痛、湿热黄疸等症。根含大叶唐松草碱（thalifaberine）等多种生物碱。（《中国药用植物志》，2016）

标本登录：湖南：新宁，罗林波 938，罗毅波 2670；永顺，北京队 1242；桑植，植化室队 291；衡山，刘瑛 296，关克俭，杨保民 73，胡忠辉 496。

江西：庐山，胡先骕 23190，关克俭 74353，聂敏祥 94313；铅山，武夷山，聂敏祥 4482。

浙江：天目山，钟观光 D247，D565，贺贤育 26366，26417，植物资源队 59−29035，王文采 s. n.，马丹丹 491；莫干山，刘慎谔 7994，Cheo & Wilson 12798；天台山，贺贤育 27982；盘山，刘慎谔 7995；泰顺，高乞 240。

江苏：溧阳，刘昉勋 2633；宜兴，刘昉勋 2353。

安徽：金寨，植物资源队 59−0319，沈显生 1435，1710；霍山，孙淼 A50192；岳西，谢中稳 97−105；黄山，刘慎谔，钟补求 2863，2990，周根生等 492；绩溪，孙淼 100059。

河南：商城，河南普查队 59−10078。

陕西：太白山，羊皮沟和大水塔，秦岭队 58−10484，58−10499。

18. **粗壮唐松草**（《秦岭植物志》） 图版 18

Thalictrum robustum Maxim. in Acta Hort. Petrop. 11: 18. 1890; Fl Tsinling. 1(2):240. 1974; S. H. Fu, Fl. Hupeh. 1: 355. 1976; W. T. Wang & S. H. Wang in Fl Reip. Pop. Sin. 27: 532, pl. 129: 1−6. 1979; W. T. Wang in Bull. Bot. Lab. N.−E. Forest. Inst. 8: 23. 1980; B. Z. Ding et al. Fl. Henan. 1: 469, fig. 601. 1981; W. T. Wang in Iconogr. Corm. Sin. Suppl. 1: 422, fig. 8607. 1982; X. W. Wang in Fl. Anhui 2: 321, fig. 623. 1987; W. T. Wang in Keys Vasc. Pl. Wuling Mount. 164. 1995; in Guihaia 17(1): 15. 1997; et in High. Pl. China 3: 466. 2000; D. Z. Fu & G. Zhu in Fl. China 6: 296. 2001; C. X. Yang et al. Keys Vasc. Pl. Chongqing 217. 2009; Q. L. Gan, Fl. Zhuxi Suppl. 242. 2011; L. Q. Li in High. Pl. Daba Mount. 118. 2014; L. Q. Li et al. Pl. Baishuijiang St. Nat. Reserve Gansu Prov. 71. 2014; W. T. Wang in High. Pl. China in Colour 3: 385. 2016; L. Xie in Med. Fl China 3: 244, with fig. & photo. 2016. Syntypes：甘肃：oppido Nan-ping, 1885−07−16, G. N. Potanin s. n. (LE, not seen)。四川：ad fl. Honton supra vicum Shidshapo, 1885−08−15, G. N. Potanin s. n. (LE, not seen).

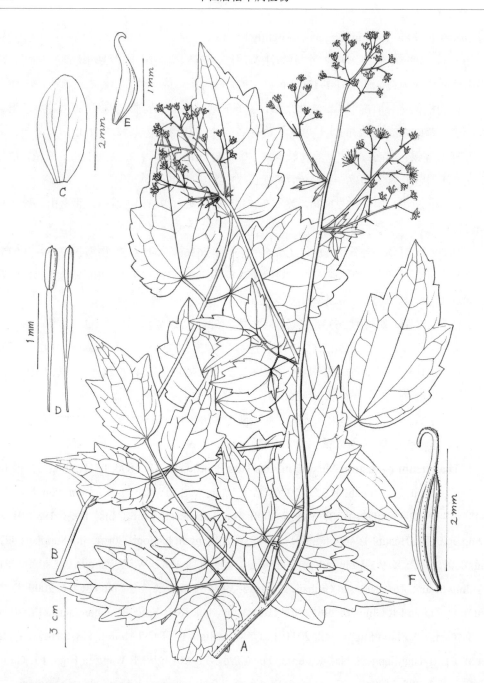

图版18. **粗壮唐松草** A. 聚伞圆锥花序，B. 中部茎生叶，C. 萼片，D. 二雄蕊，E. 心皮，F. 瘦果（据秦岭队10499）。

Plate 18. **Thalictrum robustum** A. thyrse, B. middle cauline leaf, C. sepal, D. two stamens, E. carpel, F. achene (from Qinling Exped. 10499).

T. clematidifolium Franch. in J. de Bot. 8: 273. 1894. —*T. actaeifolium* Sieb. & Zucc. var. *clematidifolium* (Franch.)Finet & Gagnep. in Bull. Soc. Bot. France 50: 611.1903; Boivin in J. Arn. Arb. 26: 112. 1945. Syntypes: 重庆：城口，R. P. Farges 4966 (P, not seen)。湖北：A. Henry 333, 6084, 7344 (P？, not seen).

T. falcatum Pamp. in Nouv. Gior. Bot. Ital., n. ser., 22: 290. 1915. Syntypes: 湖北西北部：Zan-lan-scian（沧浪山），1913−07, P. C. Silvestri 4111, 4111a,4111b (not seen).

T. laxum Ulbr. In Notizbl. Bot. Gart. Berl. 9: 225. 1925; S. H. Fu, Fl. Hupeh. 1: 354. 1976; W. T. Wang & S. H. Wang in Fl. Reip. Pop. Sin. 27: 532, 1979; D. Z. Fu & G. Zhu in Fl. China 6: 292. 2001. Isotype：湖北西部, A. Henry 6424 (US; photo. PE).

多年生草本植物。茎高（50−）80−150 cm，疏被短柔毛或无毛，上部分枝。中部茎生叶为 2−3 回三出复叶；叶片长达 25 cm；小叶纸质，卵形，(3−) 6−8.5×(1.3−)3−5 cm，基部浅心形或圆形，3 浅裂，边缘有不等大的牙齿，下面被密或疏的短柔毛，脉下面隆起，脉网明显；小叶柄长 0.6−2 cm；叶柄长 3−7 cm；托叶上部不规则分裂。聚伞圆锥花序具多数花；花梗长 1.5−3 mm，被短柔毛。花：萼片 4，早落，椭圆形，约 2.6×1.5 mm，顶端圆钝，每一萼片具 3 条基生脉（中脉和一侧生脉 2 回二歧状分枝，另一侧生脉 1 回二歧状分枝）。雄蕊 25−30，无毛；花丝长 2−3.2 mm，上部条形，宽 0.2−0.35 mm，下部丝形；花药长圆形，0.5−0.6×0.3 mm，顶端钝。心皮 6−20，无毛；子房狭倒卵形，约 1.2×0.5 mm；花柱长 1 mm，稍钩状弯曲。瘦果近纺锤形，约 2×0.8 mm，每侧有 2 条纵肋；宿存花柱长约 1 mm，顶端钩状弯曲。6−7 月开花。

Perennial herbs. Stems (50−)80−150 cm tall, above branched. Middle cauline leaves 2−3-ternate; blades up to 25 cm long; leaflets large, papery, ovate, (3−) 6−8.5×(1.3−)3−5 cm, at base subcordate or rounded, 3-lobed, at margin unequally dentate, abaxially densely or sparsely puberulous, nerves abaxially prominent, nervous nets conspicuous; petiolules 0.6-2 cm long; petioles 3−7 cm long; stipules above irregularly divided. Thyrses many-flowered; pedicels 1.5−3 mm long, puberulous. Flower: Sepals 4. caducous, elliptic, ca. 2.6× 1.5 mm, apex rounded-obtuse, each sepal with 3 basal nerves (midrib and 1 lateral nerve twice dichotomizing, and another lateral nerve once dichotomizing). Stamens 25−30, glabrous; filaments 2−3.2 mm long, above linear, 0.2−0.35 mm broad, below filiform; anthers oblong. 0.5−0.6×0.3 mm, apex obtuse. Carpels 6−20, glabrous; ovary narrow-obovate, ca. 1.2× 0.5 mm; style 1 mm long, slightly hooked. Achene subfusiform, ca. 2×0.8 mm, on each side with 2 longitudinal ribs; persistent style ca. 1 mm long, apex hooked.

分布于四川北部、重庆、湖南西北部、湖北、安徽西南部、河南、陕西和甘肃的南部。生山地林中、沟边或草坡上，海拔 940–2100 m。

根有清热解毒之效，用于痢疾、腹泻等症。(《中国药用植物志》，2016)

标本登录：四川：平武，蒋兴麐 10542；旺苍，巴山队 5042，5139，5426。

重庆：南川，金佛山，李国凤 62815；巫溪，杨光辉 58630；巫山，杨光辉 59034；奉节，三峡队 2950，3115；城口，戴天伦 101196，102605，106512，巴山队 191，458，716，1251，1379。

湖南：石门，瓶壶山队 1237，1332。

湖北：建始，王从荣 1000；宣恩，王映明 5478；巴东，鄂植考队 24528，25062，赵常明 2397；兴山，赵常明 1595。

河南：卢氏，刘继孟 4667，5034；西峡，河南队 1163，关克俭，戴天伦 1266；嵩县，河南队 2607；栾川，刘淼 40073；洛宁，刘淼 30198。

陕西：太白山，刘慎谔，钟补求 2549，王作宾 1602，秦岭队 58-10499，武建勇，武振海 173；终南山，王作宾 2123。

甘肃：天水，张珍万 222。

系 3. 小喙唐松草系

Ser. **Rostellata** W. T. Wang & S. H. Wang in Fl. Reip. Pop. Sin. 27: 534, 617. 1979. Type: *T. rostellatum* Hook. f. & Thoms.

本系在亲缘关系方面与前系糙叶唐松草系 Ser. *Scabrifolia* 甚为相近，区别为本系的瘦果具柄。

This series is closely related to the previous Ser. *Scabrifolia*, differing from the latter in its stipitate achenes.

在中国有 8 种，广布于西南至华北、海南和台湾。

19. **小喙唐松草**(《中国植物志》) 图版 19

Thalictrum rostellatum Hook. f. & Thoms. Fl. Ind. 15. 1855; et in Hook. f. Fl. Brit. Ind. 1: 12. 1872; Lecoy. in Bull. Soc. Bot. Belg. 24: 229. 1885; Finet & Gagnep. in Bull. Soc. Bot. France 50: 614. 1903; W. T. Wang & S. H. Wang in Fl. Reip. Pop. Sin. 27: 534, pl. 130: 10–12. 1979; W. T. Wang in Iconogr. Corm. Sin. Suppl. 1: 423, fig. 8608. 1982; Grierson, Fl. Bhutan 1(2): 298. 1984; W. T. Wang in Fl. Xizang. 2: 66, fig. 19: 8–10. 1985; Rau in Sharma et al. Fl. India

六、分类学处理 129

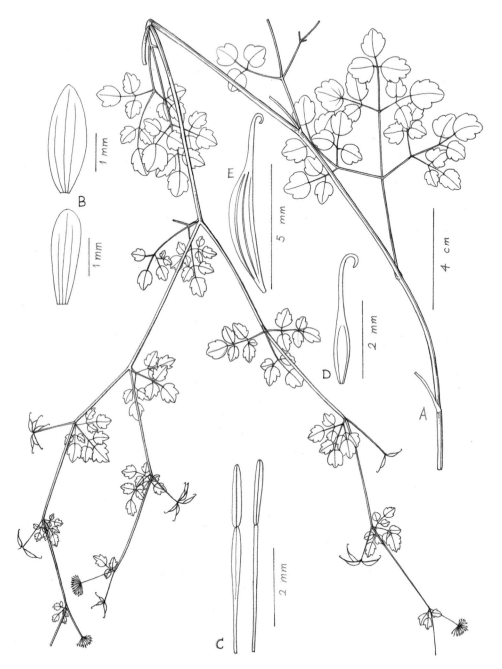

图版19. 小喙唐松草 A. 开花并结果植株，B. 二萼片，C. 二雄蕊，D. 心皮，E. 瘦果（据张永田，郎楷永 1177）。

Plate 19. **Thalictrum rostellatum** A. flowering and fruiting plant, B. two sepals, C. two stamens, D. carpel, E. achene (from Y. T. Zhang & K. Y. Lang 1177).

1: 140. 1993; W. T. Wang in Vasc. Pl. Hengduan Mount. 1: 499. 1993; in High. Pl. China 3: 466, fig. 737. 2000; et in Fl. Yunnan. 11: 162. 2003; D. Z. Fu & G. Zhu in Fl. China 6: 296. 2001. Syntypes: India: Simla, Jacquemont s. n. (K, not seen), Sikkim, J. D. Hooker s. n. (K, not seen).

多年生草本植物，全部无毛，或小叶下面疏被短柔毛。茎高 40-60 cm，上部分枝。下部和中部茎生叶为 3 回三出复叶；叶片长 6-12 cm；小叶草质，宽倒卵形，近圆形或宽卵形，0.8-1.7×0.8-1.7 cm，顶端圆形或钝，基部圆截形或宽楔形，3 浅裂，裂片全缘或有 1-2 小齿，脉平，脉网不明显；小叶柄长 0.6-1.8 cm；叶柄长 1-3.3 cm，基部有短鞘。复单歧聚花序有少数花；花梗细，长 3-12 mm。花：萼片 4，白色，近长圆形，2.8-3.2×1-1.8 mm，顶端钝，每一萼片具 3 条不分枝的基生脉。雄蕊 8-14；花丝长约 3 mm，狭条形，上部宽 0.2 mm，下部宽 0.1 mm；花药狭长圆形，1.4-1.5×0.2-0.3 mm，顶端钝。心皮 3-9；子房狭倒卵形，约 1.2×0.5 mm；花柱长约 2.2 mm，顶端钩状弯曲。瘦果扁，新月形或长圆形，4.8×1-1.5 mm，每侧有 3 条纵肋，柄长约 0.5 mm；宿存花柱长 2.5-3.5 mm，钩状弯曲。5 月开花。

Perennial herbs, totally glabrous, or leaflets abaxially sparsely puberulous. Stems 40–60 cm tall, above branched. Lower or middle leaves 3–ternate; blades 6–12 cm long; leaflets herbaceous, broad-obovate, suborbicular or broad-ovate, 0.8–1.7×0.8–1.7 cm, at apex rounded or obtuse, at base rounded-truncate or broadly cuneate, 3–lobed, lobes entire or 1–2–denticulate, nerves flat, nervous nets inconspicuous; petiolules 0.6–1.8 cm long; petioles 1–3.3 cm long, at base shortly vaginate. Compound monochasia few-flowered; pedicels slender, 3–12 mm long. Flower: Sepals 4, white, suboblong, 2.8–3.2×1–1.8 mm, apex obtuse, each sepal with 3 simple basal nerves. Stamens 8–14; filamens ca. 3 mm long, narrow-linear, above 0.2 mm broad, below 0.1 mm broad; anthers narrow-oblong, 1.4–1.5×0.2–0.3 mm, apex obtuse. Carpels 3–9; ovary narrowly obovate, ca. 1.2×0.5 mm; style ca. 2.2 mm long, apex hooked. Achene compressed, lunate or oblong, 4.8×1–1.5 mm, on each side with 3 longitudinal ribs, stipe ca. 0.5 mm long; persistent style 2.5–3.5 mm long, hooked.

分布于中国西藏南部和东南部、云南西北部、四川西部；以及不丹、尼泊尔、印度北部。生山地林中或沟边，海拔 2500-3200 m。

标本登录：西藏：察隅，王启无 65524；波密，青藏队 73-112，倪志诚等 127，D. E. Boufford et al. 29924，覃海宁 298；古乡，应俊生，洪德元 650171，650251；通麦，郎楷永等 983；易贡，应俊生，洪德元 650502；林芝，张永田，郎楷永 1177，倪志诚等 1784，罗健 9123；札木，张永田，郎楷永 462，608；米林，李渤生，程树志 5394，青藏补点队 75-0915；工布江达，D. E. Boufford et al. 30092；隆子，青藏补点队 75-0500，75-0543；亚东，傅国勋 1086，1029；吉隆，吴征镒等 75-0582，李渤生等 13395，13604。

云南：丽江，滇西北队 9771；鹤庆，秦仁昌 24076；中甸，中甸队 62-1394；维西，王启无 68109；冯国楣 4478；杨亲二，袁琼 50，2029，2273；德钦，植物研究所横断山队 3512，谢磊，毛建丰 138；贡山，青藏队 81-8337。

四川：大金，李馨 77456；绰斯甲，姜恕 1078；马尔康，郎楷永，李良千，费勇 2126。

20. 菲律宾唐松草（《海南植物志》） 图版 20

Thalictrum philippinense C. B. Rob. in Bull. Torr. Bot. Club. 35: 65. 1908; Merr. in Lingnan Sci. J. 5: 75. 1927; Boivin in Rhodora 46: 363. 1944; F. C. How & W. T. Wang in Fl. Hainan. 1: 308. 1964; W. T. Wang & S. H. Wang in Fl. Reip. Pop. Sin. 27: 551, pl. 136: 1–2. 1979; W. T. Wang in High. Pl. China 3: 470. 2000; D. Z. Fu & G. Zhu in Fl. China 6: 295. 2001; R. J. Wang in Fl. Guangdong 5: 19. 2003; X. B. Yang et al. Illustr. Fl. Hainan 2:340. 2015. Syntypes: Philippines : Luzon, Prov. Benguet, Bugulo, 1904–06–22, R. S. Williams 1137 (NY, not seen); same locality, 1904–09–18, Williams 957 (NY, not seen).

多年生小草本植物，全部无毛。茎高约 18 cm，上部分枝。基生叶为 3 回三出复叶，长约 12 cm，有长柄；叶片长约 5 cm；小叶草质，宽卵形，近圆形或宽倒卵形，0.7-1.4×0.8-1.6 cm，顶端圆形，基部圆形或浅心形，3 浅裂，裂片有 1-2 小齿或全缘，脉上面平，下面稍隆起，脉网明显；小叶柄长 1-8 mm；叶柄长约 7 cm。茎生叶长 1.7-2 cm，为 1-2 回三出复叶，具短柄。复单歧聚伞花序顶生并腋生，有少数花；花梗长 2-5 mm。花：萼片 5，长圆状椭圆形，约 1.3×1 mm，顶端圆形，每一萼片有 3 条不分枝的基生脉。雄蕊约 12；花丝丝形，长约 1.5 mm；花药宽长圆形，长约 0.5 mm，顶端钝。心皮约 5；子房狭卵形，约 0.7×0.4 mm；花柱长约 0.2 mm，直。瘦果新月形，约 1.5×0.6 mm，每侧有 3 条细纵肋，柄长约 0.4 mm；宿存花柱长约 1 mm，顶端钩状弯曲。4-5 月开花。

Small perennial herbs, totally glabrous. Stem ca. 18 cm tall, above branched. Basal leaf 3-ternate, ca. 12 cm long, long petiolate; blade ca. 5 cm long; leaflets herbaceous, broad-ovate, suborbicular or broad-obovate, 0.7–1.4×0.8–1.6 cm, at apex rounded, at base rounded or subcordate, 3-lobed, lobes 1–2-denticulate or entire, nerves adaxially flat, abaxially slightly prominent, nervous nets conspicuous; petiolules 1–8 mm long; petioles ca. 7 cm long. Cauline leaves 1.7–2 cm long, 1–2-ternate, shortly petiolate. Compound monochasia terminal and

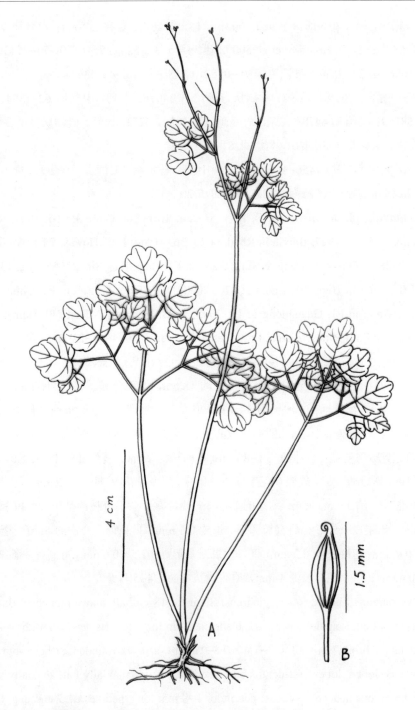

图版20. 菲律宾唐松草　A.植株全形，B.瘦果。（据F. A. McClure 2843）
Plate 20. **Thalictrum philippinense**　A. habit, B. achene. (fom F. A. McClure 2843)

axillary, few-flowered; pedicels 2–5 mm long. Flower: Sepals 5, oblong-elliptic, ca. 1.3×1 mm, apex rounded, each sepal with 3 simple basal nerves. Stamens ca. 12; filaments filiform, ca. 1.5 mm long; anthers broad-oblong, ca. 0.5 mm long, apex obtuse. Carpels ca. 5; ovary narrow-ovate, ca. 0.7×0.4 mm; style ca. 0.2 mm long, straight. Achene lunate, ca. 1.5×0.6 mm, on each side with 3 thin longitudinal ribs, and with a stipe ca. 0.4 mm long; persistent style ca. 1 mm long, apex hooked.

分布于中国海南五指山；以及菲律宾。生山地陡崖阴处，海拔 1600 m。

标本登录：海南：五指山，1922-05-01，F. A. Mc Clure 2843.

21. 察隅唐松草（《云南植物研究》） 图版 21

Thalictrum chayuense W. T. Wang in Acta Bot. Yunnan. 4(2):136, pl. 2: 7. 1982; D. Z. Fu & G. Zhu in Fl. China 6: 287. 2001. Holotype：西藏：察隅，县城附近山谷，alt. 2100 m，花白色，1980-06-29，倪志诚，汪永泽，次多，次旦 297（PE）.

多年生草本植物。茎高约 1 m，枝条疏被短柔毛（毛长约 0.1 mm）。茎生叶为 2-3 回三出复叶，长 6-27 cm；小叶草质，近圆倒卵形、椭圆形或卵形，1.2-3×1.2-2.8 cm，顶端圆形，具短尖头，基部圆形或浅心形，3 浅裂（裂片具 1-2 小钝齿或全缘），两面无毛，具五出脉，脉上面平，下面隆起，脉网明显；小叶柄长 1.5-10 mm；叶柄长 1-6 cm，基部具鞘。聚伞圆锥花序长约 15 cm，3 回稀疏分枝，约有 20 花；花序梗细，长约 10 cm，上部和花序分枝被短腺毛（毛长约 0.1 mm）；基部苞片为 2 回三出复叶，上部苞片为 1 回三出复叶或简单；花梗丝形，长 9-19 mm，无毛。花：萼片早落，未见。雄蕊约 7，无毛；花丝长 3 mm，上部狭条状棒形，宽约 0.2 mm，下部狭条形，宽 0.1 mm；花药淡黄色，狭长圆形，约 1.4×0.5 mm，顶端钝。心皮 3-4；子房斜狭倒卵形，约 1.4×0.5 mm，被短毛（毛长 0.1 mm）；花柱长约 2 mm，顶端钩状弯曲，腹面上部有长约 1 mm 的不明显柱头。瘦果扁，近新月形，3-3.5×1.2-1.5 mm，每侧有 3 条纵肋，疏被短柔毛（毛长 0.1 mm），柄长约 1.2 mm；宿存花柱长约 2 mm，顶端钩状弯曲。5-6 月开花。

Perennial herbs. Stem ca. 1 m tall, branches sparsely puberulous (hairs ca. 0.1 mm long). Cauline leaves 2–3–ternate, 6–27 cm long; leaflets herbaceous, suborbicular-obovate, elliptic or ovate, 1.2–3×1.2–2.8 cm, at apex rounded, at base rounded or subcordate, 3–lobed (lobes obtusely 1–2–denticulate or entire), on both surfaces glabrous, 5–nerved, nerves adaxially flat, abaxially prominent, nervous nets conspicuous; petiolules 1.5–10 mm long; petioles 1–6 cm long, at base vaginate. Thyrse ca. 15 cm long, thrice laxly branched, ca. 20–flowered; peduncle

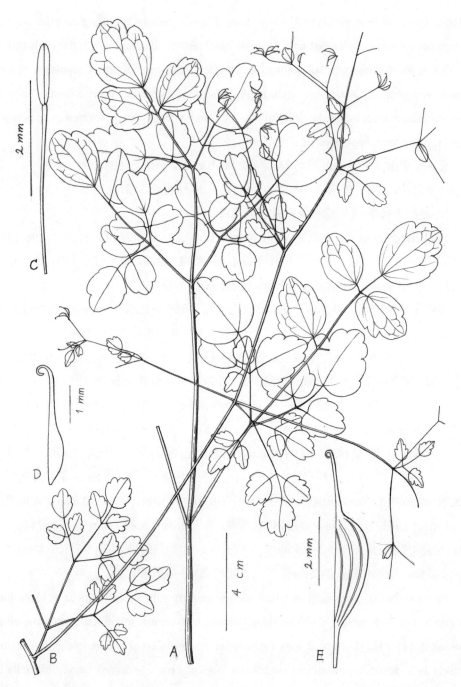

图版21. **察隅唐松草** A.茎生叶，B.腋生聚伞圆锥花序，C.雄蕊，D.心皮，E.瘦果。（据holotype）
Plate 21. **Thalictrum chayuense** A. cauline leaf, B. axillary thyrse, C. stamen, D. carpel, E. achene. (from holotype)

slender, ca. 10 cm long, above with thyrse branches glandular-puberulous (hairs 0.1 mm long); basal bract 2-ternate, upper bracts 1-ternate or simple; pedicels filiform, 9-19 mm long, glabrous. Flower: Sepals caducous, not seen. Stamens ca. 7, glabrous; filament 3 mm long, above narrow-linear-clavate, ca. 0.2 mm broad, below narrow-linear, 0.1 mm broad; anthers yellowish, narrow-oblong, ca. 1.4×0.5 mm ,apex obtuse.Carpels 3-4; vary obliguely narrow-obovate,ca 1.4×0.5 mm, puberulous (hairs 0.1 mm long); style ca. 2 mm long, at apex hoohed, adaxially above with a inconspicuous stigma ca. 1 mm long. Achene compressed, sublunate, 3-3.5×1.2-1.5 mm, on each side with 3 longitudinal ribs, sparsely puberulous (hairs 0.1 mm long), and at base with a stipe ca. 1.2 mm long; persistent style ca. 2 mm long, apex hooked.

特产西藏察隅县。生山谷林下，海拔 2700 m。

22. 纺锤唐松草（《云南植物研究》） 图版 22

Thalictrum fusiforme W. T. Wang in Acta Bot. Yunnan. 4(2): 137, pl. 2: 8. 1982; D. Z. Fu & G. Zhu in Fl. China 6: 290. 2001. Holotype：云南：察隅，上察隅，布宗，alt. 2000 m，花白色，1980-07-26，倪志诚，汪永泽，次多，次旦 837（PE）.

多年生草本植物。茎高 80-100 cm，分枝，近节处与叶柄、花序梗和花序分枝均被短柔毛和短腺毛（毛长 0.05-0.15 mm）。茎生叶为 2 回三出复叶；小叶膜质，近圆形、宽倒卵形或宽卵形，0.9-1.3×0.8-1 cm，顶端圆形，具 1 短尖头，基部圆形或浅心形，3 浅裂（裂片全缘或具 1-2 小钝齿），两面无毛，具五出脉，脉两面平；小叶柄长 2-10 mm；叶柄短。复单歧聚伞花序顶生并腋生，长约 15 cm，稀疏二歧状分枝，约有 15 花；花序梗长 7-13.5 cm；基部苞片叶状，上部苞片一回三出分裂或简单，形似小叶，或钻形，长只 1.2 mm，无毛；花梗长 7-14 mm，近无毛。花：萼片和雄蕊未见。心皮 6，无毛；子房狭卵形，约 1×0.4 mm；花柱长 2.5 mm，顶端拳卷状弯曲。瘦果扁，纺锤形，约 4×1.2 mm，无毛，每侧具 3 条纵肋，柄长 1-1.3 mm；宿存花柱长 3.6-4 mm，顶端拳卷状弯曲。6-7 月开花。

Perennial herbs. Stem 80-100 cm tall, branched, near nodes with petioles, peduncles and thyrse branches puberulous and glandular-puberulous (hairs 0.05-0.15 mm long). Cauline leaves 2-ternate; leaflets membranous, suborbicular, broad-obovate or broad-ovate, 0.9-1.3×0.8-1 cm, at apex rounded and minutely mucronate, at base rounded or subcordate, 3-lobed (lobes entire or obtusely 1-2-denticulate), on both surfaces glabrous, 5-nerved, nerves on both surfaces flat; petiolules 2-10 mm long; petioles short. Compound monochasia terminal and axillary, ca. 15 cm long, laxly dichotomously branched, ca. 15-flowered;

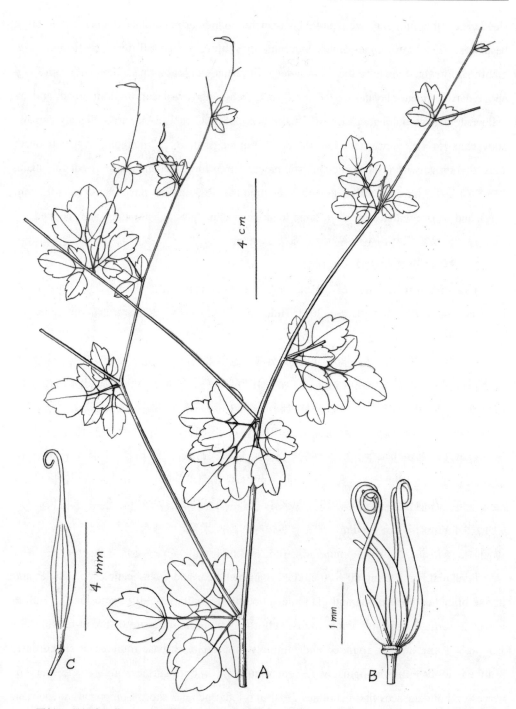

图版22. **纺锤唐松草** A. 结果植株上部，B. 雌蕊群，C. 瘦果。（据holotype）
Plate 22. **Thalictrum fusiforme** A. upper part of fruiting plant, B. gynaecium, C. achene. (from holotype)

peduncles 7–13.5 cm long; basal bract foliaceous, upper bracts 1-ternate or simple, leaflet-like, or subulate, only 1.2 mm long, glabrous; pedicels 7–14 mm long, subglabrous. Flower:Sepals and stamens not seen. Carpels 6, glabrous; ovary narrow-ovate, ca. 1×0.4 mm; style 2.5 mm long, apex circinate. Achene compressed, fusiform, ca. 4×1.2 mm, glabrous, on each side with 3 longitudinal ribs, and at base with a stipe 1–1.3 mm long; persistent style 3.6–4 mm long, apex circinate.

特产西藏察隅县。生山地河边杂木林下，海拔 2000 m。

23. **澜沧唐松草**(《植物分类学报》) 图版 23

Thalictrum lancangense Y. Y. Qian in Acta Phytotax. Sin. 35 (3): 262, fig. 1. 1997; D. Z. Fu & G. Zhu in Fl. China 6: 292. 2001. Type：云南：澜沧，alt. 2000 m, 1991-07-24, 钱义咏 2311(holotype, PE; isotype, SMAO)，3034(paratypes, PE, SMAO)。

多年生草本植物，全部无毛。茎高 17–70 cm，分枝。基生叶为 1–2 回三出复叶，具长柄，中部茎生叶为 2–3 回三出复叶，具短柄；小叶纸质，宽卵形或近圆形、稀倒卵形，0.7–2.2×0.7–2.9 cm，基部心形，近心形，或宽楔形，3 浅裂，边缘具圆牙齿，脉两面隆起，脉网明显；小叶柄长 0.3–2 cm；叶柄长达 15 cm。复单歧聚伞花序顶生并腋生，伞房状，具多数花；花梗长 4.5–15 mm。花：萼片 4，早落，白色，长椭圆形或倒卵形，约 3.2×1.6 mm，顶端钝，每一萼片具 3 条基生脉（中脉和 1 侧生脉不分枝，另一侧生脉二歧状分枝）。雄蕊 20–28；花丝长 1.5–3.5 mm，上部倒披针形或狭倒披针形，宽 0.25–0.35 mm，下部狭条形，宽 0.15 mm；花药长圆形，0.8–1×0.4–0.5 mm，顶端钝。心皮（5–）9–15；子房近纺锤形，约 1.5×0.4 mm；花柱长 1.2 mm，顶端向后弯曲。瘦果纺锤形，1.4–2.2×0.5–0.8 mm，每侧有 3 条纵肋，柄长约 0.4 mm；宿存花柱长 1.5–2 mm，顶端钩状弯曲或直。7–8 月开花。

Perennial herbs, totally glabrous. Stems 17–70 cm tall, branched. Basal leaves 1-2-ternate, long petiolate, middle leaves 2-3-ternate, shortly petiolate; leaflets papery, broad-ovate or suborbicular, rarely obovate, 0.7–2.2×0.7–2.9 cm, at base cordate, subcordate or broadly cuneate, 3-lobed, margin rotund-dentate, nerves on both surfaces prominent, nervous nets conspicuous; petiolules 0.3–2 cm long; petioles up to 15 cm long. Compound monochasia terminal and axillary, corymbiform, many-flowered; pedicels 4.5–15 mm long. Flower: Sepals 4, caducous, white, long elliptic or obovate, ca. 3.2×1.6 mm, apex obtuse, each sepal with 3 basal nerves (midrib and 1 lateral nerve simple, and the another lateral nerve dichotomizing).

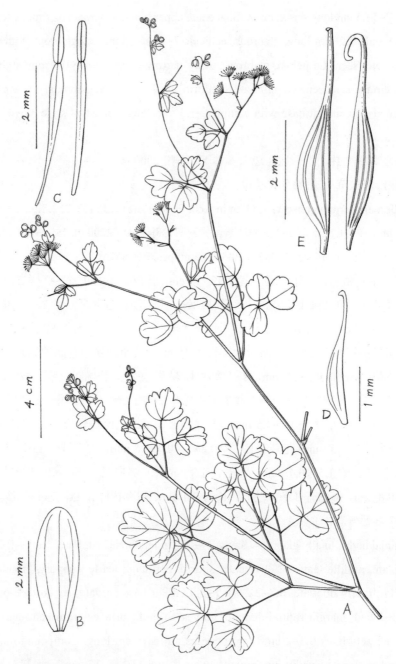

图版23. **澜沧唐松草** A. 开花植株,B. 萼片,C. 二雄蕊,D. 心皮(据holotype),E. 二瘦果(据paratype)。

Plate 23. **Thalictrum lancangense** A. flowering plant, B. sepal, C. two stamens, D. carpel (from holotype), E. two achenes (from paratype).

Stamens 20−28; filaments 1.5−3.5 mm long, above oblanceolate or narrow-oblanceolate, 0.25−0.35 mm broad, below narrow-linear, 0.15 mm broad; anthers oblong, 0.8−1×0.4−0.5 mm, apex obtuse carpels (5−)9−15; ovary subfusiform, ca. 1.5×0.4 mm; style 1.2 mm long, apex recurved. Achene fusiform, 1.4−2.2×0.5−0.8 mm, on each side with 3 longitudinal ribs and at base with a stipe 0.4 mm long; persistent style 1.5−2 mm long, apex hooked or straight.

特产云南澜沧县。生山地疏林中，海拔 2000 m。

24. **弯柱唐松草**（《秦岭植物志》） 图版 24

Thalictrum uncinulatum Franch. in Nouv. Arch. Mus. Hist. Nat. Paris, ser. 2, 8: 187. 1885; Lecoy. in Bull. Soc. Bot. Belg. 24: 169. 1885; Finet & Gagnep in Bull. Soc. Bot. France 50: 607. 1903; Finet in J. de Bot. 21: 20. 1908; Fl. Tsinling. 1(2):240. 1974; S. H. Fu, Fl, Hupeh. 1: 354, fig. 494. 1976; W. T. Wang & S. H. Wang in Fl. Reip. Pop. Sin. 27: 528, pl. 127: 5−9. 1979; W. T. Wang in Iconogr. Corm. Sin. Suppl. 1: 421, fig. 8606. 1982; Y. K. Li in Fl. Guizhou. 3: 69. 1990; W. T. Wang in Vasc. Pl. Hengduan Mount. 1: 499. 1993; in Keys Vasc. Pl. Wuling Mount. 164. 1995; in Fl. Yunnan. 11: 161. 2000, et in High. Pl. China 3: 465, fig. 735. 2000; K. M. Liu, Fl.Hunan. 2: 648, fig. 2−503. 2000; D. Z. Fu & G. Zhu in Fl. China 6: 300. 2001; Z. Y. Li & M. Ogisu, Pl. Mount Emei 255. 2007; C. X. Yang et al. Keys Vasc. Pl. Chongqing 216. 2009; Q. L. Gan, Fl. Zhuxi Suppl. 242. 2011; L. Q. Li in High. Pl. Daba Mount. 118. 2014; L. Xie in Med. Fl. China 3: 242, with fig. 2016. Holotype：四川：宝兴，1869−04，A. David s. n. (P, not seen).

多年生草本植物。茎高 60−120 cm，疏被短柔毛，上部分枝。茎生叶为 3 回三出复叶；叶片长 10−21 cm；小叶纸质，宽卵形、卵形或宽椭圆形。1.5−3×1.3−2.9 cm；基部圆形或心形，3 浅裂（裂片有 1−2 钝齿或全缘），上面近无毛，下面被短柔毛，脉下面隆起，脉网明显；小叶柄长 1−12 mm；叶柄长 2.5−7 cm，疏被短柔毛。聚伞圆锥花序长达 20 cm，有多数花；花梗长 1.5−3.5 mm，密被短柔毛。花：萼片 4，早落，白色，倒卵形或椭圆形，2.5−3×1.5 mm，顶端圆形，背面上部疏被短柔毛或近无毛，每一萼片具 3 条不分枝的基生脉。雄蕊 20−34，无毛；花丝长 3.5−6 mm，上部狭倒披针状条形，宽 0.2−0.4 mm，下部狭条形或丝形，宽约 0.1 mm；花药狭长圆形，1−1.2×0.25−0.4 mm，顶端钝。心皮 4−10；子房斜狭倒卵形，约 1.5×0.6 mm，无毛，柄长 0.4 mm；花柱长约 0.8 mm，顶端向后弯曲；柱头长约 0.2 mm。瘦果狭椭圆形，1.5−2×0.6−1 mm，每侧有 2−3 条纵肋，柄长 0.3−0.6 mm；宿存花柱长 0.5−0.8 mm，顶端钩状弯曲。7 月开花。

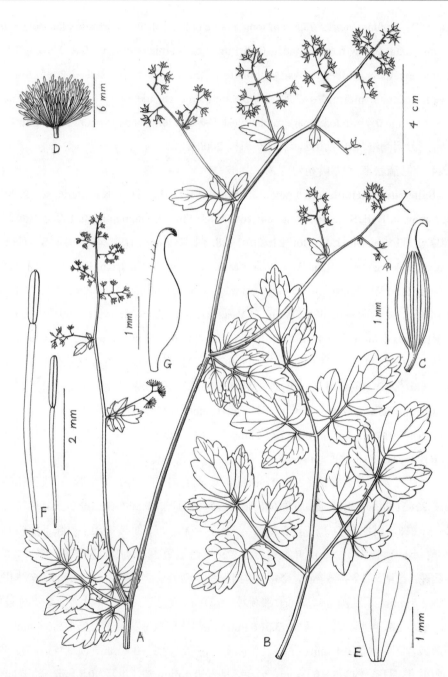

图版24. 弯柱唐松草　A. 果序，B. 中部茎生叶，C. 瘦果（据金佛山队2091），D. 花（萼片已脱落），E. 萼片，F. 二雄蕊，G. 心皮（据蒋兴麐34886）。

Plate 24. **Thalictrum uncinulatum**　A. infructescence, B. middle cauline leaf, C. achene. (from Jinfoshan Exped. 2091), D. flower. with sepals caducous, E. sepal, F. two stamens, G. carpel (from X. L. Jiang 34886).

Perennial herbs. Stems 60−120 cm tall, sparsely puberulous, above branched. Cauline leaves 3−ternate; blades 10−21 cm long; leaflets papery, broad-ovate, ovate or broad-elliptic, 1.5−3×1.3−2.9 cm, at base rounded or cordate, 3−lobed (lobes obtusely 1−2-denticulate or entire), adaxially subglabrous, abaxially puberulous, nerves abaxially prominent, nervous nets conspicuous; petiolules 1−12 mm long; petioles 2.5−7 cm long, with sparse short hairs. Thyrses up to 20 cm long, many-flowered; pedicels 1.5−3.5 mm long, densely puberulous. Flower: Sepals 4, caducous, white, obovate or elliptic, 2.5−3×1.5 mm, apex rounded, abaxially above sparsely puberulous or subglabrous, each sepal with 3 simple basal nerves. Stamens 20−34, glabrous; filaments 3.5−6 mm long, above narrowly oblanceolate-linear, 0.2−0.4 mm broad, below narrow-linear of filiform, 0.1 mm across; anthers narrow-oblong, 1−1.2×0.25−0.4 mm, apex obtuse. Carpels 4−10; ovary obliquely narrow-obovate, ca. 1.5×0.6 mm, glabrous, base with a stipe 0.4 mm long; style ca. 0.8 mm long, apex recurved; stigma ca. 0.2 mm long. Achene narrow-elliptic, 1.5−2×0.6−1 mm, on each side with 2−3 longitudinal ribs, base with a stipe 0.3−0.6 mm long; persistent style 0.5−0.8 mm long, apex hooked.

分布于云南东北部、贵州西部、重庆，湖南西北部、湖北西部、四川、甘肃和陕西的南部、河南西部。生山地草坡或林边灌丛，海拔 1350−2600 m。

根有清热凉血之效，用于胸膈饱胀、痔疮出血等症。(《中国药用植物志》，2016)

标本登录：云南：大关，滇东北队 59-361。

贵州：纳雍，毕节队 675。

重庆：南川，金佛山，熊济华，周子林 92125，92718，金佛山队 2091，2322；开县，巴山队 2185，2445；巫溪，杨光辉 58630；城口，戴天伦 101295，101337，101361。

湖南：石门，壶瓶山队 1237，1316，1332。

湖北：咸丰，王映明 6560；恩施，傅国勋，张志松 1276；宜昌，龙茹 14；巴东，傅国勋，张志松 947；神农架，神农架队 22442。

四川：雷波，俞德浚 3509；美姑，经济植物队 59-1477；越西，植物资源队 59-3839；九龙，孙石烈 60846；峨眉山，汪发缵 23477，熊济华 31865，杨光辉 56283，56656；天全，蒋兴麐 34886，应俊生 10098，关克俭，王文采 1878，2106；泸定，贡嘎山，J. I. Jeon S11466；宝兴，俞德浚 2364；崇州，冯正波等 4295；理县，米亚罗，吴中伦 33376；茂县，汪发缵 21831；黑水，李馨 74056，73989（PE），Gu Lei & Li Zhong-rong 1331，1596 (GH)；南坪，陈伟烈等 8525。

甘肃：舟曲，王作宾 14573；察岗里，夏纬瑛 6487。

陕西：商南，戴天伦 101337。

河南：嵩县，关克俭 2893。

25. 长喙唐松草（《秦岭植物志》）　图版 25

Thalictrum macrorhynchum Franch. in J. de Bot. 4: 302. 1890; Finet & Gagnep. in Bull. Soc. Bot. France 50: 608. 1903; Fl. Tsinling. 1(2): 239, fig. 203. 1974; S. H. Fu, Fl. Hupeh. 1: 355, fig. 496. 1976; W. T. Wang & S. H. Wang in Fl. Reip. Pop. Sin. 27: 526, pl. 126: 6–10. 1979; B. Z. Ding et al. Fl. Henan 1: 471. 1981; W. T. Wang in Iconogr. Corm. Sin. Suppl. 1: 420, fig. 8604. 1982; S. Y. He, Fl. Beijing, rev. ed., 1: 237, fig. 294. 1984; J. W. Wang in Fl. Hebei 1: 444, fig. 441. 1986; X. Y. Yu et al. Fl. Shanxi, 1: 594, fig–375.1992; W. T. Wang in Keys Vasc. Pl. Wuling Mount.164.1995; et in High. pl. China 3: 465, fig. 734. 2000; K. M. Liu, Fl. Hunan 2: 649, fig. 2–505. 2000; D. Z. Fu & G. Zhu in Fl. China 6: 293. 2001; C. X. Yang et al. Keys Vasc. Pl. Chongqing 216. 2009; Q. L. Gan, Fl. Zhuxi, Suppl. 243, fig. 291. 2011; L. Q. Li in High. Pl. Daba Mount. 118. 2014; W. T. Wang in High. Pl. China in Colour 3: 385. 2016; L. Xie in Med. Fl. China 3: 241, with fig. & photo. 2016. Holotype：北京西部山区，ad collum S. Michel in montibus Trappistarum, secus rivulos, jun. 1888, M. Em. Bodinier 60 (P, not seen).

多年生草本植物，全部无毛。茎高 45–65 cm，分枝。基生叶和下部茎生叶具长柄，上部茎生叶有短柄，为 2 – 3 回三出复叶；叶片长 9.5–13 cm，宽达 15 cm；小叶草质，宽倒卵形，圆菱形或宽椭圆形，1.4–4×1.2–4 cm，顶端圆形，基部圆形或浅心形，3 浅裂，裂片有 1–2 圆齿或全缘，脉上面平，下面平或稍隆起，脉网不明显；小叶柄细，长 0.9–1.6 cm；叶柄长约 8 cm，基部具狭鞘；托叶膜质，全缘。复单歧聚伞花序顶生，伞房状，宽 7–15 cm，具稀疏的花；花梗细，长 1.2–3.2 cm。花：萼片 4，早落，白色，倒卵形，3–3.5×1.5 mm，顶端圆形，每一萼片具 3 条不分枝的基生脉。雄蕊 36–42；花丝长 3–5.5 mm，上部狭倒披针形，宽 0.25–0.4 mm，下部狭条形，宽 0.1–0.15 mm；花药长圆形，0.8–0.9×0.3–0.5 mm，顶端钝。心皮 10–20；子房狭卵形，约 1.8×0.8 mm，柄长约 0.4 mm；花柱长约 1.5 mm，顶端拳卷状或螺旋状弯曲。瘦果狭卵形，4–4.5×1.6–2.2 mm，每侧有 3 条纵肋，基部的柄长 0.6–1 mm；宿存花柱长 2.2–3 mm，顶端钩状弯曲。6 月开花。

Perennial herbs, totally glabrous. Stems 45–65 cm tall, branched. Basal and lower cauline leaves long petiolate, upper cauline leaves shortly petiolate, 2–3-ternate; blades 9.5–13 cm

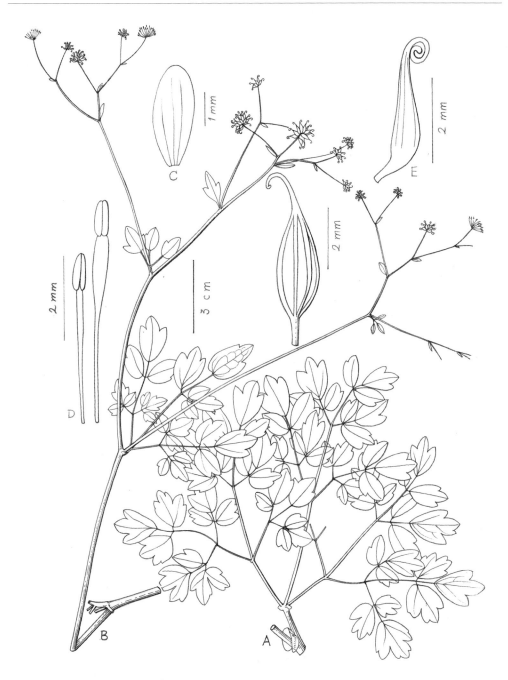

图版25. **长喙唐松草** A. 上部茎生叶，B. 聚伞圆锥花序，C. 萼片，D. 二雄蕊，E. 心皮（据巴山队4967），F. 瘦果（据巴山队1245）。

Plate 25. **Thalictrum macrorhynchum** A. upper cauline leaf, B. thyrse, C. sepal, D. two stamens, E. carpel (from Ba Shan Exped. 4967), F. achene (from Ba Shan Exped. 1245).

long, up to 15 cm broad; leaflets herbaceous, broad-obovate, orbicular-rhombic or broad-elliptic, 1.4−4×1.2−4 cm, at apex rounded, at base rounded or subcordate, 3−lobed, lobes 1−2-rotund-dentate or entire, nerves adaxially flat, abaxially flat or slightly prominent, nervous nets inconspicuous; petiolules slender, 0.9−1.6 cm long; petioles ca. 8 cm long, base narrowly vaginate; stipules membranous, entire. Compound monochasia terminal, corymbiform, 7−15 cm broad, with sparse flowers; pedicels slender, 1.2−3.2 cm long. Flower: Sepals 4, caducous, white, obovate, 3−3.5×1.5 mm, apex rounded, each sepal with 3 simple basal nerves. Stamens 36−42; filaments 3−5.5 mm long, above narrow-oblanceolate, 0.25−0.4 mm broad, below narrow-linear, 0.1−0.15 mm broad; anthers oblong, 0.8−0.9×0.3−0.5 mm, apex obtuse. Carpels 10−20; ovary narrow-ovate, ca. 1.8×0.8 mm, base with a stipe ca. 0.4 mm long; style ca. 1.5 mm long, apex circinate or spirally curved. Achene narrow-ovate, 4−4.5×1.6−2.2 mm, on each side with 3 longitudinal ribs, at base with a stipe 0.6−1 mm long; persistent style 2.2−3 mm long, apex hooked.

分布于湖南西北部、湖北西部、重庆、四川东北部、甘肃和陕西的南部、河南西部、山西南部、河北西部、北京。生山地林中或灌丛中，海拔850-2900 m。

根和根状茎有清热利湿、消肿之效，在甘肃、湖北等地作"马尾莲"用。(《中国药用植物志》，2016)

标本登录：湖南：石门，壶瓶山队 1397。

湖北：恩始，王国荣 1466；建始，王国荣 875；兴山，刘瑛 599；神农架，236 部队 6 所 2634。

重庆：奉节，张泽荣 25092, 25275；巫溪，杨光辉 58731；城口，戴天伦 105442, 105660，巴山队 1245。

四川：平武，四川大学川东队 10553；旺苍，巴山队 4967, 5231。

甘肃：夏河，王作宾 7138；兰州，崔友文 10300，何业祺 5425；连城，孙学刚 6451；天水，李全喜 347；平凉，崔友文 10268。

陕西：佛坪，傅坤俊 4597, 4805；太白山，秦岭队 2646, 10156；户县，郭本兆 279。

河南：内乡，D. E. Boufford et al. 26113；西峡，关克俭，戴天伦 1293；栾川，？119。

山西：垣曲，包世英 105；翼城，黄河队 57-63，刘心源 20453；运城，黄河队 57-468；霍州，黄河队 57-801；灵空山，关克俭，陈艺林 418；介休，刘继孟 1241。

河北：武安，关克俭 5839；内丘，刘心源 686, 1259；阜平，阜平队 778；涞水，刘继孟 2213，2531。

北京：门头沟区，西灵山，杨朝广 734，1302。

26. 南湖唐松草（*Flora of Taiwan*） 淡红唐松草（《中国植物志》） 图版 26

Thalictrum rubescens Ohwi in Acta Phytotax. Geobot. 2: 156. 1933; Liu & Hsieh in Fl. Taiwan 2: 512. 1976; W. T Wang & S. H. Wang in Fl. Reip. Pop. Sin. 27: 526. 1979; T. Y. Yang & T. C. Huang in Taiwania 34: 62, fig. 4. 1989; et in Fl. Taiwan, 2nd ed., 2: 567, pl. 272. 1996; D. Z. Fu & G. Zhu in Fl. China 6: 297. 2001. Type：台湾：宜兰，南湖大山，1933-07，J. Ohwi 4152 (holotype, KYO, not seen; isotype P, photo PE).

多年生草本植物，全部无毛。茎高 10–30 cm，中部之上有 1–2 叶，不分枝或有少数分枝。基生叶约 3，为 2–3 回三出复叶，具长柄；叶片长 5–6 cm；小叶宽卵形，宽椭圆形，近圆形或卵形，5–15×3–14 mm，顶端通常急尖，基部圆形或截形，3 浅裂，裂片通常全缘，或有 1–2 小齿，脉两面稍隆起；小叶柄长 1–7 mm；叶柄长 4.5–5.5 cm；托叶膜质，全缘。单歧聚伞花序顶生，通常简单，具 2 花；花梗细，长 0.8–2 cm。花：萼片 4–6，白色，椭圆形，4.5–5.8×1.8–3.2 mm，顶端钝，每一萼片具 3 基生脉（中脉 2 回二歧状分枝，2 侧生脉 3 回二歧状分枝）。雄蕊多数，长 4–4.5 mm；花丝上部倒披针形，下部丝形；花药宽长圆形，顶端钝。心皮 40–60，长 3–3.5 mm；子房狭椭圆形，具短柄；花柱短，顶端钩状弯曲，瘦果具短柄，纺锤形，长 3.5–5 mm，每侧有 3 条纵肋；宿存花柱短，顶端钩状。6–7 月开花。

Perennial herbs, totally glabrous. Stems 10–30 cm tall, above the middle 1–2-leaved, simple or few-branched. Basal leaves ca. 3, 2–3-ternate, long petiolate; blades 5–6 cm long; leaflets broad-ovate, broad-elliptic, suborbicular or ovate, 5–15×3–14 mm, at apex usually acute, at base rounded or truncate, 3-lobed, lobes usually entire or 1–2-denticulate, nerves on both surfaces slightly prominent; petiolules 1–7 mm long; petioles 4.5–5.5 cm long; stipules membranous, entire. Monocbasia terminal, usually simple, 2–flowered; pedicels slender, 0.8–2 cm long. Flower: Sepals 4–6, white, elliptic, 4.5–5.8×1.8–3.2 mm, apex obtuse, each sepal with 3 basal nerves (midrib twice dichotomizing, and 2 lateral nerves thrice dichotomizing). Stamens numerous, 4–4.5 mm long; filaments above oblanceolate, below filiform; anthers broad-oblong, apex obtuse. Carpels 40–60, 3–3.5 mm long; ovary narrow-elliptic, base shortly stipitate; style short, apex hooked. Achene shortly stipitate, fusiform, 3.5–5 mm long, on each side with 3 longitudinal ribs; persistent style short, apex hooked.

特产台湾宜兰县南湖大山和雪山。生山区高海拔阴湿处。

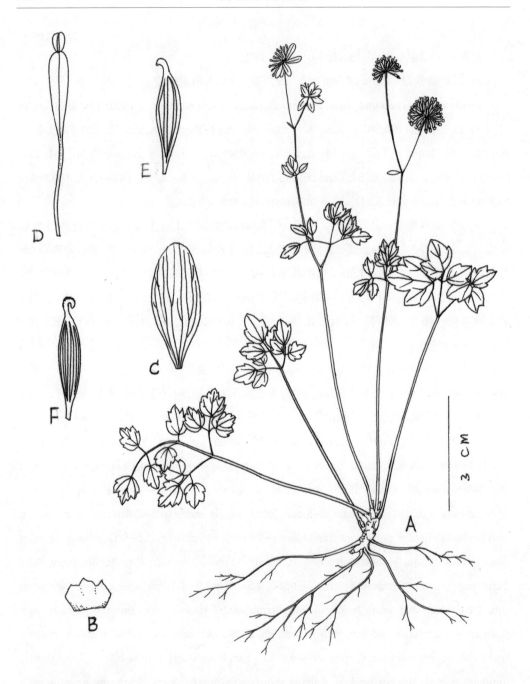

图版26. **南湖唐松草** A. 开花、结果植株全形, B. 托叶, C. 萼片, D. 雄蕊, E. 心皮, F. 瘦果。（仿*Flora of Taiwan*, 2nd ed.）

Plate 26. **Thalictrum rubescens** A. flowering and fruting plant, B. stipule, C. sepal, D. stamen, E. carpel, F. achene. (redrawn from *Flora of Taiwan*, 2nd ed.)

27. 台湾唐松草(《中国植物志》) 图版 27

Thalictrum urbainii Hayata, Icon. Pl. Formos. 1: 25. 1911; Iconogr. Corm. Sin. 1: 677, fig. 1353. 1972; W. T. Wang & S. H. Wang in Fl. Reip. Pop. Sin. 27: 550, pl. 125: 4-6. 1979; T. Y. Yang & T. C. Huang in Taiwania 34: 65, fig. 6. 1989; et in Fl. Taiwan, 2nd ed., 2: 570, pl. 274. 1996; W. T. Wang in High. Pl. China 3: 470, fig. 746. 2000; D. Z. Fu & G. Zhu in Fl. China 6: 301. 2001.——*T. fauriei* Hayata in J. Coll. Sci. Univ. Tokyo 22: 7, t. 1. Sep. 1906, non Lévl. & Van., Jul. 1906; Liu & Hsieh in Fl. Taiwan 2: 310, pl. 393. 1976.——*T. hayatanum* Koidz. In Bot. Mag. Tokyo 30: 20. 1925. Holotype: 台湾：台北，Mt. Tatunshan, Faurie 197(T1, not seen)

T. micrandrum Hayata, Icon. Pl. Formos. 3: 4, pl. 1. 1913; W. T. Wang & S. H. Wang in Fl. Reip. Pop. Sin. 27: 551. 1979. Holotype: 台湾：Kinkwaseki, 1912-05, Y. Shimada s. n. (T1, not seen; photo of isotype, PE).

T. morii Hayata, I. C. 5; W. T. Wang & S. H. Wang in l. c. Holotype: Tonkurankei, 1910-04, U. Mori s. n. (T1, not seen).

27a. var. **urbainii**

多年生草本植物，全部无毛。茎 2-3 条，高 13-30 cm，不分枝或上部有少数分枝。基生叶约 3，为 2 回二出复叶；小叶近革质，菱状圆形，倒卵形或椭圆形，4-10×4-10 mm，顶端圆形，基部宽楔形或圆形，3 浅裂，裂片全缘或有 1-2 小钝齿，脉下面稍隆起；小叶柄长 3-15 mm；叶柄长约 5 cm。单歧聚伞花序顶生，有少数花；花梗长 1-1.6 cm。花：萼片 4，早落，椭圆形，3-4×1.5-2 mm，顶端圆钝，每一萼片具 3 条不分枝的基生脉。雄蕊 28-40；花丝长 3-3.5 mm，上部倒披针形，宽 0.4-0.8 mm，下部丝形，粗约 0.1 mm；花药宽长圆形，0.6-0.8×0.3-04 mm，顶端钝。心皮 5-17；子房纺锤形，约 1.8×0.8 mm，柄长约 0.8 mm；花柱长约 1 mm，顶端钩状弯曲。瘦果扁纺锤形，4-6×0.7-1 mm，每侧有 3 条纵肋，基部的柄长 1.5-2 mm；宿存花柱长 0.7-1 mm，顶端钩状弯曲。5 月开花。

Perennial herbs, totally glabrous. Stems 2-3, 13-30 cm tall, simple or above few-branched. Basal leaves ca. 3, 2-ternate; leaflets subcoriaceous, rhombic-orbicular, obovate or elliptic, 4-10×4-10 mm, at apex rounded, at base broadly cuneate or rounded, 3-lobed, lobes entire or obtusely 1-2-denticulate, nerves abaxially slightly prominent; petiolules 3-15 mm long; petioles ca. 5 cm long. Monochasia terminal, few-flowered; pedicels 1-1.6 cm long. Flower: Sepals 4, caducous, elliptic, 3-4×1.5-2 mm, apex rounded-obtuse, each sepal

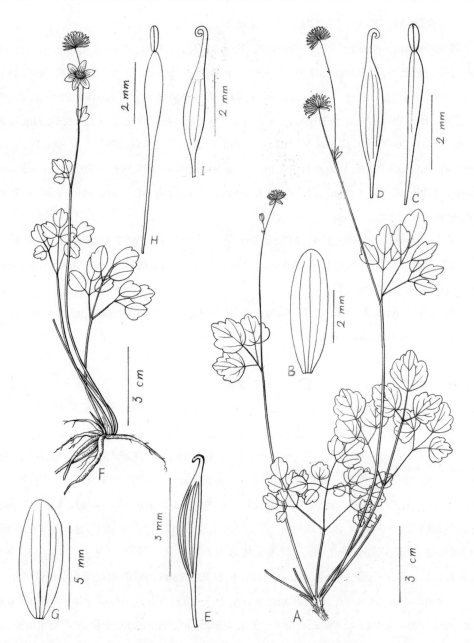

图版27. 台湾唐松草 A-E. 模式变种 A. 植株全形，E. 瘦果（据C. M. Wang & Y. H. Tsai 02974），B. 萼片，C. 雄蕊，D. 心皮。（据C. M. Wang & Y. H. Tsai 03154） F-I. 大花台湾唐松草 F. 植株全形，G. 萼片，H. 雄蕊，I. 心皮。（据T. Y. Yang. C. F. Hsieh et al. 5460）

Plate 27. **Thalictrum urbainii** A-E. var. **urbainii** A. habit, E. achene, (from C. M. Wang & Y. H. Tsai 02974) B. sepal, C. stamen, D. carpel. (from C. M, Wang & Y. H. Tsai 03154) F-I. var. **majus** F. habit, G. sepal, H. stamen. I. carpel. (from T. Y. Yang, C. F. Hsieh et al 5460)

with 3 simple basal nerves. Stamens 28–40; filaments 3–3.5 mm long, above oblanceolate, 0.4–0.8 mm broad, below filiform, 0.1 mm across; anthers broad-oblong, 0.6–0.8×0.3–0.4 mm, apex obtuse. Carpels 5–17; ovary fusiform, ca. 1.8×0.8 mm, base with a stipe ca. 0.8 mm long; style ca. 1 mm long, a pex hooked. Achene compressed, fusiform. 4–6×0.7–1 mm, on each side with 3 longitudinal ribs, and base with a stipe 1.5–2 mm long; persistent style 0.7–1 mm long. apex hooked.

特产台湾，广布台湾山区。生山地林下或林边阴湿处，海拔 550–2680 m。

标本登录：台湾：南投，alt. 2680 m，王秋美等 3154；花莲，alt. 800–1460 m，S. T. Chiu 等 6173，陈添财 12276；台北，alt. 550 m，王秋美等 2974。

27b. 大花台湾唐松草（《中国植物志》）

Var. **majus** T. Shimizu in J. Fac. Tex. Sci. Technol., Shinchu Univ., ser. A, 10: 27. 1961; W. T. Wang & S. H. Wang in Fl. Reip. Pop. Sin. 27: 551. 1979; T. Y. Yang & T. C. Huang in Taiwania 34: 67, fig. 6. 1989; et in Fl. Taiwan, 2nd ed., 2: 573. 1996; D. Z. Fu & G. Zhu in Fl. China 6: 301. 2001. Holotype：台湾：花莲，Tienchang cliff, T. Shimizu 11271（KYO, not seen; photo, PE）.

与台湾唐松草的区别在于本变种的花较大，具 6 枚萼片。茎高（10–）20–70 cm。花直径 2–3 cm；萼片 6, 8–15×3–8 mm.

This variety differs from var. *urbainii* in its larger 6-sepalous flowers. Stems (10–)20–70 cm tall. Flower 2–3 cm in diam.; sepals 6, 8–15×3–8 mm.

特产台湾东部，生海拔 1150–1600 m 一带山地阴湿处。

标本登录：台湾：花莲，杨宗愈等 5460。

系 4. 思茅唐松草系

Ser. **Simaoensia** W. T. Wang, ser. nov. Type: *T. simaoense* W. T. Wang & G. Zhu.

Series nova haec est affinis Ser. *Brevisericeis* W. T. Wang, a qua differt staminum filamentis filiformibus. Compound monochasia corymbiform or paniculiform, few- to many-flowered. Achenes sessile.

本系与绢毛唐松草系近缘，区别在于本系的雄蕊花丝呈丝形。单歧聚伞花序伞房状或圆锥状，具少数或多数花。瘦果无柄。

2 种。

28. 思茅唐松草（*Flora of China*） 图版 28

Thalictrum simaoense W. T. Wang & G. Zhu in Phytologia 79(5): 385, fig. 1. 1995; D. Z. Fu & G. Zhu in Fl. China 6: 297. 2001. Type: 云南：思茅, alt. 2000 m, A. Henry 13096 (holotype, MO; isotype, NY).

多年生小草本植物，全部无毛。茎数根丛生，纤细，高 9–15 cm，在基部之上或中部二歧状分枝。基生叶为 2 回三出复叶，具长柄；叶片 5–9×6–8 cm；小叶薄纸质，圆卵形、肾形或扁圆形，0.9–1.8×1–2.8 cm，顶端圆形或近截形，基部心形或浅心形，微 3 裂，边缘具少数小齿，脉下面隆起，脉网明显；叶柄细，长 1.8–3.8 cm。茎生叶 1–2，与基生叶相似，但较小，1.5–5.5×1.5–3.4 cm；小叶多为扁圆形，0.4–1×0.45–1.2 cm，基部浅心形，微 3 裂，裂片全缘；叶柄长 0.2–3 cm。单歧聚伞花序顶生，具 2 (–3) 花；苞片具短柄，为三出复叶，长约 7 mm；花梗丝形，长 1.7–2.2 cm。花：萼片 4，白色，宽椭圆形，4.2–5×3–5.5 mm，顶端圆形或钝，每一萼片具 3 基生脉（中脉和 1 侧生脉 2 回二歧状分枝，另一侧生脉 1 回二歧状分枝）。雄蕊 9–15；花丝丝形，长 2.2–3 mm，近顶端稍增粗；花药长圆形，长 1–1.2 mm，顶端有不明显的小尖头。心皮 10–13；子房纺锤形，长 1.4 mm；花柱钻形，与子房等长，顶端钩状弯曲，腹面具不明显柱头组织。瘦果纺锤形，长 1.9 mm，每侧有 3 条细纵肋；宿存花柱顶端钩状弯曲。7 月开花。

Small perennial herbs, totally glabrous. Stems several caespitose, slender, 9–15 cm tall, above base or from the middle dichotomously branched. Basal leaves 2–ternate, long petiolate; blades 5–9×6–8 cm; leaflets thinly papery, orbicular-ovate, reniform or depressed-orbicular, 0.9–1.8×1–2.8 cm, at apex rounded or subtruncate, at base cordate or subcordate, 3–lobulate, margin few-denticulate, nerves abaxially prominent, nervous nets conspicuous; petioles slender, 1.8–3.8 cm long. Cauline leaves 1–2, similar to basal leaves, but smaller, 1.5–5.5×1.5–3.4 cm; leaflets mostly depressed-orbicular, 0.4–1×0.45–1.2 cm, at base subcordate, 3–lobulate, lobes entire; petioles 0.2–3 cm long. Monochasia terminal, 2(–3)–flowered; bracts shortly petiolate, ternate, ca. 7 mm long; pedicels filiform, 1.7–2.2 cm long. Flower: Sepals 4, white, broad-elliptic, 4.2–5×3–3.5 mm, apex rounded-obtuse, each sepal with 3 basal nerves (midrib and 1 lateral nerve twice dichotomizing, and the another lateral nerve once dichotomizing). Stamens 9–15; filaments filiform, 2.2–3 mm long, near apex slightly thickened; anthers oblong, 1–1.2 mm long, apex inconspicuously mucronate, Carpels 10–13; ovary fusiform, 1.4 mm long; style subulate, as long as ovary, apex hooked, adaxially

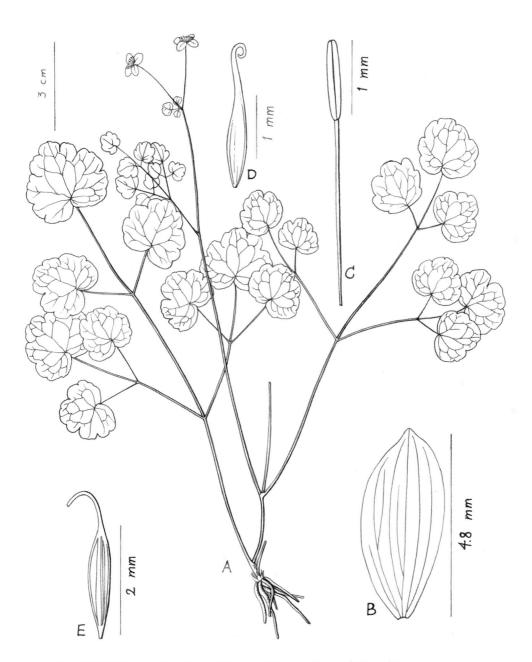

图版28. **思茅唐松草** A. 植株全形, B. 萼片, C. 雄蕊, D. 心皮, E. 瘦果。(据isotype)
Plate 28. **Thalictrum simaoense** A. habit, B. sepal, C. stamen, D. carpel, E. achene. (from isotype)

with inconspicuous stigmatic tissue. Achene fusiform, 1.9 mm long, on each side with 3 thin longitudinal ribs; persistent style at apex hooked.

特产云南思茅。生山地陡崖阴湿处。

29. 短蕊唐松草(《广西植物》) 图版 29

Thalictrum brachyandrum W. T. Wang in Guihaia 37(4): 411, fig. 3. 2017. Type: 河南: 商城县, 朝阳洞, 1984-06-20, 河南植物资源队 D0543(fl., holotype and isotypes, PE); 同地, 1954, 河南植物普查队 1087(fr., paratype, PE).

多年生草本植物。茎高 80-85 cm, 与叶柄和小叶柄均疏被短柔毛(毛长 0.2-0.5 mm), 中部以上分枝。茎生叶为 2-3 回三出复叶; 叶片长 4-6.5 cm; 小叶纸质, 宽卵形、卵形或椭圆形, 1.6-3×1-2.6 cm, 顶端急尖, 基部圆形、浅心形或截形, 3 浅裂(裂片具 1-2 小齿或全缘), 两面无毛, 具三出或五出脉, 脉上面平, 下面隆起, 脉网明显; 小叶柄长 0.5-1.5 mm; 叶柄长 0.3-5 cm, 基部具鞘。复单歧聚伞花序顶生或腋生, 长 3-7 cm, 形成顶生聚伞圆锥花序, 具多数花, 无毛; 下部苞片为三出复叶, 上部苞片简单, 长 1-6 mm; 花梗长 2-3 mm。花无毛: 萼片 4, 淡紫色, 2 枚较小, 狭椭圆形, 2×1 mm, 具 3 条基生脉, 另 2 枚较大, 宽椭圆形或近圆形, 2×1.5-2 mm, 具 4 条基生脉。雄蕊 40-50; 花丝粗丝形, 长 1-1.2 mm; 花药淡黄色, 长圆形, 长 0.8-1 mm, 顶端钝。心皮 4-6; 子房长 1 mm; 花柱长 0.8-1 mm, 直或顶端钩状弯曲, 腹面顶端具不明显柱头。瘦果扁, 狭椭圆形, 2.5-3×1-1.2 mm, 无毛, 每侧具 2 条纵肋; 宿存花柱长约 1.2 mm, 顶端钩状弯曲。6-7 月开花。

Perennial herbs. Stems 80-85 cm tall, with petioles and petiolules sparsely puberulous (hairs 0.2-0.5 mm long), above the middle branched. Cauline leaves 2-3-ternate; blades 4-6.5 cm long; leaflets papery, broad-ovate, ovate or elliptic, 1.6-3×1-2.6 cm, at apex acute, at base rounded, subcordate or truncate, 3-lobed (lobes 2-3-denticulate or entire), on both surfaces glabrous, 3- or 5-nerved, nerves adaxially flat, abaxially prominent, nervous nets conspicuous; petiolules 0.5-15 mm long; petioles 0.3-5 cm long, at base vaginate. Compound monochasia terminal and axillary, 3-7 cm long and forming a terminal many-flowered thyrse, glabrous; lower bracts ternate, upper bracts simple, 1-6 mm long; pedicels 2-3 mm long. Flower glabrous: Sepals 4, purplish, 2 smaller, narrow-elliptic, 2×1 mm, with 3 basal nerves, another 2 larger, broad-elliptic or suborbicular, 2×1.5-2 mm, with 4 basal nerves. Stamens 40-50; filaments thickly filiform, 1-1.2 mm long; anthers yellowish, oblong, 0.8-1 mm long,

153

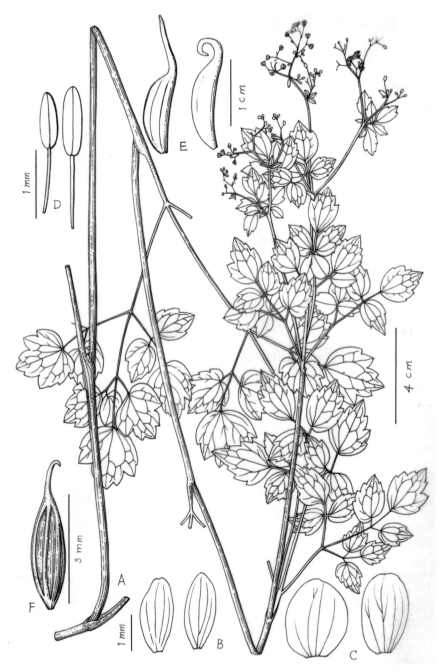

图版29. 短蕊唐松草 A. 开花植株，B. 二萼片（具3条基生脉），C. 二萼片（具4条基生脉），D. 二雄蕊，E. 二心皮，F. 瘦果。（据holotype）

Plate 29. **Thalicrtum brachyandrum1** A. flowering plant, B. two sepals with three basal nerves, C. two sepals with four basal nerves, D. two stamens, E. two carpels, F. achene. (from holotype)

apex obtuse. Carpels 4–6; ovary 1 mm long; style 0.8–1 mm long, straight or apex hooked, adaxially on apex with an inconspicuous stigma. Achene compressed, narrow-elliptic, 2.5–3×1–1.2 mm, glabrous, on each side with 2 longitudinal ribs; persistent style ca. 1.2 mm long, apex hooked.

特产河南商城县。

系 5. 钩柱唐松草系

Ser. **Uncata** W. T. Wang & S. H. Wang in Fl. Reip. Pop. Sin. 27: 536, 617. 1979. Type: *T. uncatum* Maxim.

复单歧聚伞花序狭，似总状花序。每朵花的雄蕊 5–15；花丝条形，棒状狭条形或丝形。

Compound monochasia narrow, raceme-like. Stamens 5–15 per flower; filaments linear, narrowly clavate-linear or filiform.

3 种．分布于西南部。

30. 钩柱唐松草（《中国高等植物图鉴》） 图版 30

Thalictrum uncatum Maxim. In Acta Hort. Petrop. 11: 14. 1890; Iconogr. Corm. Sin. 1: 675, fig. 1349. 1972; Fl. Tsinling. 1(2): 241, fig. 205. 1974; W. T. Wang & S. H. Wang in Fl. Reip. Pop. Sin. 27: 536, pl. 130: 1–4. 1979; W. T. Wang in Bull. Bot. Lab. N.-E. Forest. Inst. 8: 24. 1980; in Fl. Xizang. 2: 66, fig. 19: 1–3. 1985; in Vasc. Pl. Hengduan Mount. 1: 499. 1993; in Fl. Yunnan. 11: 163. 2000; et in High. Pl. China 3: 466, fig. 738. 2000; D. Z. Fu & G. Zhu in Fl. China 6: 300. 2001; H. Y. Fang & J. B. Pan, Field Guid. Wild Pl. China: Qilian Mount. 364. 2016; W. T. Wang in High. Pl. China in Colour 3: 386. 2016; X. T. Ma et al. Wild Flow. Xizang 29, photo 57. 2016. Type: 甘肃：trajectu inter vicos Morping et Wuping, 1885–06–28, G. N. Potanin s. n. (syntype, LE, not seen); 同地，1885–07–04, G. N. Potanin s. n. (syntype, LE, not seen; isosyntype, PE).

T. hamatum Maxim. in Acta Hort. Petrop. 11: 14. 1890. Type: 甘肃：Lumbu 河流[①]，1885–07–11, G. H. Potanin s. n. (syntype, LE, not seen; isosyntype, PE). 四川：valle fl. Honton infra vicum Sanshei, 1885–08–13, G. N. Potanin s. n.(syntype, LE, not seen).

30a. var. **uncatum**

多年生草本植物，全部无毛。茎高 45–90 cm，上部分枝。下部茎生叶为 4–5 回三出复叶，有长柄；叶片长达 15 cm；小叶草质，倒卵形，宽菱形或宽卵形，0.9–1.3×0.5–

[①] Lumbu 为拉丁文的一条河的名字，无法考证其中文名。——编辑注

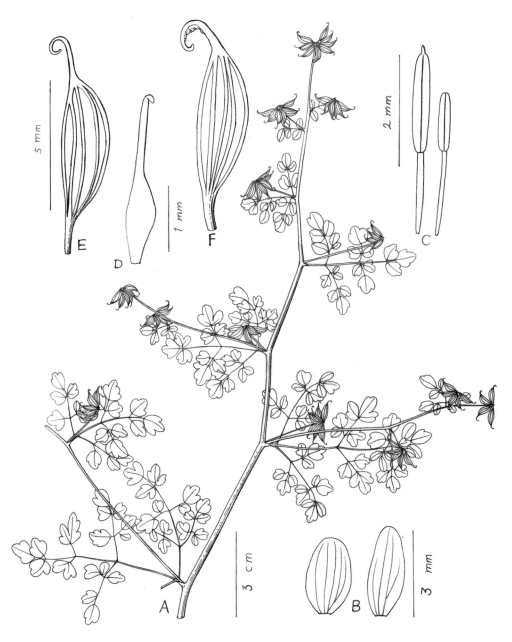

图版30. **钩桂唐松草** A. 开花茎上部，B. 二萼片，C. 二雄蕊，D. 心皮（据姜恕等1069），E、F. 瘦果（据横断山植被组4862）。

Plate 30. **Thalictrum uncatum** var. **uncatum** A. upper part of flowering stem, B. two sepals, C. two stamens, D. carpel (from S. Jiang et al. 4862). E, F. achene (from HengduanShan Vegetation Group 4862).

1 cm，基部宽楔形或圆形，3 浅裂，裂片多全缘或有 1–2 小齿，脉平，脉网不明显；叶柄长约 7 cm，基部有鞘。复单歧聚伞花序顶生并腋生，狭，似总状花序；花梗细，长 2–4 mm。花：萼片 4–5，淡紫色，宽椭圆形或狭卵形，3×1.5–2 mm，顶端圆钝，每一萼片具 3 条基生脉（中脉和 1 侧脉不分枝，另 1 侧脉 1–2 回二歧状分枝）。雄蕊 10–15；花丝狭条形，1.2–1.5×0.15 mm；花药狭长圆形，1–1.8×0.4 mm，顶端具长 0.1–0.2 mm 的短尖头，稀钝。心皮 5–12；子房近菱形，约 1.2×0.6 mm，基部的柄长约 0.8 mm；花柱长约 1.8 mm，顶端向后弯曲。瘦果扁平，新月形，4–5×1.5–2.2 mm，每侧有 3 条纵肋，基部的柄长 1–1.2 mm；宿存花柱长约 2 mm，顶端拳卷状弯曲。5–7 月开花。

Perennial herbs, totally glabrous. Stems 45–90 cm tall, above branched. Lower cauline leaves 4–5-ternate, long petiolate; blades up to 15 cm long; leaflets herbaceous, obovate, broad-rhombic or broad-ovate, 0.9–1.3×0.5–1 cm, at base broad-cuneate or rounded, 3-lobed, lobes mostly entire or 1–2-denticulate, nerves flat, nervous nets inconspicuous; petioles ca. 7 cm long, base vaginate. Compound monochasia terminal or axillary, narrow, raceme-like; pedicels slender, 2–4 mm long. Flower: Sepals 4–5, purplish, broad-elliptic or narrow-ovate, 3×1.5–2 mm, apex rounded-obtuse, each sepal with 3 basal nerves (midrib and 1 lateral nerve simple, and another lateral nerve 1–twice dichotomizing). Stamens 10–15; filaments narrow-linear, 1.2–1.5×0.15 mm; anthers narrow-oblong, 1–1.8×0.4 mm, apex with mucrones 0.1–0.2 mm long, rarely obtuse. Carpels 5–12; ovary subrhombic, ca. 1.2×0.6 mm, base with a stipe ca. 0.8 mm long; style ca. 1.8 mm long, apex recurved. Achene flattened, lunate, 4–5×1.5–2.2 mm, on each side with 3 longitudinal ribs, base with a stipe 1–1.2 mm long; persistent style ca. 2 mm long, apex circinate.

分布于西藏东南部、云南西北部、四川西部、青海东部、甘肃南部。生山地草坡上或灌丛中，海拔 2400–4000 m。

标本登录：西藏：江达，青藏队 76–12415；察雅，青藏队 76–12282；索县，青藏队 76–11017；米林，西藏中草药队 3706；林芝，色季拉山，罗建 9113。

云南：中甸，俞德浚 11686，中甸队 62–809，杨亲二，袁琼 1818，谢磊，毛建丰 93；德钦，俞德浚 10825，青藏队 81–2318。

四川：德昌，四川植物队 11994；美姑，经济植物队 59–1057，四川植物队 13088，13126；昭觉，四川植物队 12823；木里，俞德浚 14746，应俊生 4405；乡城，横断山植被组 4862；郎楷永，李良千，费勇 2649，D. E. Boufford, B. Bartholomew et al. 28389；稻城，四川植被队 2060；理塘，汪发缵 21579，郎楷永，李良千，费勇 2807；雅江，郎楷永，李良千，费勇 2890，D. E. Boufford, B. Bortholomew et al. 35950；康定，蒋兴鏖，熊济华 35620，关克俭，王文采 837，1325，赵振渠 114557，郎楷永，李良千，费勇 684；小金，经济植物队 59-135；姜恕 1418；米亚罗，姜恕 903；汶川，杨亲二，袁琼 214；绰斯甲，姜恕 1069，第八森林队 2438；马尔康，李馨 71148，郎楷永，李良千，费勇 2132，2192，谢磊 2001-9019；松潘，植物资源队 59-1590；阿坝，李馨 1198；若尔盖，郎楷永，李良千，费勇 1958，D. E. Boufford et al. 40313；郎木寺，姜恕 1418。

青海：湟中，何建农 939；互助，郭本兆 9158；门源，崔鸿宾 80-230。

甘肃：岷县，王作宾 4782；Radja and Yellow River gorges, J. F. Rock 14126；康乐，廉永善 94267；连城，孙学刚 6212，6363；天祝，何业祺 4844。

30b. 狭翅钩柱唐松草（《中国植物志》，变种）

Var. angustialatum W. T. Wang in Fl. Reip Pop. Sin. 27: 536, 618. 1979; D. Z. Fu & G. Zhu in Fl. China 6: 300. 2001. —— *T. uncatum* Maxim. SSP. *angustialatum* (W. T. Wang) Luferov in Novit. Syst. Pl Vasc. 45:147.2014. Holotype：贵州：威宁，戛利乡，alt. 1550 m, 1959-07-09，毕节队 125(PE)。

与钩柱唐松草的区别在于本变种的心皮柱头有 2 侧生狭翅。

This variety differs from var. *uncatum* in its stigma with 2 lateral narrow wings.

特产贵州威宁县。生山谷沟边，海拔 1550 m。

31. 狭序唐松草（《中国高等植物图鉴》） 图版 31

Thalictrum atriplex Finet & Gagnep. in Bull. Soc. Bot. France 50: 613, pl. 19: B. 1903; Hand.-Mazz, Symb. Sin. 7: 311. 1931; Iconogr. Corm. Sin. 1: 675, fig. 1350. 1972; W. T. Wang & S. H. Wang in Fl. Reip. Pop. Sin. 27: 537, pl. 130: 5-9. 1979; W. T. Wang in Fl. Xizang. 2: 66, fig. 19: 4-7. 1985; in Vasc. Pl. Hengduan Mount. 1: 499. 1993; in Fl. Yunnan. 11: 163. 2000; et in High. Pl. China 3: 467, fig. 789. 2000; D. 2. Fu & G. Zhu in Fl.China 6: 286. 2001; W. T. Wang in High. Pl. China in Colour 3: 386. 2016; L. Xie in Med. Fl. China 3: 245, with fig. & photo. 2016. Holotype：四川：康定，东俄洛，1893-08-05，J. A. Soulie 391 (P, not seen; photo, PE)。

多年生草本植物，全部无毛。茎高 40-80 cm，上部分枝。下部茎生叶为 4 回三出复叶，具长柄，长约 25 cm；叶片长约 15 cm；小叶草质，倒卵形、宽菱形或宽卵形，长 0.8-2.2×0.8-3 cm，顶端圆或钝，基部宽楔形、圆形或浅心形，3 浅裂或 3 中裂，裂片有 1-2 齿，脉近平，脉网不明显；叶柄长约 12 cm，基部有狭鞘。上部茎生叶较小。复单歧聚伞花序

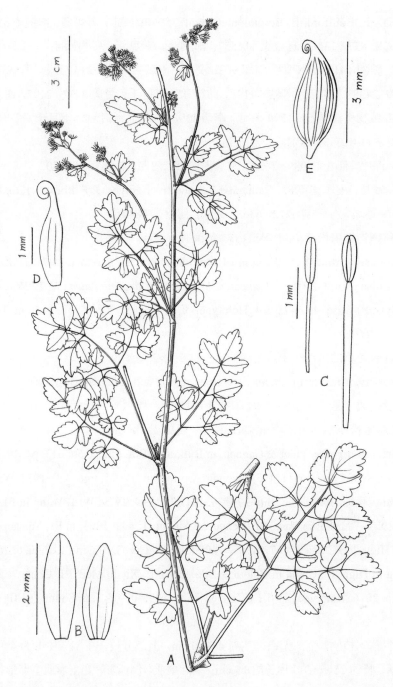

图版31. **狭序唐松草** A. 开花茎上部，B. 二萼片，C. 二雄蕊，D. 心皮，E. 瘦果。（据关克俭，王文采69）
Plate 31. **Thalictrum atriplex** A. upper part of flowering stem, B. two sepals, C. two stamens, D. carpel, E. achene. (from K. C. Kuan & W. T. Wang 69)

顶生和腋生，狭，似总状花序；花梗长 1-5 mm。花：萼片 4，早落，白色或带黄绿色，狭卵形或长圆形，2-2.2×0.8 mm，顶端钝，每一萼片具 1-2 条基生脉。雄蕊 6-12；花丝狭条形，长 1.8-3 mm，宽 0.1 mm 或上部宽 0.2 mm，下部宽 0.1 mm；花药狭长圆形，0.7-0.8×0.3-0.4 mm，顶端有不明显短尖头。心皮 3-5（-8）；子房近长圆形，约 2×1 mm；花柱长约 1.2 mm，顶端钩状弯曲。瘦果宽椭圆形，2.2-2.8×1.2-2 mm，每侧有 2 条纵肋，基部的柄长 0.1-0.3 mm；宿存花柱长 1-2 mm，顶端钩状弯曲。6-7 月开花。

Perennial herbs, totally glabrous. Stems 40-80 cm tall, above branched. Lower cauline leaves 4-ternate, long petiolate, ca. 25 cm long; blades ca. 15 cm long; leaflets herbaceous, obovate, broad-rhombic or broad-ovate, 0.8-2.2×0.8-3 cm, at apex rounded or obtuse, at base broad-cuneate, rounded or subcordate, 3-lobed or 3-fid, lobes 1-2-dentate, nerves nearly flat, nervous nets inconspicuous; petioles ca. 12 cm long, base narrowly vaginate. Compound monochasia terminal and axillary, narrow, raceme-like; pedicels 1-5 mm long. Flower: Sepals 4, caducous, white or tinged with yellowish-greenish, narrow-ovate or oblong, 2-2.2×0.8 mm, apex obtuse, each sepal with 1-2 basal nerves. Stamens 6-12; filaments narrow-linear, 1.8-3 mm long, 0.1 mm broad or above 0.2 mm broad, below 0.1 mm broad; anthers oblong, 0.7-0.8×0.3-0.4 mm, apex inconspicuously mucronate. Carpels 3-5(-8); ovary suboblong, ca. 2×1 mm; style ca. 1.2 mm long, apex hooked. Achene broad-elliptic, 2.2-2.8×1.2-2 mm, on each side with 2 longitudinal ribs; persistent style 1-2 mm long, apex hooked.

分布于西藏东南部、云南西北部、四川西部、青海南部。生山地草坡、林边或疏林中，海拔 2300-4000 m。

根和根状茎有清热解毒之效，用于痢疾、肠炎等症。根含大叶唐松草碱（thalifaberine）等生物碱。（《中国药用植物志》，2016）

标本登录：西藏：察隅，察瓦龙，王启无 65517；松多，张永田，郎楷永 1374，2754；冈沱，贾慎修 341；波密，古乡，应俊生，洪德元 65-0211；林芝，倪志诚等 1771；色季拉山，罗建 09-261；米林，西藏中草药队 3806，青藏补点队 75-0911，倪志诚等 2870；工布江达，青藏队 74-3704；隆子，青藏补点队 75-0421；错那，吴征镒等 75-1100；吉隆，PE-西藏队 4153。

云南：丽江，王启无 70849，冯国楣 21352；中甸，俞德浚 12346，杨亲二，袁琼 1821，谢磊 72；维西，王启无 63686；德钦，俞德浚 9515，10511，10568，青藏队 81-2318，横断山植被组 8648。

四川：九龙，俞德浚 6787，D. E. Boufford et al. 33109,33271；乡城，青藏队 81-4704；稻城，D.E.Boufford et al. 28108；巴塘，郎楷永，李良千，费勇 2307、2328；白玉，D. E. Boufford et al. 36969；甘孜，崔友文 4248；德格，崔友文 5063；道孚，D. E. Boufford et al. 27697；乾宁，四川植物队 5558；雅江，D. E. Boufford et al. 35843；康定，刘振书 695，蒋兴麐 36027，崔友文 4194，姜恕 2060，关克俭，王文采 69、656，四川植物队 4823，C. R. Lancaster941A；绰斯甲，第八森林队 2336；马尔康，张泽民，周洪富 22837。

青海：玉树，何廷农等 2594。

32. **螺柱唐松草**（《广西植物》） 图版 32

Thalictrum spiristylum W. T. Wang in Guihaia 37(6): 682, fig. 5. 2017. Type：云南：宁蒗县，宁蒗林业局，水草坝，alt. 3100 m，1981-07-10，韩裕丰，邓坤梅，陈永瑞 81-1070（holotype and isotypes, PE）。

多年生草本植物。茎高约 60 cm，近无毛，分枝。中部和上部茎生叶为 2－3 回三出羽状复叶，无毛；叶片 10-15×10-18 cm；小叶薄纸质，宽菱形，宽倒卵形、倒卵形或宽卵形，1-3.5×0.7-3.6 cm，基部圆形、宽楔形或近截形，3 浅裂或 3 中裂（裂片在顶端急尖，具 1-2 尖齿或全缘），具 3 或 5 出脉，脉平；小叶柄长 0.1-2 cm；叶柄长 0.3-3.5 cm，基部具鞘。顶生复单歧聚伞花序狭，似总状花序，长约 4 cm，密生 15 花，腋生复单歧花序约 3，也似总状花序，长 3.5-8.5 cm，约有 10 花，基部有一具 1-4 花的短分枝，近无毛；苞片叶状，只叶柄疏被短柔毛，其他部分无毛；花梗粗壮，长约 1 mm，无毛。花无毛：萼片 4，白色，卵状长圆形，约 2.2×0.8 mm，顶端微尖，每一萼片具 1 条脉。雄蕊 5-6；花丝丝形，长 1.5-3 mm；花药长圆形或狭长圆形，0.8×0.3-0.45 mm，顶端具或不具小短尖头。心皮 4-5；子房长圆状椭圆形，2×0.6-0.7 mm；花柱长约 2 mm，顶端螺旋状弯曲，腹面具不明显柱头。7 月开花。

Perennial herbs. Stems ca. 60 cm tall, subglabrous, branched. Middle and upper cauline leaves 2–3-ternate-pinnate, glabrous; blades 10–15×10–18 cm; leaflets thinly papery, broad-rhombic, broad-obovate, obovate or broad-ovate, 1–3.5×0.7–3.6 cm, at base rounded, broad-cuneate or subtruncate, 3–lobed or 3–fid (lobes at apex acute, acutely 1–2-dentate or entire), 3– or 5–nerved, nerves flat; petiolules 0.1–2 cm long; petiole 0.3–3.5 cm long, base vaginate. Terminal compound monochasium narrow, raceme-like, ca. 4 cm long, densely 15–flowered, axillary compound monochasia ca. 3, also raceme-like, 3.5–8.5 cm long, ca. 10–flowered, at base with a short branch 1–4-flowered, subglabrous; bracts foliaceous, with only petioles

161

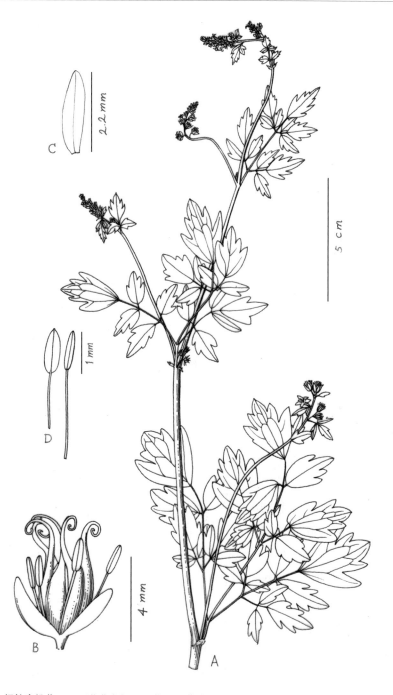

图版32. **螺柱唐松草**　A. 开花茎上部，B. 花，C. 萼片，D. 2雄蕊。（据holotype）
Plate 32. **Thalictrum spiristylum**　A. upper part of flowering stem, B. flower, C. sepal, D. two stamens. (from holotype)

sparsely puberulous, elsewhere glabrous; pedicels short and robust, ca. 1 mm long. Flower glabrous: Sepals 4, white, ovate-oblong, ca. 2.2×0.8 mm, apex slightly acute, each sepal with 1 basal nerve. Stamens 5–6; filaments filiform, 1.5–3 mm long; anthers oblong or narrow-oblong, 0.8×0.3–0.45 mm, apex minutely mucronate or not mucronate. Carpels 4–5; ovary oblong-elliptic, 2×0.6–0.7 mm; style ca. 2 mm long, at apex spirally curved, adaxially with an inconspicuous stigma.

特产云南宁蒗县。生山谷坡地，海拔 3100 m。

组 2. 钻柱唐松草组

Sect. **Dolichostylus** W. T. Wang, sect. nov. Type: *T. tenuisubulatum* W. T. Wang

Ob carpelli stylo apice uncato vel circinato sectio nova haec est fortasse affinis Sect. *Leptostigmati* Boivin, a qua carpello stylo elongato anguste subulato-lineari ovario multo longiore praedito et stigmate carente differt. In Sect. *Leptostigmate*, stylus plerumque ovario brevior, raro ei aequilongus, ventro superne stigmate in conspicuo praeditus.

由于心皮花顶端钩状或拳卷状弯曲，本组可能与叉枝唐松草组 Sect. *Leptostigma* Boivin 有亲缘关系。与后者的区别在于本组心皮的花柱伸长，膜质，呈狭钻状条形，比子房长得多，同时心皮无柱头。在叉枝唐松草组，花柱在多数情况下比子房短，稀与子房等长，在腹面上部具一不明显柱头。

1 种，特产云南西部。

33. 钻柱唐松草（《云南植物研究》） 图版 33

Thalictrum tenuisubulatum W. T. Wang in Acta. Bot. Yunnan. 4 (3): 135, pl. 2: 6. 1982; et in Fl. Yunnan. 11: 162. 2000; D. Z. Fu & G. Zhu in Fl. China 6: 299. 2001. Holotype: 云南：腾冲，狼牙山，青草岭，alt. 3350 m，1964-06-10，武素功 7130 (LBG).

多年生草本植物，全部无毛。茎高约 40 cm，分枝。茎生叶为 3 回三出复叶，长 2–14 cm；小叶薄草质，倒卵形、宽倒卵形或近圆形，0.6–1.8×0.7–1.4 cm，顶端圆形，基部圆形或浅心形，3 浅裂，裂片多全缘，有时具 1–2 小钝齿，脉近平；小叶柄长 1.2–8 mm；叶柄长 1–2.5 cm，基部具短鞘。复单歧聚伞花序顶生并侧生，长约 2 cm，有 3 花；花梗细，长 0.2–1.6 cm。花：萼片 4，白色，椭圆形，6–8×4–5 mm，顶端圆形或钝，每一萼片具 5 条基生脉（中脉不分枝，侧生脉 1–3 回二歧状分枝）。雄蕊约 50；花丝狭条形，长

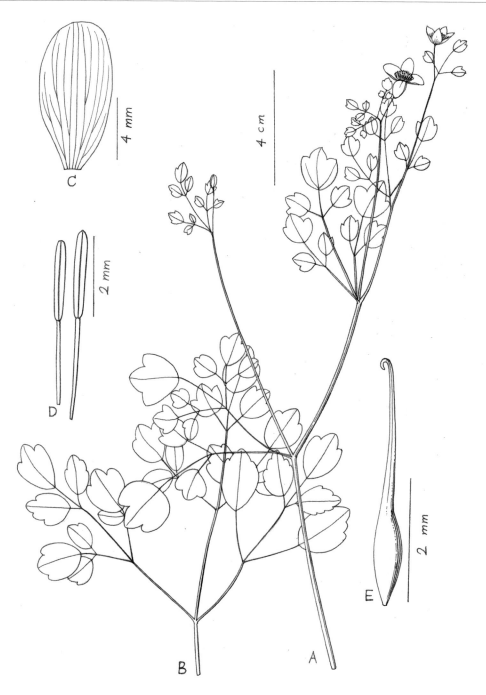

图版33. 钻柱唐松草　A. 开花茎上部，B. 上部茎生叶，C. 萼片，D. 二雄蕊，E. 心皮。（据holotype）
Plate 33. **Thalictrum tenuisubulatum**　A. upper part of flowering stem, B. upper cauline leaf, C. sepal, D. two stamens, E. carpel. (from holotype)

1.5–2 mm，上部宽约 0.1 mm，下部宽约 0.08 mm；花药狭长圆形或条形，1.5–1.6×0.25–0.4 mm，顶端钝。心皮约 7；子房纺锤形，约 1.6×0.4 mm；花柱比子房长许多，膜质，狭钻状条形，长约 2.4 mm，下部宽 0.35 mm，上部宽 0.15 mm，具 1 条纵脉，顶端钩状或拳卷状弯曲。5–6 月开花。

Perennial herbs, totally glabrous. Stem ca. 40 cm tall, branched. Cauline leaves 3-ternate, 2–14 cm long; leaflets thinly herbaceous, obovate, broad-obovate or suborbicular, 0.6–1.8×0.7–1.4 cm, at apex rounded, at base rounded or subcordate, 3-lobed, lobes mostly entire, sometimes obtusely 1–2-denticulate, nerves nearly flat; petiolules 1.2–8 mm long; petioles 1–2.5 cm long, at base shortly vaginate. Compound monochasia terminal or lateral, ca. 2 cm long, 3-flowered; pedicels slender, 0.2–1.6 cm long. Flower: Sepals 4, white, elliptic, 6–8×4–5 mm, apex rounded or obtuse, each sepal with 5 basal nervs (midrib simple, and lateral nerves 1–thrice dichotomizing). Stamens ca. 50; filaments narrow-linear, 1.5–2 mm long, above ca. 0.1 mm broad, below ca. 0.08 mm broad; anthers narrow-oblong or linear, 1.5–1.6×0.25–0.4 mm, apex obtuse. Carpels ca. 7; ovary fusiform, ca. 1.6×0.4 mm; style much longer than ovary, membranous, narrowly subulate-linear, ca. 2.4 mm long, below 0.35 mm broad, above 0.15 mm broad, longitudinally 1-nerved, apex hooked or circinate.

特产云南腾冲县，生山地针阔混交林中，海拔 3350 m。

组 3. 宽柱唐松草组

Sect. **Platystylus** W. T. Wang, sect. nov. Type: *T. latistylum* W. T. Wang.

Sectio nova haec carpello stylo membranaceo complanato oblongo praedito et stigmate carente insignis est, et cum charateribus duobus his a generis *Thalictri* sectionibus sinensibus ecteris distinguitur. In sectionibus ceteris, carpellum stylo vulgo tenuiter terete et stigmate inconspicuo vel conspicuo praeditus est.

本组的特点是心皮具膜质、扁平、长圆形花柱，同时心皮无柱头，据此二特征即可与中国唐松草属的其他组相区别。在其他组，花柱通常呈细圆柱形，并在腹面上部具不明显或明显柱头。

1 种，特产甘肃南部。

34. 宽柱唐松草（《广西植物》） 图版 34

Thalictrum latistylum W. T. Wang in Guihaia 37(6): 676, fig. 2. 2017. Holotype：甘肃：文县，白水江自然保护区，邱家坝，alt. 2900 m，2006-06-27，白水江考察队 0454（PE）.

多年生草本植物。茎高约 50 cm，下部有 5 条浅纵沟，上部分枝。中部和上部茎生叶为 2-3 回三出复叶，具短柄；叶片 2-7×4-10 cm；小叶薄纸质，宽倒卵形、近圆形或宽卵形，0.4-1.6×0.4-2 cm，基部圆形或浅心形，3 浅裂或 3 中裂（裂片在顶端钝或圆形，全缘或有 1-2 钝齿），稀 3 微裂，上面无毛，下面被短腺毛（毛长 0.1-0.2 mm），具 3 出或 5 出脉，脉上面平，下面稍隆起；小叶柄丝形，长 2-4 mm，无毛；叶柄长 0.3-2 cm，无毛或被短柔毛。复单歧聚伞花序顶生，长 3.5-4.5 cm，有 4-6 花；苞片叶状，为三出复叶，长 0.5-3 cm；花梗细，长 0.2-1.5 cm，无毛。花无毛：萼片 4，白色，长圆状椭圆形，约 4×2 mm，顶端圆形，每一萼片具 3 条基生脉（中脉和 1 侧生脉 2 回二歧状分枝，另一侧生脉不分枝）。雄蕊 35-50；花丝狭条形，长 2.5-3 mm，上部宽 0.2-0.25 mm，下部宽 0.15 mm；花药长圆形，0.8-1×0.4 mm，顶端钝。心皮 4；子房斜狭倒卵形，约 1×0.5 mm，上部与花柱具 1 纵脉；花柱膜质，扁平，长圆形，约 0.8×0.35 mm，顶端钝；柱头不存在。6-7 月开花。

Perennial herbs. Stem ca. 50 cm tall, below longitudinally shallowly 6-sulcate, above branched. Middle and upper cauline leaves 2–3-ternate, shortly petiolate; blades 2–7×4–10 cm; leaflets thinly papery, broad-obovate, suborbicular or broad-ovate, 0.4–1.6×0.4–2 cm, at base rounded or subcordate, 3-lobed or 3-fid (lobes at apex obtuse or rounded, entire or obtusely 1–2-dentate), rarely 3-lobulate, adaxially glabrous, abaxially glandular-puberulous (hairs 0.1–0.2 mm long), 3- or 5-nerved, nerves adaxially flat, abaxially slightly prominent; petiolules filiform, 2–4 mm long; petioles 0.3–2 cm long, glabrous or puberulous. Compound monochasia terminal, 3.5–4.5 cm long, 4–6-flowered; bracts foliaceous, ternate, 0.5–3 cm long; pedicels slender, 0.5–1.5 cm long, glabrous. Flower glabrous: Sepals 4, white, oblong-elliptic, ca. 4×2 mm, apex rounded, each sepal with 3 basal nerves (midrib and 1 lateral nerve twice dichotomizing, and another lateral nerve simple). Stamens 35–50; filaments narrow-linear, 2.5–3 mm long, above 0.2–0.25 mm broad, below 0.15 mm broad; anthers oblong, 0.8–1×0.4 mm, apex obtuse. Carpels 4; ovary obliquely narrow-obovate, ca. 1×0.5 mm, above with style longitudinally 1-nerved; style membranous, flattened, oblong, ca. 0.8×0.35 mm, apex obtuse; stigma absent.

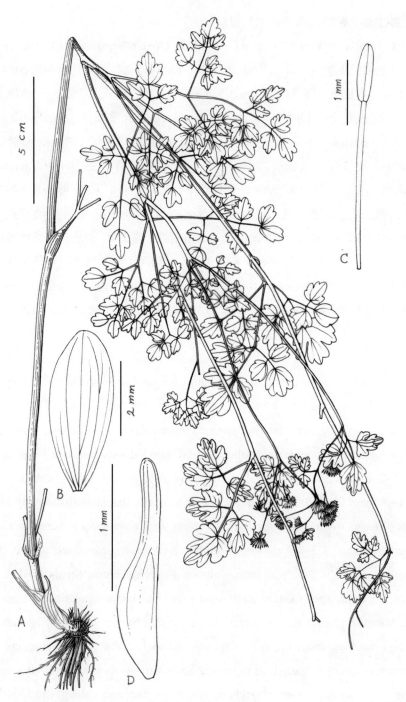

图版34. 宽柱唐松草 A. 植株全形，B. 萼片，C. 雄蕊，D. 心皮。（据holotype）
Plate 34. **Thalictrum latistylum** A. habit, B. sepal, C. stamen, D. carpel. (from holotype)

特产甘肃文县。生山顶山坡草丛中，海拔 2900 m。

组 4. 瓣蕊唐松草组

Sect. 3. **Erythrandra** B. Boivin in Rhodora 46: 360. 1944; Tamura in Sci. Rep. Osaka Univ. 17: 50. 1968; in Acta Phytotax. Geobot. 43: 56. 1992; et in Hiepko, Nat. Pflanzenfam., Zwei. Aufl., 17a Ⅳ : 482. 1995; Emura in J. Fac. Sci. Univ. Tohyo. Sect. 3, Bot. 9: 126. 1972. Type: *T. petaloideum* L.

花组成伞房状单歧聚伞花序。萼片具 3 基生脉。雄蕊花丝稀呈狭条形，通常分化成在形状和大小方面不同的二部分。心皮具 1 直的花柱；花柱腹面具 1 明显柱头或密集柱头组织。瘦果无翅。

Flowers in corymbiform monochasia. Sepal with 3 basal nerves. Stamen filament rarely narrow-linear, usually differentiated into two parts different in shape and size. Carpel with a straight style, on adaxial face of which there is a conspicuous stigma or a fascicle of dense stigmatic tissue. Achenes not winged.

10 种．隶属 5 系

系 1. 细柱唐松草系

Ser. **Megalostigmata** W. T. Wang, ser. nov. Type: *T. megalostigma* (Boivin) W. T. Wang.

Secies nova haec est affinis Ser. *Baicalensibus* (Tamura) W. T. Wang & S. H. Wang, a qua staminum filamentis et stigmatibus ambabus anguste linearibus differt. Achene broad-ellipsoid, on each side with 3(–4) longitudinal ribs and at base with a short stipe.

本系与贝加尔唐松草系 Ser. *Baicalensia* 近缘，与后者的区别在于本系的雄蕊花丝以及柱头都呈狭条形。瘦果圆椭圆形，在每侧具 3 (–4) 条纵肋，基部具 1 短柄。

1 种，分布于四川西部、甘肃南部。

35. 细柱唐松草 长柱贝加尔唐松草（《中国植物志》） 图版 35

Thalictrum megalostigma (B. Boivin) W. T. Wang in Bull. Bot. Lab. N.-E. Forest. Inst. 8: 27. fig. 3: 1 et 3. 1980; in Vasc. Pl. Hengduan Mount. 1: 500. 1993; et in High. Pl. China 3: 468. 2000.—*T. baicalense* Turcz. var. *megalostigma* Boivin in Rhodora 46: 363, fig. 9. 1944; W. T. Wang & S. H. Wang in Fl. Reip. Pop. Sin. 27: 542. pl. 120:16a. 1979; D. Z. Fu & G. Zhu in Fl. China 6: 287. 2001; L. Xie in Med. Fl. China 3: 250. 2016. Type：四川：康定，9500 ft., 1928-09-27 (holotype, GH, not seen); 抚边, alt. 3400 m, 1930-06-19, 汪发缵 21377 (paratypes, GH, PE).

多年生草本植物，全部无毛。茎高 40-100 cm，分枝，稀不分枝。茎生叶为 2-3 回三出复叶或 2-3 回三出羽状复叶；叶片卵形或宽卵形，7-16×7-26 cm；小叶纸质，宽倒卵形、倒卵形、近圆形、宽椭圆形或卵形，1.5-3.5×0.7-4 cm，顶端微尖，常具短尖头，基部圆形、宽楔形或近截形，3 浅裂或 2-3 中裂（裂片具 1-2 牙齿，有时全缘），具 5 出脉，脉两面平；小叶柄长 0.1-2 cm；叶柄长 2-15 cm。聚伞圆锥花序顶生并腋生，长 5-20 cm，有少数或多数花；下部苞片叶状，为 1-2 回三出复叶，长 2-8 cm，上部苞片简单，宽卵形、椭圆形或条形，长 3-10 mm，3 中裂或不分裂；花梗长 0.3-1.8 cm。花：萼片 4，浅绿色带紫色，椭圆状卵形，约 2.6×1.2 mm，顶端圆形，每一萼片具 3 条基生脉（中脉和 1 侧生脉不分枝，另 1 侧生脉二歧状分枝）。雄蕊 10-14；花丝狭条形，长 4.5-5 mm，上部宽 0.2 mm，下部宽 0.1 mm；花药长圆形，约 1×0.3-0.4 mm，顶端钝。心皮 5-6；子房宽椭圆体形，约 3×1.5 mm；花柱长约 1.2 mm；柱头狭条形，长约 1 mm，顶端稍向后弯曲。瘦果宽椭圆体形，2.8-3×1.6-2.2 mm，每侧有 3(-4) 条纵肋，基部有长 0.5-1 mm 的柄；宿存花柱长 1-1.5 mm，顶端稍向后弯曲。5-6 月开花，6-8 月结果。

Perennial herbs, totally glabrous. Stems 40-100 cm tall, branched, rarely simple. Cauline leaves 2-3-ternate or 2-3-ternate-pinnate; blades ovate or broad-ovate, 7-16×7-26 cm; leaflets papery, broad-obovate, obovate, suborbicular, broad-elliptic or ovate, 1.5-3.5 ×0.7-4 cm, at apex slightly acute and often mucronate, at base rounded, broad-cuneate or subtruncate, 3-lobed or 2-3-fid (lobes 1-2-dentate, sometimes entire), 5-nerved, nerves on both surfaces flat; petiolules 0.1-2 cm long, petioles 2-15 cm long. Thyrses terminal and axillary, 5-20 cm long, few- or many-flowered; lower bracts foliaceous, 1-2-ternate, 2-8 cm long, upper bracts simple, broad-ovate, elliptic or linear, 3-10 mm long, 3-fid or undivided; pedicels 0.3-1.8 cm long. Flower: Sepals 4, greenish and tinged with purple, elliptic-ovate, ca. 2.6×1.2 mm, apex rounded, each sepal with 3 basal nerves (midrib and 1 lateral nerve simple, and the another lateral nerve dichotomizing). Stamens 10-14; filaments narrow-linear, 4.5-5 mm long, above 0.2 mm broad, below 0.1 mm broad; anthers oblong. ca. 1×0.3-0.4 mm, apex obtuse. Carpels 5-6; ovary broad-ellipsoid, ca. 3×1.5 mm; style ca. 1.2 mm long; stigma narrow-linear, ca. 1 mm long., apex slightly recurved. Achene broad-ellipsoid, 2.8-3×1.6-2.2 mm, on

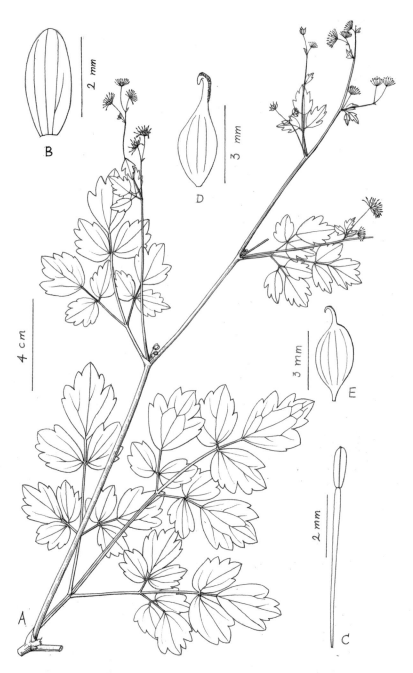

图版35. **细柱唐松草** A. 开花茎上部, B. 萼片, C. 雄蕊, D. 心皮, （据李馨77675）, E. 瘦果。（据D. E. Boufford等37958）

Plate 35. **Thalictrum megalostigma** A. upper part of flowering stem, B. sepal, C. stamen, D. carpel. (from X. Li 77675) E. achene. (from D. E. Boufford et al. 37958)

each side with 3(–4) longitud inal ribs, at base with a stipe 0.5–1 mm long; persistent style 1–1.5 mm long, apex slightly recurved.

分布于四川西部和甘肃南部，生山地沟边、灌丛中或林中，海拔 2000–3000 m。

根和根状茎有清热解毒之效，用于痢疾、传染性肝炎、麻疹等症。在四川、甘肃等地作"马尾连"药材使用。(《中国药用植物志》, 2016)

标本登录：**四川**：康定，刘振书 134，蒋兴麐 36784，胡文光，何铸 11289；贡嘎山，alt. 3100 m，郎楷永，李良千，费勇 1073；宝兴，硗碛，关克俭，王文采 3002；丹八，第八森林队 5503，姜恕 2210，D. E. Boufford et al. 37958；理县，吴中伦 33545；小金，alt. 3200 m，张秀实 7054，郎楷永，李良千，费勇 1519；金川，alt. 2230–2900 m，李馨 77671，77675，D. E. Boufford et al. 38806；绰斯甲，第八森林队 2447；马尔康，D. E. Boufford et al. 39874；刷经寺，张泽云，周洪富 22825；若尔盖，李文华，韩裕丰 H82–511。

甘肃：舟曲，alt. 2500–2800 m，王作宾 14344，姜恕 421；莲花山自然保护区，alt. 2050 m，孙学刚 2280。

系 2. 贝加尔唐松草系

Ser. **Baicalensia** (Tamura) W. T. Wang & S. H. Wang in Fl. Reip. Pop. Sin. 27: 541. 1979.——Sect. *Erythrandra* Boivin subsect. *Baicalensia* Tamura in Sci Rep. Osaka Univ. 17: 51. 1968; et in Hiepko, Nat. Pflanzenfam. Zwei. Aufl., 17a Ⅳ : 482. 1995——Sect. *Baicalensia* (Tamura) Emura in J. Fac. Sci. Univ. Tokyo, Sect. 3, Bot. 9: 107. 1972. Type: *T. baicalense* Turcz.

雄蕊花丝分化成在形状和大小都不同的上、下二部分。心皮具 1 直花柱；花柱腹面顶端具一椭圆体形小柱头。瘦果扁球形，具细纵肋和一短柄。

Stamen filament differentiated into upper and lower two parts different in shape and size. Carpel with a straight style, and style adaxially on apex with a small ellipsoid stigma. Achene depressed-globose, with thin longitudinal ribs and a short stipe at base.

1 种，广布东亚。

36. 贝加尔唐松草（《东北植物检索表》） 图 36

Thalictrum baicalense Turcz. In Bull. Soc. Nat. Mosc. 11: 85. 1838; Lecoy. in Bull. Soc. Bot. Belg. 24: 164. 1885; Maxim, Fl. Tangut. 5. 1889; et in Acta Hort. Petrop. 11: 14. 1890; Finet & Gagnep. in Bull. Soc. Bot. France 50: 607. 1903; Nevski in Fl. URSS 7: 515, t. 517.

1937; Kitagawa, Lineam. Fl. Mansh. 226. 1939; Boivin in Rhodora 46: 362. 1944; Kom. & K. Alisova, Key Pl. Far East. USSR 1: 554. 1961; C. Pei & T. Y. Chou, Icon. Chin. Med. Pl. 7: fig. 326. 1964; Iconogr. Corm. Sin. 1: 676, fig. 1352. 1972; Fl. Tsinling. 1(2):242, fig. 207. 1974; S. H. Li in Fl. Pl. Herb. Chinae Bor. –Or. 3: 206, fig. 89. 1975; W. T. Wang & S. H. Wang in Fl. Reip. Pop. Sin. 27: 542, pl. 132. 1979; W. T. Wang in Bull. Bot. Lab. N.-E. Forest. Inst. 8: 25. 1980; B. Z. Ding et al. Fl. Henan. 1: 473, fig. 606. 1981; S. Y. He, Fl. Beijing, rev. ed., 1: 238, fig. 296. 1984; Tamura & Shimizu in Satake, Wild Flow. Pl. Japan 2: 86, Pl. 86: 1–2. 1984; W. T. Wang in Fl. Xizang. 2: 68. 1985; J. W. Wang in Fl. Hebei 1: 443, fig. 439. 1986; D. Z. Ma, Fl. Ningxia. 1: 174. 1986; Y. Z. Zhao in Fl. Intramongol., 2nd ed., 2: 451, pl. 182: 4–6. 1990; X. Y. Yu et al. in Fl. Shanxi. 1: 596, fig. 376. 1992; W. T. Wang in Vasc. Pl. Hengduan Mount. 1: 501. 1993; S. H. Li in Clav. Pl. China Bor.-Or:, ed. 2: 205, pl. 96: 1. 1995; L. H. Zhou in Fl. Qinghai. 1: 329. 1997; W. T .Wang in High. Pl. China 3: 468, fig. 742. 2000; D. Z. Fu & G. Zhu in Fl. China 6: 286. 2001; Y. Kadota in Iwatsuki et al. Fl. Japan 2a:337. 2006; F. H. Li et al. Pl. Yanqing 142. 2015; W. T. Wang in High. Pl. China in Colour 3: 387. 2016; L. Xie in Med. Fl. China 3: 249, with fig. & photo. 2016. Described from the shores of Lake Baikal.

T. giraldii Ulbr. In Notizbl. Bot. Gart. Berl. 9: 224. 1925. Holotype: 陕西: Ngo-san, 1899-08, G. Giraldi 4889（not seen）.

36a. f. **baicalense**

多年生草本植物,全部无毛。茎高 40-80 cm,不分枝或分枝。茎生叶为 3 回三出复叶;叶片长 9-16 cm;小叶草质,宽菱形,宽倒卵形或宽卵形,1.8-4.5×1.5-5 cm,基部宽楔形或圆形,3 浅裂,裂片有 1-4 钝齿,脉下面隆起,脉网稍明显;小叶柄长 0.2-3 cm;叶柄长 1-2.5 cm,基部有狭鞘。复单歧聚伞花序伞房状,长 2.5-4.5 cm,有 8-20 花;苞片狭倒卵形、狭卵形或钻形,长 2-8 mm;花梗细,长 4-9 mm。花:萼片 4,早落,绿白色,椭圆形,约 3×1.5 mm,顶端圆形,每一萼片具 3 条不分枝的基生脉。雄蕊 10-20;花丝长 2.2-2.4 mm,上部狭倒披针形或近条形,宽 0.2-0.4 mm,下部丝形,粗 0.1 mm,花药宽长圆形,0.8-1×0.3-0.4 mm,顶端短。心皮 3-7;子房椭圆体形,约 1.4×0.6 mm;花柱长约 0.5 mm,腹面顶端具椭圆体形,长 0.2-0.3 mm 的柱头。瘦果宽椭圆体形,约 3.5×2.5 mm,有 8 条细纵肋,基部的柄长 0.2-0.7 mm;宿存花柱长约 0.8 mm。5-6 月开花。

Perennial herbs, totally glabrous. Stems 40–80 cm tall, simple or branched. Cauline

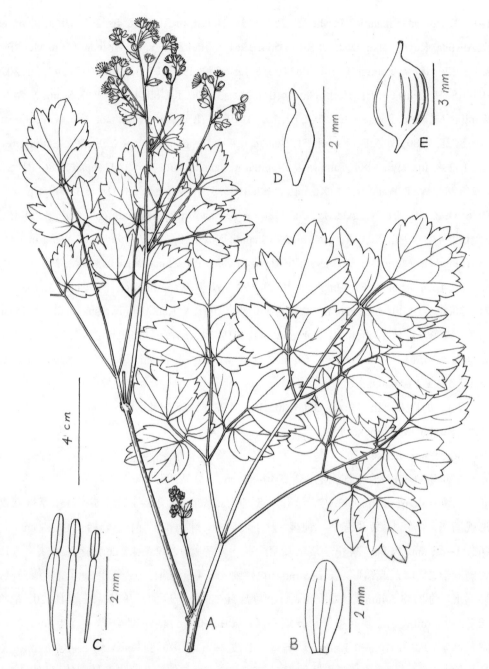

图版36. 贝加尔唐松草　A. 开花茎上部，B. 萼片，C. 三雄蕊，D. 心皮（据刘继孟180），E. 瘦果。（据刘瑛，刘心源13096）

Plate 36. **Thalictrum baicalense**　A. upper part of flowering stem, B. sepal, C. three stamens, D. carpel (from K. M. Liou 180), E. achene, (from Y. Liu and X. Y. Liu 13096)

leaves 3-ternate; blades 9–16 cm long; leaflets herbaceous, broad-rhombic, broad-obovate or broad-ovate, 1.8–4.5×1.5–5 cm, at base broad-cuneate or rounded, 3-lobed, lobes obtusely 1–4-dentate, nerves abaxially slightly prominent, nervous nets slightly conspicuous; petiolules 0.2–3 mm long; petioles 1–2.5 cm long, at base narrowly vaginate. Compound monochasia terminal, corymbiform, 2.5–4.5 cm long, 8–20-flowered; bracts narrow-obovate, narrow-ovate or subalate, 2–8 mm long; pedicels slender, 4–9 mm long. Flower: Sepals 4, caducous, greenish-white, elliptic, ca. 3×1.5 mm, apex rounded, each sepal with 3 simple basal nerves. Stamens 10–20; filaments 2.2–2.4 mm long, above narrow-oblanceolate or linear, 0.2–0.4 mm broad, below filiform, 0.1 mm across; anthers broad-oblong, 0.8–1×0.3–0.4 mm, apex obtuse. Carpels 3–7; ovary ellipsoid, ca. 1.4×0.6 mm; style ca. 0.5 mm long, adaxially on apex with a ellipsoid stigma 0.2–0.3 mm long. Achene broadly ellipsoid, ca. 3.5×2.5 mm, with 8 thin longitudinal ribs, at base with a stipe 0.2–0.7 mm long; persistent style ca. 0.8 mm long.

分布于西藏东南部、四川西北部、青海东部、甘肃和陕西的南部、河南西部、山西南部和中部、河北西部和北部、内蒙古、吉林和黑龙江的东部；以及俄国西伯利亚东部和远东地区、朝鲜、日本。生山地林下或湿润草坡；海拔分布：自西藏东南部至甘肃南部为2400–2800 m，自青海东部至山西南部为1600–1900 m，自河北至东北为900–1500 m。

根和根状茎有清热解毒之效，用于痢疾、传染肝炎、麻疹等症。根含小檗碱（berberine）等生物碱。（《中国药用植物志》，2016）

标本登录：西藏：察隅，察瓦龙，1960-06-03，? 8900。

四川：若尔盖，朱大海等4428；昭化，alt. 900 m, 1930-05，郝景盛306。

青海：湟中，刘尚武2972；湟源，刘更喜 s. n.；互助，郭本兆9156；民和，何廷农608，654；大通，刘继孟5958，5992。

甘肃：舟曲，夏纬瑛6783，王作宾15267；迭部，杨亲二，孔宏智98-87；岷县，王作宾5009；临潭，莲花山，廉永善96-816，孙学刚2720；夏河，王作宾6896，傅坤俊875，939；和政，孙学刚2001-0433；卓尼，刘继孟5409；榆中，何业祺5369，崔鸿宾等367，李捷91；兰州，何业祺5051；连城，孙学刚6015；天水，刘继孟10483。

陕西：佛坪，郭本兆4599；周至，郭本兆1231，秦岭队53；太白山，刘慎谔，钟补求1846；终南山，孔宪武2680；渭南，王作宾15582；南五台山，刘慎谔11054；华山，夏纬瑛4268。

河南：西峡，关克俭1042；卢氏，刘继孟4438，4953；栾川，黄土高原队2793；嵩县，植物资源队 L0504。

山西：垣曲，包世英 261；翼城，刘心源 20362；隰县，王作宾 3180；洪洞，陈艺林 806，包世英 806；介休，刘继孟 1241。

河北：武安，关克俭 5794；邢台，刘瑛，刘心源 13096；内丘，刘心源 252；赞皇，石家庄队 535；涞源，刘继孟 2393；涞水，刘继孟 2232，2532；小五台山，黄秀兰等 1992；赤城，黄秀兰等 7687；遵化，唐进 2081；雾灵山，刘继孟 180，王忠涛等 14-0224；围场，承德队 32。

北京：百花山，刘冰，刘夙 682；延庆，北京中草药队 130；密云，北京中草药队 597。

内蒙古：乌盟，西公旗，乌拉山，郎学忠 221；巴林左旗，1962-07-24，？12；牙克石，牙资南队 237。

吉林：长白山，钱家驹 132，314，王薇等 2011，2046；和龙，延边 2 组 820。

黑龙江：带岭，宋朝枢 88，中德队 7078，7487；石道河，郝景盛 16037。

36b. 光果唐松草（变型）

f. **levicarpum** Tamura in Acta Phytotax. Geobot. 15: 85. 1953; S. H. Li in Fl. Pl. Herb. Chinae Bor.-Or. 3: 206. 1975; et in Clav. Pl. Chinae Bor.-Or., ed. 2: 205. 1995; W. T. Wang in Bull. Bot. Lab. N.-E. Forest. Inst. 8: 25. 1980. Holotype：日本：Hondo med.: Prov. Shimotsuke: Niko, Matsuda s. n. (KYO, not seen).

瘦果光滑，具不明细肋。

Achenes smooth, with inconspicuous thin ribs.

分布于中国吉林（和龙县，安图县）和黑龙江（伊春）（据李书馨，1975）；以及日本。生山地林下。

系 3. 瓣蕊唐松草系

Ser. **Petaloidea** (Prantl) W. T. Wang & S. H. Wang in Fl. Reip. Pop. Sin. 27: 542. 1979. — Sect. *Camptonota* Prantl subsect. *Petaloidea* Prantl in Bot. Jahrb. 9: 271. 1887, p.p.; Tamura in Hiepko, Nat. Pflanzenfam. 17a Ⅳ: 483. 1995. Lectotype: *T. petaloldeum* L. (W. T. Wang & S. H. Wang, 1979).

雄蕊花丝分化成在形状和大小方面不同的上、下两部分。心皮具直花柱和狭条形柱头。瘦果无柄，具强烈隆起的纵肋。

Stamen filament differentiated into upper and lower two parts different in shape and size. Carpel with a straight style and a narrow-linear stigma. Achene sessile, with strongly elevated longitudinal ribs.

1 种，广布东亚北部。

37. **瓣蕊唐松草**（《北京植物志》） 图版 37

Thalictrum petaloideum L. Sp. Pl., ed. 2, 771. 1762; DC. Syst. 1: 175. 1818; et Prodr. 1: 12. 1824; Lecoy. in Bull. Soc. Bot. Belg. 24: 165. 1885; Forbes & Hemsl. In J. Linn. Soc. Bot.23: 9. 1886; Maxim in Acta Hort. Petrop.11: 11. 1890; Finet & Gagnep. in Bull. Soc. Bot. France 50: 610. 1903; Nevski in Fl. URSS 7: 518, t. 5: 5. 1937; Kitagawa, lineam. Fl. Mansh. 227. 1939; Fl. Beijing 1: 293, fig. 241. 1962; C. Pei & T. Y. Chou, Icon. Chin. Med. Pl. 7: fig. 327. 1964; Iconogr. Corm. Sin. 1: 676, fig. 1351. 1972; Fl. Tsinling. 1(2): 242. 1974; S. H. Li in Fl. Pl. Herb. Chinae Bor.-Or. 3: 211, pl. 92: 1–5. 1975; W. T. Wang & S. H. Wang in Fl. Reip. Pop. Sin. 27: 544, pl. 133. 1979; B. Z. Ding et al. Fl. Henan. 1: 472, fig. 604. 1981; S. Y. He, Fl. Beijing, rev. ed., 1: 237, fig. 295. 1984; W. T. Wang in Acta Bot. Yunnan. 6(4): 377. 1984; J. W. Wang in Fl. Hebei 1: 492, fig. 438. 1986; D. Z. Ma, Fl. Ningxia. 1: 173. 1986; X. W. Wang in Fl. Anhui 2: 323, fig. 626. 1987; Y. Z. Zhao in Fl. Intramongol. ed, 2, 2: 453, pl. 182: 1–6. 1990; X. Y. Yu et al. in Fl. Shanxi. 1: 596, fig. 377. 1992; W. T. Wang in Vasc. Pl. Hengduan Mount. 1: 501. 1993; J. G. Liu in Fl. Xinjiang. 2(1): 271. 1994; S. H. Li in Clav. Pl. Chinae Bor.-Or., ed. 2: 207. 1995; W. T. Wang in Bull. Res. Bot. Harbin 16(2):158. 1996; L. H. Zhou in Fl. Qinghai. 1: 329, pl. 73: 7–10. 1997; W. T. Wang in High. Pl. China 3: 468, fig. 743. 2000; D. Z. Fu & G. Zhu in Fl. China 6: 295. 2001; Q. Lu et al. Desert Pl. China 162, with photo. 2012; M. B. Deng & K. Ye in Fl. Jiangsu 2: 85, fig. 2–138. 2013; W. T. Wang in High. Pl. China in Colour 3: 387. 2016; L. Xie in Med. Fl. China 3: 250, with fig. & photo. 2016. Described from Siberia.

T. petaloideum L. var. *latifoliolatum* Kitag. in Rep. Inst. Sci. Res. Manch. 4(7): 82. 1940; et in J. Jap. Bot. 36: 18. 1961.

37a. var. petaloideum

多年生草本植物，全部无毛。茎高 20-80 cm，上部分枝。基生叶数个，为 3-4 回三出或羽状复叶；叶片长 5-15 cm；小叶草质，倒卵形、宽倒卵形、菱形、椭圆形或近圆形，3-12×2-15 mm，顶端钝，基部圆形、宽楔形或楔形，3 浅裂至 3 深裂，有时不分裂，裂片全缘，脉平，脉网不明显；小叶柄长 5-7 mm；叶柄长达 10 cm，基部有鞘。茎生叶较小。复单歧聚花序顶生，伞房状，有少数至多数花；花梗细，长 0.5-2.5 cm。花：萼片 4，早

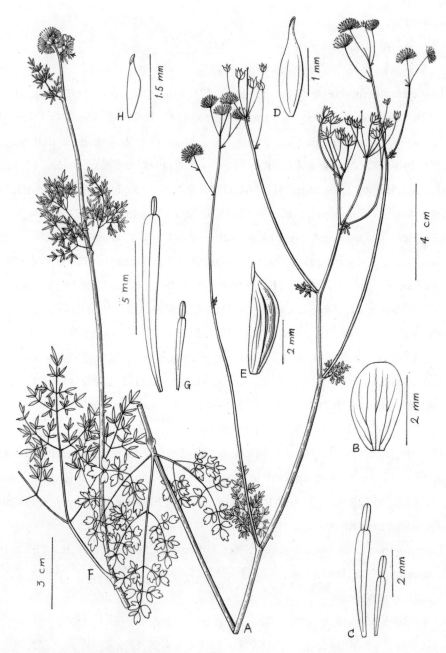

图版37. 瓣蕊唐松草 A–E. 模式变种 A. 开花茎上部，B. 萼片，C. 二雄蕊，D. 心皮，E. 瘦果。（据刘瑛，刘心源12931） F–H. 细裂瓣蕊唐松草 F. 开花茎上部，G. 二雄蕊，H. 心皮。（据刘慎谔等3205）

Plate 37. **Thalictrum petaloideum** A–E. var. **petaloideun** A. upper part of flowering and fruiting stem, B. sepal, C. two stamens, D. carpel, E. achene (from Y. Liu and X. Y. Liu 12931) F–H. var. **supradecompositum** F. upper part of flowering stem, G. two stamens, H. carpel. (from T. N. Liou et al. 3205)

落，白色，倒卵形，2.6–5×1.4–2 mm，顶端圆形，每一萼片具3条基生脉（中脉1回二歧状分枝，2侧生脉2回二歧状分枝）。雄蕊25–45；花丝长1.8–2.5 mm，上部狭倒披针形，宽1–1.2 mm，下部狭条形，宽0.1–0.2 mm；花药长圆形，0.7–1×0.3–0.5 mm，顶端钝。心皮4–13；子房长圆形，约1.5×0.5 mm；花柱长0.5 mm；柱头长0.3 mm。瘦果椭圆状卵形，4–6×1.4–1.6 mm，有8条明显隆起的纵肋；宿存花柱长0.8–1 mm。6–7月开花。

Perennial herbs, totally glabrous. Stems 20–80 cm tall, above branched. Basal leaves several, 3–4-ternate or 3–4-pinnate; blades 5–15 cm long; leaflets herbaceous, obovate, broad-obovate, rhombic or suborbicular, 3–12×2–15 mm, at apex obtuse, at base rounded, broad-cuneate or cuneate, 3-lobed to 3-parted with lobes entire, sometimes undivided, nerves flat, nervous nets inconspicuous; petiolules 0.5–7 mm long; petioles up to 10 cm long, at base vaginate. Cauline leaves smaller. Compound monochasia terminal, corymbiform, few-to many-flowered; pedicels slender, 0.5–2.5 cm long. Flower: Sepals 4, caducous, white, obovate, 2.6–5×1.4–2 mm, apex rounded, each sepal with 3 basal nerves (midrib once dichotomizing, and 2 lateral nerves twice dichotomizing), Stamens 25–45; filaments 1.8–5.5 mm long, above oblanceolate, 1–1.2 mm broad, below narrow-linear, 0.1–0.2 mm broad; anthers oblong, 0.7–1×0.3–0.5 mm, apex obtuse. Carpels 4–13; ovary oblong, ca. 1.5×0.5 mm; style 0.5 mm long; stigma 0.3 mm long. Achene elliptic-ovate, 4–6×1.4–1.6 mm, with 8 conspicuously elevated longitudinal ribs; persistent style 0.8–1 mm long.

分布于中国的四川西部和北部、湖北、浙江（舟山岛）、江苏南部（南京）、安徽、山东、河南、陕西、甘肃、青海、新疆、宁夏、山西、河北、内蒙古、辽宁、吉林、黑龙江；以及朝鲜、俄国西伯利亚地区。生山坡草地；海拔分布：在四川、青海、甘肃一带为1800–3900 m，在山西、河北一带为800–1800 m，在东北为700 m以下。

根有清热解毒之效，用于黄疸性肝炎、赤白痢疾等症。（《中国药用植物志》，2016）

标本登录：四川：新龙，alt. 3900 m，王晋峰6395；若尔盖，alt. 2500 m，郎楷永，李良千，费勇1851；南坪，综考会队H82-601。

湖北：枣阳，木德义37。

浙江：舟山岛：定海，大菜花山，王国民1614。

安徽：全椒，王德群5665；萧县，刘晓龙569。

山东：午山，F. H. Sha 227，241。

河南：伊阳，河南普查队6071；林县，河南普查队4303。

陕西：华阴，夏纬瑛 4478；陕北盆地，乐天宇 s. n.。

甘肃：舟曲，王作宾 15197，姜恕 349，白龙江队 1261；迭部，白水江队 1652，1848；岷县，王作宾 15254；夏河，王作宾 5771；临兆，傅坤俊 829；漳县，黄河队 56-4633；天水，王作宾 4336，黄河队 56-4056；通渭，黄河队 56-5035；会宁，黄河队 56-5205；榆中，何业祺 5906；天祝，何业祺 4640；山丹，1964-07，? 852。

青海：贵德，何廷农 982；共和，张珍万 1976；湟源，郑斯绪 247；乐都，吴玉虎 3813；西宁，郝景盛 813，张永和 4289；互助，刘更喜 s. n.；大通，刘继孟 6334，关克俭 77286；门源，郭本兆 7259；Sincheng，刘继孟 5953。

新疆：皮山，李渤生，郑度，郭柯 11680。

宁夏：隆德，王作宾 13002；六盘山，刘继孟 5647，黄河队 56-2250，黄河队甘一队 56-2198；海原，黄河队 56-5565。

山西：垣曲，H. Smith 6710，包世英 334；阳城，包世英 2163；陵川，刘继孟 7560，包世英 1290；平顺，刘继孟 8012；沁水，段长虹 99；沁源，刘继孟 1407；隰县，王作宾 3244；介休，刘继孟 1266；中阳，中阳甲队 168；离石，黄河队 55-2386；灵空山，关克俭，陈艺林 710；交城，夏纬瑛 1054；关帝山，黄河队 57-890；临县，黄河队 55-1851；兴县，黄河队 55-2506；岢岚，黄河队 55-2965；五寨，黄河队 55-2704；宁武，刘继孟 1921；五台，关克俭，陈艺林 2246；繁峙，段长虹 125。

河北：涉县，关克俭 5573；武安，关克俭 5728；邢台，刘瑛，刘心源 12931；内丘，刘心源 635；正定，5 分队 609；阜平，阜平队 66；涞源，刘继孟 2390，2494，2739；易县，黄秀兰等 2728；涞水，刘继孟 2213；小五台山，孔宪武 174，王作宾 450，王启无 61512，刘瑛 11293，黄秀兰等 420；怀来，黄秀兰等 2178；张家口，崔友文 1745，黄秀兰等 3585；雾灵山，王忠涛等 14-082；平泉，王忠涛等 14-412；围场，北师大队 19239，李良千等 42。

北京：百花山，王启无 60005；昌平，? 942；居庸关，傅德志等 85-002；延庆，延庆队 157。

内蒙古：乌审旗，吴佐祺 64-002；凉城，蛮汉山，内蒙古大学队 24；克什克腾旗，黄岗梁林场，陈又生等 140573；阿尔山，张玉良等 359；海拉尔，李学文 s. n.；牙克石，东北林学院队 170。

辽宁：沈阳，鸡冠山，孔宪武 650；公主岭，中德队 506。

吉林：长春，王薇等 2008。

黑龙江：哈尔滨，王光正 49；牡丹江市，J. Sato 7726。

37b. 狭裂瓣蕊唐松草（《中国植物志》，变种）

var. **sapradecompositum** (Nakai) Kitag. Lineam. Fl. Mansh. 227. 1939; et in Rep. Inst. Sci. Res. Manch. 4(7): 82, 1940; S. H. Li in Fl. Pl. Herb. Chinae Bor.-Or. 3: 211, pl. 92: 6. 1975; W. T. Wang & S. H. Wang in Fl. Reip. Pop. Sin. 27: 544. 1979; J. W. Wang in Fl. Hebei 1: 443. 1986; Y. Z. Zhao in Fl. Intramongol. 2nd ed., 2: 455, pl. 183: 7. 1990; S. H. Li in Clav. Pl. Chinae Bor-Or., ed. 2: 207, pl. 97: 7. 1995; D. Z. Fu & G. Zhu in Fl. China 6: 295. 2001; L. Xie in Med. Fl. China 3: 251, with photo. 2016. — *T. supradecompositum* Nakai in Bot. Mag. Tokyo 46: 54. 1932. Holotype：北京：明陵，A. Kikuchi s. n. (T1, not seen; photo, PE).

本变种与瓣蕊唐松草的区别在于其小叶或全裂小叶的裂片狭长，呈狭长椭圆形或狭披针形。

This variety differs from var. *petaloideum* in its long narrow-elliptic leaflets or segments of leaflets.

分布于河北北部、内蒙古、辽宁和吉林的西部，黑龙江西南部。生低山干燥山坡或草原多沙草地或林边。

标本登录：**河北**：崇礼，崔友文 1969；沙岭子，察哈尔农业处 60。

内蒙古：锡盟，西乌旗，内蒙古大学普查队 14 组 3；东乌旗，内蒙古大学普查队 3 组 17；昭盟，克旗，蒙宁队 1108；兴安盟，科右中旗，孔令韶 64；东科治旗，刘慎谔等 3205。

吉林：通榆，叶居新 63；镇赉，白城组 91。

黑龙江：洋尔庙，第一线之二 s. n.

系 4. 网脉唐松草系

Ser. **Reticulata** W. T. Wang & S. H. Wang in Fl. Reip. Pop. Sin. 27: 539,618. 1979. Type: *T. reticulatem* Franch.

雄蕊花丝分化成在形状和大小方面不同的上、下二部分。心皮具极短花柱，在花柱腹面有密集的柱头组织。瘦果无柄，具明显 6 或 8 条纵肋。

Stamen filament differentiated into upper and lower two parts different in shape and size. Carpel with a very short inconspicuous style, and on vetral surface of style there are dense stigmatis tissues. Achene sessile, with 8 or 6 conspicuous longitudinal ribs.

2 种，1 种分布于中国西南部，另 1 种分布于东南部武夷山区。

38. 网脉唐松草(《药学学报》)　　图版 38

Thalictrum reticulatum Franch. in Bull. Soc. Bot. France 33: 371. 1886; et Pl. Delav. 17, pl. 4. 1889; Finet & Gagnep. in Bull. Soc. Bot. France 50: 604. 1903; Hand-Mazz. Symb. Sin. 7: 310. 1931; W. T. Wang & S. H. Wang in Fl. Reip. Pop. Sin. 27: 539, pl. 131: 7–9. 1979; W. T. Wang in Iconogr. Corm. Sin. Suppl. 1: 424, fig. 8610. 1982; in Vasc. Pl. Hengduan Mount. 1: 500. 1993; et in Fl. Yunnan. 11: 164. 2000; D. Z. Fu & G. Zhu in Fl. China 6: 296. 2001; L. Xie in Med. Fl. China 3: 248, with fig. 2016. Holotype: 云南：大理，Ta-pin-tze 之北，1882-08-06，Delavay 17 (P; photo, PE).

38a. var. **reticulatum**

多年生草本植物，全部无毛。茎高 26–40 cm，分枝。基生叶为 3 回三出复叶，长约 12 cm；叶片长 3.5–5 cm；小叶纸质或薄革质，宽卵形或宽菱形，(0.5–)1–1.8× (0.5–)1.3–2 cm，顶端圆形或钝，基部浅心形或圆形，不明显 3 浅裂，边缘有少数圆齿，脉上面近平，下面隆起，脉网明显；小叶柄长 0.4–2.5 cm；叶柄长 6.5–7.5 cm，基部具狭鞘。茎生叶 2–3，较小。复单歧聚伞花序顶生，伞房状，宽达 12 cm；花梗长 3–7 mm。花：萼片 4，早落，白色。椭圆形或宽倒卵形，约 3×2 mm，顶端圆钝，每一萼片具 3 条不分枝的基生脉。雄蕊 24–45；花丝长 2.8–3.8 mm，上部狭倒披针形，宽 0.3–0.5 mm，下部丝形，粗 0.1 mm；花药长圆形，0.8×0.3–0.4 mm，顶端钝。心皮 4–12；子房狭卵形，约 1×0.4 mm；花柱长约 0.4 mm，腹面具密集柱头组织、瘦果稍扁，狭椭圆体形，约 1.6×0.8 mm，每侧有 3 条明显纵肋；宿存花柱长约 0.4 mm。6 月开花。

Perennial herbs, totally glabrous. Stems 26–40 cm, branched. Basal leaves 3-ternate, ca. 12 cm long; blades 3.5–5 cm long; leaflets papery or subcoriaceous, broad-ovate or broad-rhombic, (0.5–)1–1.8× (0.5–)1.3–2 cm, at apex rounded or obtuse, at base subcordate or rounded, inconspicuously 3-lobed, margin few-rotund-dentate, nerves adaxially nearly flat, abaxially prominent, nervous nets conspicuous; petiolules 0.4–2.5 cm long; petioles 6.5–7.5 cm long, at base narrowly vaginate. Cauline leaves 2–3, smaller. Compound monochasia terminal, corymbiform, up to 12 cm broad; pedicels 3–7 mm long. Flower: Sepals 4, caducous, white, elliptic or broad-obovate, ca. 3×2 mm, apex rounded-obtuse, each sepal with 3 simple basal nerves. Stamens 24–45; filaments 2.8–3.8 mm long, above narrow-oblanceolate, 0.3–0.5 mm broad, below filiform, 0.1 mm across; anthers oblong, 0.8×0.3–0.4 mm, apex obtuse. Carpels 4–12; ovary narrow-ovate, ca. 1×0.4 mm; style ca. 0.4 mm long, abaxially with dense

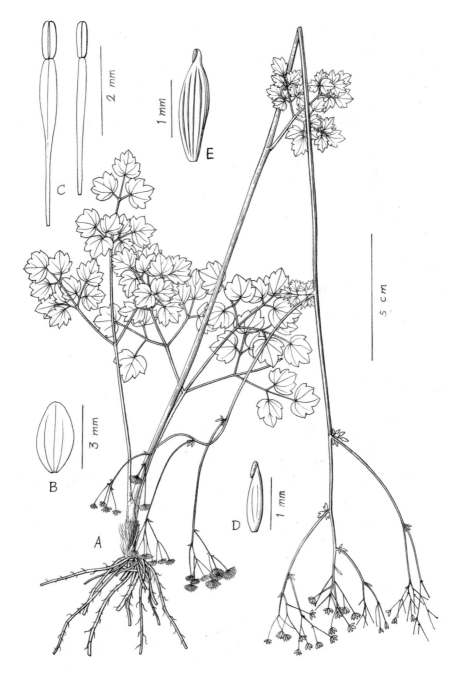

图版38. **网脉唐松草** A. 植株全形，B. 萼片（据滇西北金沙江队4672），C. 二雄蕊，D. 心皮，E. 瘦果，（据滇西北金沙江队6517）。

Plate 38. **Thalictrum reticulatum** var. **reticulatum** A. habit, B. sepal [from Jinshajiang (NW Yunnan) Exped. 4672], C. two stamens, D. carpel, E. achene, (from Jinshajiang (NW Yunnan) Exped. 6517).

stigmatic tissues. Achene slightly compressed, narrow-ellipsoid, ca. 1.6×0.8 mm, on each side with 3 conspicuous longitudinal ribs; persistent style ca. 0.4 mm long.

分布于云南西北部和四川西部．生山地草坡或疏松林中，海拔 2200-2500 m。

根有清热解毒之效，用于感冒、咳嗽、痢疾等症。(《中国药用植物志》，2016)

标本登录：云南：鹤庆，马耳山，滇西北金沙江队 4672，6517；丽江，大坝，中甸队 62-1233；中甸，冯国楣 1240，中甸队 62-1066，D. E. Boufford et al. 35251；南涧，杨云辉等 160。

四川：盐源，青藏队 83-12516；木里，武素功 2561，3666，青藏队 83-13625，D. E. Boufford et al. 42802；雅江，黄治平等 690，姜恕 2683，5198；康定，？691；泰宁和牦牛之间，H. Smith 12576；丹八，姜恕 2192，D. E. Boufford et al. 38010；理县，姜恕 788。

38b. 毛叶网脉唐松草(《中国植物志》)

var. hirtellum W. T. Wang & S. H. Wang in Fl. Reip. Pop. Sin. 27: 541, 618. 1979; W. T. Wang in Vasc. Pl. Hengduan Mount. 1: 500. 1993; D. Z. Fu & G. Zhu in Fl. China 6: 296. 2001. Holotype：四川：木里，麦地龙，alt. 2140 m，1960-06-28，应俊生 4181(PE)．

与网脉唐松草的区别在于本变种的小叶下面脉上被短硬毛。

This variety differs from var. *reticulatum* in its leaflets abaxially on nerves hirtellous.

特产四川木里县。生山地灌丛中，海拔 2140 m。

39. 武夷唐松草(《中国植物志》)　图版 39

Thalictrum wuyishanicum W. T. Wang & S. H. Wang in Fl. Reip. Pop. Sin. 27: 541, 618. Pl. 131: 1-3. 1979; S. R. Lin & x. Z. Zhao in Fujian 2: 19, fig. 14. 1985; W. T. Wang in High. Pl. China 3: 467, fig. 741: 1-6. 2000; D. Z. Fu & G. Zhu in Fl. China 6: 301. 2001; W. T. Wang in Fl. Jiangxi 2: 175, fig. 182. 2004; et in High. Pl. China in Colour 3: 386. 2016. Type：福建：武夷山，alt. 2000 m，1956-04-06，王名金 3101(holotype, PE)。江西：南丰，军峰山，1953-04-06，聂敏祥 2174(paratype, PE)．

多年生草本植物，全部无毛。茎高 17-29 cm，中部分枝或不分枝。基生叶 1-3，为 2 回三出复叶，有长柄；叶片长 4.5-9 cm；小叶纸质，圆菱形，倒卵状菱形或宽卵形，1.8-2.5×1.5-2.5 cm，两端钝或圆形，3 浅裂，裂片有 1-2 浅圆齿或全缘，脉在两面隆起，脉网明显；小叶柄长 1.5-2 cm；叶柄长 8-12 cm。茎生叶约 1，小，为 1 回三出复叶。单歧聚伞花序简单或复杂，顶生或腋生，有 2－3(-5)花；花梗长 1.2-3 cm。花：萼片 4，早落，淡粉红色，狭椭圆形或长圆形，4-4.2×0.8-1.8 mm，顶端圆钝，每一萼片具 3 条不分

六、分类学处理   183

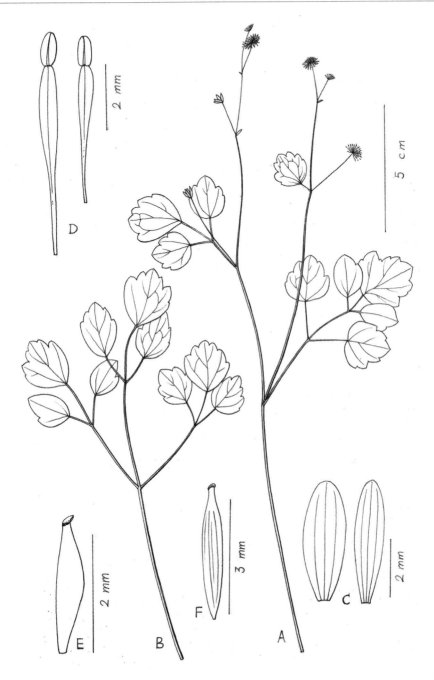

图版39. **武夷唐松草** A. 开花茎上部，B. 下部茎生叶，C. 二萼片，D. 二雄蕊，E. 心皮（据王名金3101），F. 瘦果（据聂敏祥2174）。

Plate 39. **Thalictrum wuyishanicum** A. upper part of flowering stem, B. lower cauline leaf, C. two sepals, D. two stamens, E. carpel (from M. J. Wang 3101), F. achene (from M. X. Nie 2174).

枝的基生脉。雄蕊 12-16；花丝长 3.2-4 mm，上部倒披针形，宽 0.3-0.8 mm，下部丝形，粗 0.1-0.15 mm；花药宽长圆形，0.6×0.45 mm，顶端钝。心皮（4-）6-10；子房斜狭倒卵形，约 2×0.5 mm；花柱长 0.1-0.3 mm，腹面有密集的柱头组织。瘦果纺锤形，约 3×0.5 mm，每侧有 3 条纵肋；宿存柱头长约 0.3 mm。4 月开花。

Perennial herbs, totally glabrous. Stems 17-29 cm tall, at the middle branched or simpl. Basal leaves 1-3, 2-ternate, long petiolate; blades 4.5-9 cm long; leaflets papery, orbicular-rhombic, obovate-rhombic or broad-ovate, 1.8-2.5×1.5-2.5 cm, at both ends obtuse or rounded, 3-lobed, lobes 1-2-crenate or entire, nerves on both surfaces prominent, nervous nets conspicuous; petiolules 1.5-2 cm long; petioles 8-12 cm long. Cauline leaf ca. 1, small, 1-ternate. Monochasia simple or compound, terminal or axillary, 2-4(-5)-flowered; pedicels 1.2-3 cm long. Flower: Sepals 4, cadacous, pinkish, narrow-elliptic or oblong, 4-4.2×0.8-1.8 mm, apex rounded-obtuse, each sepal with 3 simple basal nerves. Stamens 12-16; filaments 3.2-4 mm long, above oblanceolate, 0.3-0.8 mm broad, below filiform, 0.1-0.15 mm across; anthers broad-oblong, 0.6×0.45 mm, apex obtuse. Carpels (4-)6-10; ovary obliquely narrow-obovate, ca. 2×0.5 mm; style 0.1-0.3 mm long, ventrally with dense stigmatic tissues. Achene fusiform, ca. 3×0.5 mm, on each side with 3 longitudinal ribs; persistent stigma ca. 0.3 mm long.

分布于江西东部（南丰）和福建西北部（崇安）。生山地阴处石上，海拔 2000 m。

系 5. 西南唐松草系

Ser. **Fargesiana** W. T. Wang & S. H. Wang in Fl. Reip. Pop. Sin. 27: 537, 618. 1979. Type: *T. fargesii* Franch. ex Finet & Gagnep.

雄蕊花丝分化成形状和大小不同的上、下两部分，稀呈狭条形（察瓦龙唐松草），心皮具直花柱和狭条形柱头。瘦果多少强烈扁压，具短柄。

Stamen filament differentiated inta upper and lower two parts different in shape and size, rarely narrow-linear (*T. tsawarungense*). Carpel with a straight style and a narrow-linear stigma. Achene more or less strongly compressed, shortly stipitate.

5 种，多分布于西南部，1 种向北达秦岭山区。

40. 西南唐松草（《秦岭植物志》） 图版 40

Thalictrum fargesii Franch. ex Finet & Gagnep. in Bull. Soc. Bot. Franch. 50: 608, pl.

19: c. 1903; Fl. Tsinling. 1(2): 241, fig. 206. 1974; S. H. Fu, Fl. Hupeh. 1: 357, fig. 499. 1976; W. T. Wang & S. H. Wang in Fl. Reip. Pop. Sin. 27: 537, pl. 137: 1–4. 1979; W. T. Wang in Bull. Bot. Lab. N.–E. Forest. Inst. 8: 23. 1980; B. Z. Ding et al. Fl. Henan. 1: 472, fig. 605. 1981; W. T .Wang in Iconogr. Corm. Sin. Suppl. 1: 423, fig. 8609. 1982; Y. K. Li in Fl. Guizhou. 3: 73. 1990; W. T. Wang in Vasc. Pl. Hengduan Mount. 1: 500. 1993; in Keys Vasc. Pl. Wuling Mount. 164. 1995; et in High. Pl. China 3: 467, fig. 740: 1–4. 2000; D. Z. Fu & G. Zhu in Fl. China 6: 289. 2001; C. X. Yang et al. Keys Vasc. Pl. Chongqing 216. 2009; Q. L. Gan, Fl. Zhuxi Suppl. 244, fig. 293. 2011; L. Q. Li in High. Pl. Daba Mount. 117. 2014; L. Q. Li et al. Pl. Baishuijiang St. Nat. Reserve Gansu Prov. 70. 2014; W. T. Wang in High. Pl. China in Colour 3: 386. 2016; L. Xie in Med. Fl. China 3: 246, with fig. & photo. 2016. Holotype：重庆：城口, alt. 1400 m, R. P. Farges 552（P, photo, PE）.

T. pallidum Franch. in Nouv. Arch. Mus. Hist. Nat. Paris, ser. 2, 8: 187. 1885, non Fisch., 1841; et Pl. David. 2: 5. 1888; Finet & Gagnep. in Bull. Soc. Bot. France 50: 610. 1903. Holotype：四川：宝兴，1867–06, A. David s. n. (P, not seen).

T. silvestrii Pamp. in Nouv. Giorn. Bot. Ital. n. s. 22: 291. 1915. Syntypes：湖北西北部：Nang-tciang, 1910, P. C. Silvestri 3956; U-tan-scian（武当山），1912–03, Silvestri 3788; Jun-Fang, 1912–03, Silvestri 3790; Zan-lan-scian（苍浪山），1913–10, Silvestri 3789（not seen）.

多年生草本植物，通常全部无毛，稀在茎上有少数短毛（四川西部的一些居群）。茎高达 50 cm，纤细，分枝。中部茎生叶为 3-4 回三出复叶；叶片长 8-14 cm；小叶草质，菱状倒卵形、菱形或卵形，0.9-3×0.6-2.5 cm，顶端钝，基部宽楔形、圆形或浅心形，3 浅裂或不分裂，裂片全缘或有 1-3 小齿，脉下面隆起，脉网明显；小叶柄长 0.3-2 cm；叶柄长 3.5-5 cm。复单歧聚伞花序顶生，伞房状，宽 4-12 cm，有少数至多数花；花梗细，长 3-3.5 cm。花：萼片 4，白色或带淡紫色，宽卵形或椭圆形，3-6×2-4.4 mm，顶端圆钝，每一萼片具 3 条基生脉（中脉不分枝，2 侧生脉二歧状分枝）。雄蕊 14-18；花丝长 2.2-4.8 mm，上部狭倒披针形或条形，宽 0.2-0.3 mm，下部丝形，粗 0.1 mm；花药狭长圆形，0.8-1.5×0.2-0.25 mm。顶端钝。心皮 2-5；子房狭卵形，约 2×0.6 mm；花柱长约 1 mm；柱头长约 0.4 mm。瘦果近新月形或狭倒卵形，3-3.6×1-1.2 mm，每侧有 3 条纵肋，基部的柄长约 0.4 mm；宿存花柱长 1 mm；宿存柱头长 0.8 mm。5-6 月开花。

Perennial herbs, usually totally glabrous, rarely on stems with sparse short hairs (in some populations found in western Sichuan Province). Stems up to 50 cm tall, slender, branched.

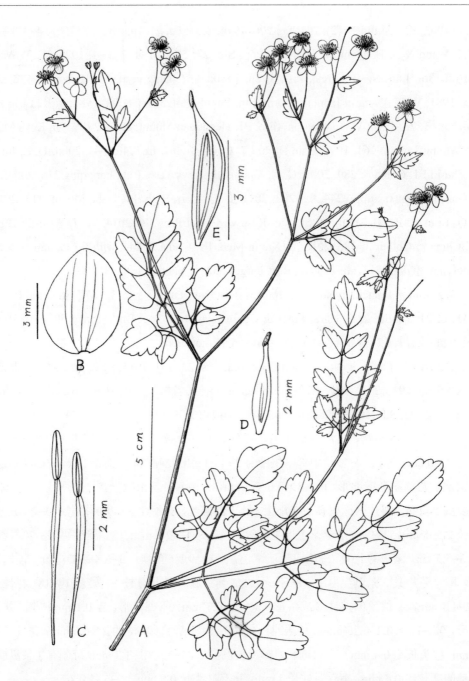

图版40. **西南唐松草** A. 开花茎上部，B. 萼片，C. 二雄蕊，D. 心皮（据杨光辉57743），E. 瘦果（据戴天伦105332）。

Plate 40. **Thalictrum fargesii** A. upper part of flowering stem, B. sepal, C. two stamens, D. carpel (from G. H. Yang 5773) E. achene (from T. L. Dai 105332).

Middle cauline leaves 3–4-ternate; blades 8–14 cm long; leaflets herbaceous, rhombic-obovate, rhombic or ovate, 0.9–3×0.6–2.5 cm, at apex obtuse, at base broad-cuneate, rounded or subcordate, 3–lobed or undivided, lobes entire or 1–3-denticulate, nerves abaxially prominent, nervous nets conspicuous; petiolules 0.3–2 cm long; petioles 3.5–5 cm long. Compound monochasia terminal, corymbiform, 4–12 cm broad, few- to many-flowered; pedicels slender, 3–3.5 cm long. Flower: Sepals 4, white or tinged with purplish, broad-ovate or elliptic, 3–6×2–4.4 mm, apex rounded-obtuse, each sepal with 3 basal nerves (midrib simple, and 2 lateral nerves dichotomizing). Stamens 14–18; filaments 2.2–4.8 mm long, above narrow-oblanceolate or linear, 0.2–0.3 mm broad, below filiform, 0.1 mm across; anthers narrow-oblong, 0.8–1.5×0.2–0.25 mm, apex obtuse. Carpels 2–5; ovary narrow-ovate, ca. 2×0.6 mm; style ca. 1 mm long; stigma ca. 0.4 mm long. Achene nearly lunate or narrow-obovate, 3–3.6×1–1.2 mm, on each side with 3 longitudinal ribs, base with a stipe 0.4 mm long; persistent style 1 mm long; persistent stigma 0.8 mm long.

分布于贵州北部、四川、重庆、湖北、河南西南部和南部、陕西和甘肃的南部。生山地林中、草地、陡崖旁或沟边，海拔1080–2400 m。

根状茎有清热解毒之效，用于目赤肿痛、咽喉痛、痢疾等症。根含香唐松草碱（Thalfoetidine）等生物碱。（《中国药用植物志》，2016）

标本登录：贵州：正安，西部科学院队3322。

四川：天全，曲桂龄2586，蒋兴麐34431，胡文光，何铸10108；宝兴，杜大华4347，曲桂龄2932，宋滋圃38214，张秀实，任有铣4765，4941，5332；彭州，冯正波，朱大海2007-486；都江堰，何铸12194；理县，姜恕1590；汶川，四川植物队8109，8293，19757，郎楷永，李良千，费勇1227；茂县，何铸，周子林14289，12468，李永江77；平武，蒋兴麐10221，姜恕7082。

重庆：长寿，刘正宇183805；巫山，杨光辉57743；城口，戴天伦100598，105332，巴山队477。

湖北：秭归，周鹤昌355；兴山，罗毅波5，赵常明Ex2416；神农架，神农架队20665；房县，杨奥生s.n.；竹溪，刘继孟8876。

河南：内乡，D. E. Boufford et al. 26417；信阳，南部山区队D0064。

陕西：洋县，朱、陈、徐、王1059，2327；佛坪，傅坤俊4706，朱、陈、徐、王1165；山阳，杨金祥2908；宁陕，田光华93-5064；终南山，孔宪武2675；南五台山，刘慎谔11055；

太白山，傅坤俊2523，王作宾6460，6596；眉县，王作宾6595；周至，郭本兆1221；户县，郭本兆59，248。

甘肃：文县，陈学林91-1134，王新180，D. E. Boufford et al. 37682，37813；武都，张志英2814；玛曲，朱格麟1785。

41. 兴山唐松草（《植物分类学报》） 图版41

Thalictrum xingshanicum G. F. Tao in Acta Phytotax. Sin. 22(5): 423, fig. 1. 1984; D. Z. Fu & G. Zhu in Fl. China 6: 302. 2001. Type: 湖北：兴山, alt. 1700 m, 1982-05-08, 蒋祖德，陶光复167(Holotype, WUBI; isotype. PE).

多年生草本植物，全部无毛。茎高40-60 cm，纤细，常自下部分枝。茎生叶为1-3回三出复叶；小叶草质，卵形、圆卵形、倒卵形或椭圆形，2-6×1.5-5 cm，顶端圆形，基部圆形、浅心形或宽楔形，不分裂或不明显3微裂，边缘具浅钝齿，脉下面隆起，脉网明显；小叶柄长0.4-3 cm；叶柄长2-5 cm。复单歧聚伞花序顶生，伞房状，具3-9花；花梗细，长1-3.5 cm。花：萼片4，白色或淡紫色，椭圆形，长3-6 mm，顶端圆形，每一萼片具3条二歧状分枝的基生脉。雄蕊约30；花丝长4-5 mm，上部棒状，宽约0.25 mm，下部丝形，粗0.1 mm；花药狭长圆形，1.2-1.5×0.25-0.3 mm，顶端钝。心皮5-13；子房狭倒卵形，约1.2×0.4 mm；花柱长约1 mm，顶端稍弧状弯曲；柱头长约0.8 mm。瘦果纺锤形，约4×1.2 mm，每侧有3条纵肋，基部具短柄；宿存花柱长约1 mm。4-5月开花。

Perennial herbs, totally glabrous. Stems 40-60 cm tall, slender, often from below branched. Cauline leaves 1-3-ternate; leaflets herbaceous, ovate, orbicular-ovate, obovate or elliptic, 2-6×1.5-5 cm, at apex rounded, at base rounded, subcordate or broadly cuneate, undivided or inconspicuously 3-lobulate, margin crenate, nerves abaxially prominent, nervous net conspicuous; petiolules 0.4-3 cm long; petioles 2-5 cm long. Compound monochasia terminal, corymbiform, 3-9-flwoered; pedicels slender, 1-3.5 cm long. Flower: Sepals 4, white or purplish, elliptic, 3-6 mm long, apex rounded, each sepal with 3 dichotomizing basal nerves. Stamens ca. 30; filaments 4-5 mm long, above clavate, ca. 0.25 mm broad, below filiform, 0.1 mm across; anthers narrow-oblong, 1.2-1.5×0.25-0.3 mm, apex obtuse, Carpels 5-13; ovary narrow-obovate, ca. 1.2×0.4 mm; style ca. 1 mm long, apex slightly arcuate; stigma ca. 0.8 mm long. Achene fusiform, ca. 4×1.2 mm, on each side with 3 longitudinal ribs, at base with a short stipe; persistent style ca. 1 mm long.

特产湖北兴山县。生海拔1700 m山地。

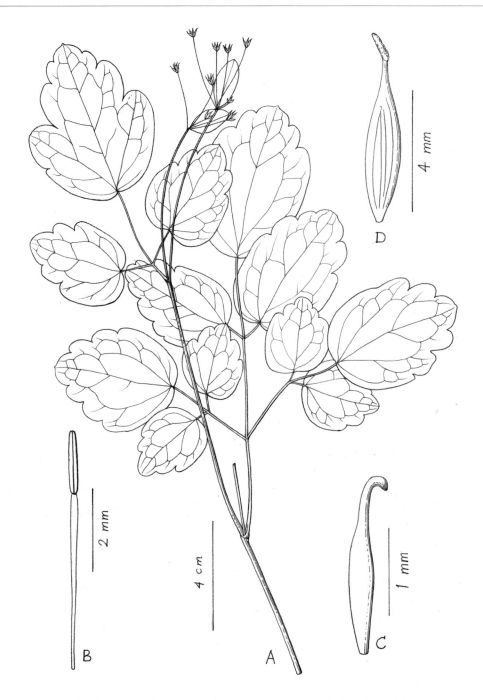

图版41. 兴山唐松草　A. 开花茎上部，B. 雄蕊，C. 心皮，D. 瘦果。（据蒋祖德，陶光复167）
Plate 41. **Thalictrum xingshanicum**　A. upper part of flowering stem, B. stamen, C. carpel, D. achene. (from Z. D. Jiang & G. F. Tao 167)

42. 丽江唐松草（《中国植物志》） 图版 42

Thalictrum wangii B. Boivin in J. Arn. Arb. 26: 116, pl. 1: 8–11. 1945, p. p. excl. paratypo; W. T. Wang & S. H. Wang in Fl. Reip. Pop. Sin. 27: 538, pl. 142: 1–3. 1979; W. T. Wang in Vasc. Pl. Hengduan Mount. 1: 500. 1993; et in Fl. Yunnan. 11: 164, pl. 49: 1–3. 2000; D. Z. Fu & G. Zhu in Fl. China 6: 301. 2001; L. Xie in Med. Fl. China 3: 247, with fig. & photo. 2016. Type: 云南：丽江, alt. 2700 m, 1935-07, 王启无 71546 (holotype, GH; isotype, PE).

多年生草本植物。茎高 30-58 cm，与小叶下面和花序均被极短腺毛，常自下部分枝。中部茎生叶为 3 回三出复叶；叶片长 5.5-9.5 cm；小叶膜质，圆卵形、圆菱形或宽倒卵形，4-11×3-13 mm，基部宽楔形或圆形，3 浅裂或不分裂，裂片多全缘，有时具 1-2 钝齿，脉上面平，下面稍隆起；小叶柄长 0.5-6 mm；叶柄长 0.8-3 cm。单歧聚伞花序顶生，有 2-3 花；花梗长 7-14 mm。花：萼片 4，早落，白色，狭椭圆形，约 3.5×1.8 mm，顶端圆钝，每一萼片具 3 条基生脉（中脉不分枝，2 侧生脉 1-2 回二歧状分枝）。雄蕊 20-25；花丝长 3-4 mm，上部条形或狭倒披针形，宽 0.15-2 mm，下部丝形，粗 0.1 mm；花药狭长圆形，1-1.5×0.15 mm，顶端钝。心皮 4-7；子房纺锤形，长约 2 mm，具极短柄；花柱长 1.5 mm；柱头长 1 mm。瘦果扁平，新月形，3.5-4.5×1 mm，每侧有 3 条细纵肋，基部具长 0.5 mm 的柄；宿存花柱长约 2 mm。6-8 月开花。

Perennial herbs. Stems 30–58 cm tall, with abaxial surfaces of leaflets and monochasia all covered with very short glandular hairs, often from below branched. Middle cauline leaves 3-ternate; blades 5.5–9.5 cm long; leaflets membranous, orbicular-ovate, orbicular-rhombic or broad-ovate, 4–11×3–13 mm, at base broadly cuneate or rounded, 3-lobed or undivided, lobes mostly entire, sometimes obtusely 1–2-dentate, nerves adaxially flat, abaxially slightly prominent; petiolules 0.5–6 mm long; petides 0.8–3 cm long. Monochasia terminal, 2–3-flowered; pedicels 7–14 mm long. Flower: Sepals 4, caducous, white, narrow-elliptic, ca. 3.5×1.8 mm, apex obtuse, each sepal with 3 basal nerves (midrib simple, and 2 lateral nerves 1–twice dichotomizing). Stamens 20–25; filaments 3–4 mm long, above linear or narrow-oblanceolate, 0.15–2 mm broad, below filiform, 0.1 mm across; anthers narrow-oblong, 1–1.5×0.15 mm, apex obtuse. Carpels 4–7; ovary fusiform, ca. 2 mm long, base with a very short stipe; style 1.5 mm long; stigma 1 mm long. Achene flattend, lunate, 3.5–4.5×1 mm, on each side with 3 thin longitudinal ribs, base with a stipe 0.5 mm long; persistent style ca. 2 mm

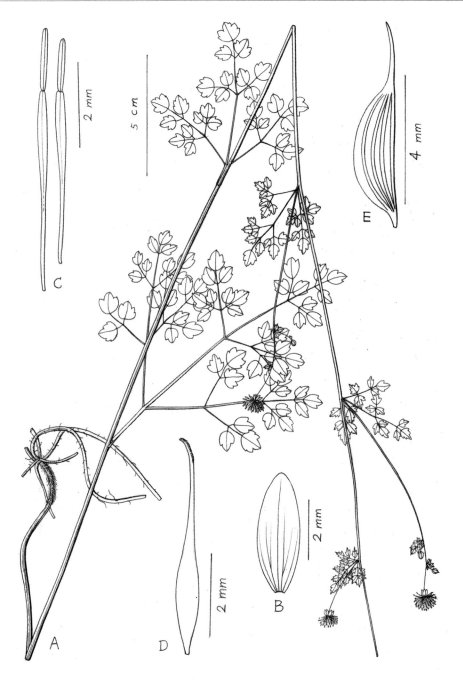

图版42. **丽江唐松草** A. 植株全形，B. 萼片，C. 二雄蕊，D. 心皮（据冯国楣1041），E. 瘦果（据谢磊，毛建丰59）。

Plate 42. **Thalictrum wangii** A. habit, B. sepal, C. two stamens, D. carpel (from K. M. Feng 1041), E. achene (from L. Xie. & J. F. Mao 59).

long.

分布于云南西北部和西藏东南部（察隅）。生山地云南松林中、沟边或草坡上，海拔2500–3100 m。

全草用于关节炎、泻痢等症，根和根状茎用于炭疽病等症。(《中国药用植物志》, 2016)

标本登录：云南：宁蒗，泸沽湖，谢磊58；中甸，冯国楣1041，1150，中甸队62–1016；德钦，俞德浚8264，冯国楣5214，青藏队81–1846，81–2589。

西藏：察隅，青藏队73–230。

43. 察瓦龙唐松草（《中国植物志》） 图版43

Thalictrum tsawarungense W. T. Wang & S. H. Wang in Fl. Reip. Pop. Sin. 27: 539, 618, pl. 142: 4. 1979; W. T. Wang in Fl. Xizang. 2: 68. 1985; et in Vasc. Pl. Hengduan Mount. 1: 500. 1993; D. Z. Fu & G. Zhu in Fl. China 6: 300. 2001.——*T. wangii* Boivin in J. Arn. Arb. 26: 116. 1945, p.p. quoad paratypum. Holotype：西藏：察隅，察瓦龙，折那，alt. 3000 m, 1935–09，王启无66523（PE; *T. wangii* Boivin 的paratype）.

多年生草本植物。茎高25–30 cm，细，上部具短分枝，与叶柄、小叶下面和花梗均被极短的腺毛。下部茎生叶为3回三出复叶；叶片长约5 cm；小叶膜质，宽卵形、圆卵形或宽菱形，0.5–1×0.6–1.3 cm，基部圆截形、宽楔形或浅心形，3浅裂，裂片全缘或有1–2小钝齿，脉在两面不明显隆起；小叶柄长2–6 mm；叶柄长3–4.2 cm，基部有狭鞘。茎的顶生单歧聚伞花序有3花，分枝的顶生单歧聚伞花序有2花；花梗细，长4–8 mm。花：萼片4，早落，白色，长椭圆形，约3.6×2 mm，顶端圆钝，每一萼片具3条不分枝的基生脉。雄蕊约15；花丝狭条形，1.2×0.15 mm；花药狭长圆形，约1.6×0.3 mm，顶端钝。心皮约2；子房狭卵形，约0.8×0.4 mm；花柱长约1.2 mm；柱头狭长圆形，长约0.6 mm。瘦果斜椭圆形，两面凸，约3×1.8–2 mm，每侧有3条不明显钝纵肋；宿存花柱长约1.2 mm。8–9月开花。

Perennial herbs. Stems 25–30 cm tall, slender, above shortly branched, with petioles, abaxial surfaces of leaflets and pedicels all covered with very short glandular hairs. Lower cauline leaves 3-ternate; blades ca. 5 cm long; leaflets membranous, broad-ovate, orbicular-ovate or broad-rhombic, 0.5–1×0.6–1.3 cm, at base rounded-truncate, broadly cuneate or subcordate, 3-lobed, lobes entire or obtusely 1–2-denticulate, nerves on both surfaces slightly prominent; petiolules 2–6 mm long; petioles 3–4.2 cm long, at base narrowly vaginate.

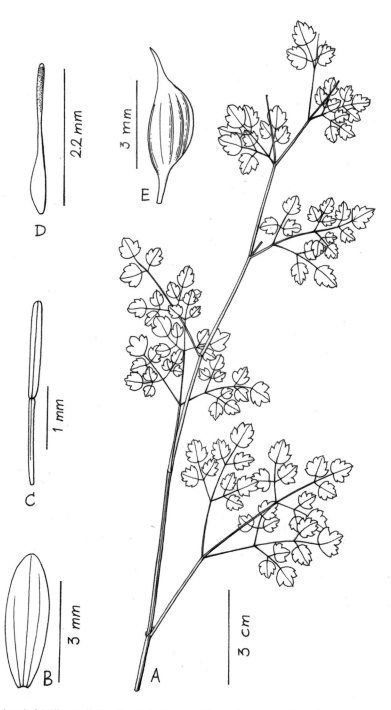

图版43. **察瓦龙唐松草** A. 植株全形，B. 萼片，C. 雄蕊，D. 心皮，E. 瘦果。（据holotype）
Plate 43. **Thalictrum tsawarungense** A. habit, B. sepal, C. stamen, D. carpel, E. achene. (from holotype)

Terminal monochasia of stem 3-flowered, and terminal monochasia of branches 2-flowered; pedicels slender, 4–8 mm long. Flower: Sepals 4, caducous, white, long elliptic, ca. 3.6×2 mm, apex rounded-obtuse, each sepal with 3 simple basal nerves. Stamens ca. 15; filaments narrow-linear, 1.2×0.15 mm; anthers narrow-oblang, ca. 1.6×0.3 mm, apex obtuse. Carpels ca. 2; ovary narrow-ovate, ca. 0.8×0.4 mm; style ca. 1.2 mm long; stigma narrow-oblong, ca. 0.6 mm long. Achene obliquely elliptic, biconvex, ca. 3×1.8–2 mm, on each side obtusely inconspicuously longitudinally 3-ribbed; persistent style ca. 1.2 mm long.

特产西藏察隅县察瓦龙。生山地林边，海拔 3000 m。

44. 矮唐松草（《云南植物志》） 图版 44

Thalictrum pumilum Ulbr. In Bot. Jabrb. 48: 23. 1913; W. T. Wang in Fl. Yunnan. 11: 162. 2000. Syntypes：云南：东川,Bonati 2657, E. E. Maire 采（B, not seen; isosyntype, NY）; Pe-long-tsin, alt. 3300 m, 1910-08,Bonati 2662, E. E. Maire 采（B. not seen）.

多年生草本植物。茎细，高 12–20 cm，有少数分枝，下部无毛，上部和花梗被短柔毛（毛长约 0.05 mm）。茎生叶为 2 回三出复叶；叶片长 3–5 cm；小叶薄纸质或草质，近圆形或宽倒卵形，0.3–1×0.4–1.4 cm，基部近心形或宽楔形，3 浅裂（裂片顶端圆形，全缘，稀具 1 小齿），上面被短柔毛（毛长约 0.05 mm），下面脉上有短而粗的小毛（毛长约 0.05 mm），脉两面平，脉网不明显；小叶柄丝形，长 1–7 mm，无毛；叶柄长 0.2–1.8 cm，无毛。单歧聚花序顶生，有 2–3 花；花梗长 3–12 mm。花：萼片 5，白色或带粉红色，长圆形，3×1.2–1.5 mm，顶端圆形，背面被短柔毛，每一萼片具 3 条不分枝的基生脉。雄蕊 10–12，无毛；花丝长 1.4–2.2 mm，上部狭倒披针形或条形，宽 0.2–0.3 mm，下部狭条形，宽 0.15 mm；花药狭长圆形，1–1.2×0.3 mm，顶端钝。心皮 5–6；子房纺锤形或狭倒卵形，约 1.2×0.8 mm，密被长约 0.05 mm 的小毛；花柱长约 1.5 mm；柱头狭长圆形，长约 1 mm。瘦果具柄，斜长圆状卵形，长约 5 mm，每侧有 3 条细纵肋，疏被短毛；宿存花柱长约 2 mm，稍弯曲；柄长约 1 mm。7–8 月开花。

Perennial herbs. Stems slender, 12–20 cm tall, few-branched., below glabrous, above with pedicels minutely puberulous (hairs ca. 0.05 mm long). Cauline leaves 2-ternate; blades 3–5 cm long; leaflets thinly papery or herbaceous, suborbicular or broad–obovate, 0.3–1×0.4–1.4 cm, at base subcordate or broadly cuneate, 3-lobed (lobes at apex rounded, entire, rarely 1-denticulate), adaxially puberulous (hairs ca. 0.05 mm long), abaxially on nerves with short and thick minute hairs also ca. 0.05 mm long, nerves on both surfaces flat, nervous

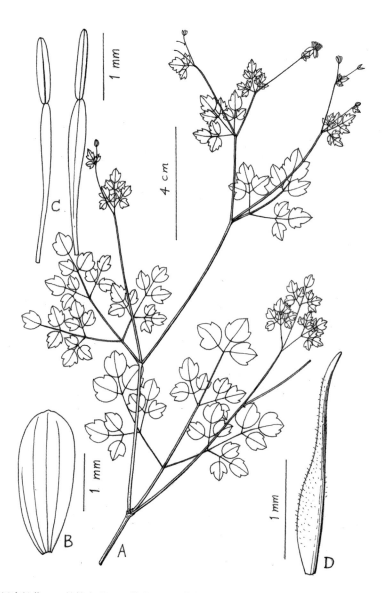

图版44. **矮唐松草** A. 植株全形,B. 萼片,C. 二雄蕊,D. 心皮。（据isosyntype）
Plate 44. **Thalictrum pumilum** A. habit, B. sepal, C. two stamens, D. carpel. (from isosyntype)

nets inconspicuous; petiolules filiform, 1–7 mm long; petioles 0.2–1.8 cm long, glabrous. Monochasia terminal, 2–3-flowered; pedicels 3–12 mm long. Flower: Sepals 5, white or tinged with pink, oblong, 3×1.2–1.5 mm, at apex rounded, abaxially puberulous, each sepal with 3 simple basal nerves. Stamens 10–12; filaments 1.4–2.2 mm long, above narrow-oblanceolate or linear, 0.2–0.3 mm broad, below narrow-linear, 0.15 mm broad; anthers narrow-obong, 1–1.2×0.3 mm, apex obtuse. Carpels 5–6; ovary fusiform or narrow-obovate, ca. 1.2×0.8 mm, densely covered with minute hairs ca. 0.05 mm long; style ca. 1.5 mm long; stigma narrow-oblong, ca. 1 mm long. Achene stipitate, obliquely oblong-ovate, ca. 5 mm long, on each side with 3 thin longitudinal ribs; persistent style ca. 2 mm long, slightly curved; stipe ca. 1 mm long.

特产云南东川和 Pe-long-tsin 一带。生于海拔 3300 m 山地。

组 5. 散花唐松草组

Sect. **Omalophysa** Turcz. ex Fisch. & Mey. in Ind. Sem. Hort. Petrop. 1: 40, 1835; Emura in J. Fac. Sci. Univ. Tokyo, sect. 3, Bot. 9: 106. 1972; Tamura in Acta Phytotax. Geobot. 43: 57. 1992; et in Hiepko, Nat. Pflanzenfam., ZWei. Aufl, 17a Ⅳ : 487. 1995; Park & Festerlin Jr. in Fl. N. Amer. 3: 264. 1997—Subsect. *Sparsiflora* Prantl in Bot. Jrhrb. 9: 272. 1887—Subsect. *Omalophysa* (Turcz. ex Fisch. & Mey.) Tamura in Sci. Rep. Osaka Univ. 17: 51. 1968.—Ser. *Sparsiflora* (Prant) W. T. Wang & S. H. Wang in Fl. Reip Pop. Sin. 27: 548. 1979. Type: *T. sparsiflorum* Turcz. ex Fisch. & Mey.

花组成聚伞圆锥花序。萼片具 3 或 5 条基出脉。雄蕊花丝多少分化成上、下两部分。心皮具直花柱和狭条形柱头，在基部具柄。瘦果具长柄，每侧具 3 条纵肋，无翅。

Flowers in thyrses. Sepal with 3 or 5 basal nerves. Stamen filament more or less differentiated to upper and lower two parts. Carpel with 1 straight style and 1 narrow-linear stigma, and at base with a stipe. Achene long stipitate, on each side with 3 longitudinal ribs, not winged.

3 种，2 种特产中国，另一种分布于中国北部和东北部，以及俄罗斯远东地区和北美洲西部。

45. 毛蕊唐松草（《广西植物》） 图版 45

Thalictrum lasiogynum W. T Wang in Guihaia 37(4): 414. fig. 4. 2017. Holotype：四川：平武县，王朗自然保护区，alt. 3110 m, 2008-07-19, 鲁丽敏 2008-337（PE）.

多年生草本植物。茎高约 90 cm，在节上密被、在节间疏被短柔毛（毛长 0.1-0.25 mm）。上部茎生叶为 3 回三出羽状复叶；叶轴和小叶柄疏被短毛或近无毛；叶片长 5-8 cm；小叶纸质，卵形或椭圆形，0.4-1.2×0.2-1 cm，顶端急尖或钝，基部圆形或宽楔形，3 浅裂（裂片具 1-2 小齿或全缘），两面疏被短柔毛（毛长 0.1-0.2 mm），具 3 出脉，脉上面平，下面稍隆起；小叶柄长 0.5-7 mm；叶柄长 2 mm，具宽耳。聚伞圆锥花序顶生并腋生，包括花序梗长 12-21 cm，具多数花；花序梗长 4.5-8.5 cm，近无毛；下部苞片叶状，长 1-2 cm，上部苞片简单，钻形，长 0.6-1 mm；花梗丝形，长 5-14 mm，疏被短柔毛。花：萼片 4，椭圆形或狭倒卵形，2.5-2.8×1-1.2 mm，上部被缘毛，每一萼片具 3 条不分枝的基生脉。雄蕊约 40，无毛；花丝长 5-6 mm，上部狭倒披针形，宽 0.25-0.4 mm，下部丝形，粗 0.1 mm；花药长圆形，0.8-1×0.25-0.45 mm，顶端钝。心皮 7-11；子房斜狭倒卵形，长 1-1.5 mm，上部密被短柔毛，基部柄长 0.5-0.7 mm；花柱长 0.5-0.7 mm；柱头长 0.2-0.3 mm。瘦果扁平，半圆形，长约 1.5 mm，每侧具 3 条纵肋，基部柄长约 1.2 mm；宿存花柱长约 0.6 mm。6-7 月开花。

Perennial herbs. Stems ca. 90 cm tall, on nodes densely and on internodes sparsely puberulous (hairs 0.1-0.25 mm long). Upper cauline leaves 3-ternate-pinnate; blades 5-8 cm long; leaflets papery, ovate or elliptic, 0.4-1.2×0.2-1 cm, at apex acute or obtuse, at base rounded or broadly cuneate, 3-lobed (lobes 1-2-denticulate or entire), on both surfaces sparsely puberulous (hairs 0.1-0.2 mm long), 3-nerved, nerves adaxially flat, abaxially slightly prominent; petiolules 0.5-7 mm long; petioles 2 mm long, broadly auriculate. Thyrses terminal and axillary, including peduncles 12-21 cm long, many-flowered; peduncles 4.5-8.5 mm long, subglabrous; lower bracts foliaceous, 1-2 cm long, upper bracts simple, subulate, 0.6-1 mm long; pedicels filiform, 5-14 mm long, sparsely paberulous. Flower: Sepals 4, elliptic. or narrow-obovate, 2.5-2.8×1-1.2 mm, above ciliate, each sepal with 3 simple basal nerves. Stamens ca. 40, glabrous; filaments 5-6 mm long, above narrow-oblanceolate, 0.25-0.4 mm broad, below filiform, 0.1 mm across; anthers oblong, 0.8-1×0.25-0.45 mm, apex obtuse. Carpels 7-11; ovary obliquely narrow-obovate, 1-1.5 mm long, above densely puberulous, base with a stipe 0.5-0.7 mm long; style 0.5-0.7 mm long; stigma 0.2-0.3 mm long. Achene

图版45. 毛蕊唐松草　A. 开花茎上部，B. 二萼片，C. 雄蕊，D. 雌蕊群，E. 幼瘦果。（据holotype）
Plate 45. **Thalictrum lasiogynum**　A. upper part of flowering stem, B. two sepals, C. stamen, D. gynoecium, E. young achene. (from holotype)

flattened, semi-orbicular, ca. 1.5 mm long, on each side with 3 longitudinal ribs, at base with a stipe ca. 1.2 mm long; persistent style ca. 0.6 mm long.

特产四川平武县，王郎自然保护区。生山地路边，海拔 3110 m。

46. **长柄唐松草**(《北京植物志》) 图版 46

Thalictrum przewalskii Maxim. in Bull. Acad. Sci. st.-Petersb. 23: 305. 1877; et Fl. Tangut. 4, t. 1: 1–6. 1889; et in Acta Hort. Petrop. 11: 11. 1890; Lecoy. in Bull. Soc. Belg. 24: 157. 1885; Forbes & Hemsl. in J. Soc. Linn. Bot. 23: 9. 1886; Finet & Gagnep. in Bull. Soc. Bot. France 50: 609. 1903; Hand.-Mazz. Symb. Sin. 7: 310. 1931; Fl. Beijing 1: 293. 1962; Iconogr. Corm. Sin. 1: 678, fig. 1356. 1972; Fl. Tsinling. 1c(2):243. 1974; S. H. Li in Fl. Pl. Herb. Chinae Bor-Or. 3: 204, pl. 88: 1–3. 1975; S. H. Fu, Fl. Hupeh. 1: 357, fig. 500. 1976; W. T. Wang & S. H. Wang in Fl. Reip. Pop. Sin. 27: 548, pl. 135: 1–4. 1979; W. T. Wang in Bull. Bot. Lab. N.-E. Forest. Inst. 8: 28. 1980; B. Z. Ding et al. Fl. Henan. 1: 467, fig. 598. 1981; S. Y. He, Fl. Beijing, rev. ed., 1: 238, fig. 297: 1. 1984; W. T. Wang in Fl. Xizang. 2: 68. 1985; J. W. Wang in Fl. Hebei 1: 443, fig. 440. 1986; D. Z. Ma. Fl. Ningxia. 1: 174, fig. 152. 1986; Y. Z. Zhao in Fl. Intramongol., ed. 2, 2: 453, pl. 182: 1–3. 1990; W. T. Wang in Vasc. Pl. Hengduan Mount. 1: 501. 1993; et Keys Vasc. Pl. Wuling Mount. 164. 1995; S. H. Li in Clav. Pl. Chinae Bor.-Or., ed. 2: 205, pl. 96: 6. 1995; L. H. Zhou in Fl. Qinghai. 1: 330, pl. 73: 7–10. 1997; K. M. Liu, Fl. Hunan 2: 654. 2000; W. T. Wang in High. Pl. China 3: 469, fig. 145: 1–4. 2000; D. Z. Fu & G. Zhu in Fl. China 6: 295. 2001; C. X. Yang et al. Keys Vasc. Pl. Chongqing 216. 2009; L. Q. Li in High. Pl. Daba. Mount. 118. 2014; L. Q. Li et al. Pl. Baishuijiang St. Nat. Reserve Gansu Prov. 71. 2014; F. H. Li et al. Pl. Yanqing 142. 2015; W. T. Wang in High. Pl. China in Colour 3: 388. 2016; L. Xie in Med. Fl. China 3: 253, with fig. & photo. 2016. Holotype：甘肃，1872–06, N. M. Przewalski s. n.（LE, not seen）.

T. rockii Boivin in J. Arn. Arb. 26: 115, pl. 1: 28–30. 1945. Type：甘肃：Upper Tebbu, alt. 11000 ft., 1925–07, J. F. Rock 13054 (holotype, GH, not seen); valley of Tayüku, alt. 8500 ft., 1925–07, J. F. Rock 12835 (paratype, GH, not seen). 青海：Ba valley, alt. 9900 ft., 1926–06–23, J. F. Rock 14271 (paratype, GH). 河北：涿鹿，Ling-shan-kou, alt. 1600 m, 1930–09–06，夏纬瑛 2400 (paratypes, NY, PE).

多年生草本植物。茎高 50–120 cm，变无毛，分枝。下部茎生叶为 4 回三出复叶，长达 25 cm；叶片长 7–25 cm；小叶草质，倒卵形，菱状椭圆形，卵形或近圆形，0.6–3×0.5–

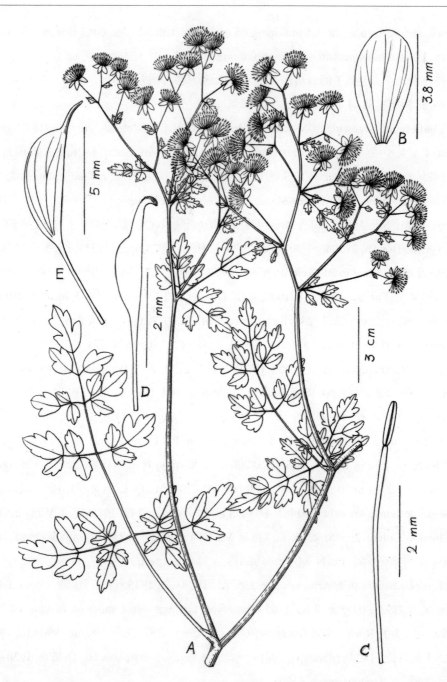

图版46. **长柄唐松草** A. 开花茎上部，B. 萼片，C. 雄蕊，D. 心皮（据孙学刚等2174），E. 瘦果（据黄河队3353）。

Plate 46. **Thalictrum przewalskii** A. upper part of flowering stem, B. sepal, C. stamen, D. carpel. (from X. G. Sun et al. 2174), E. achene (from Huanghe Exped. 3353).

2.5 cm，顶端钝或圆形，基部圆形、浅心形或宽楔形，3 浅裂或 3 中裂，裂片有 1-2 钝齿或全缘，脉下面隆起，有短毛（毛长 0.1-0.4 mm），常变无毛；小叶柄长 0.5-9 mm；叶柄长约 6 cm，基部具鞘。聚伞圆锥花序顶生，长 20-30 cm，有多数花，无毛；花梗长 3-5 mm。花无毛：萼片 4，早落，白色或带黄绿色，倒卵形或近长圆形，2.5-5×1.5-2.2 mm，顶端圆形，每一萼片具 5 条基生脉（中脉和 3 侧生脉二歧状分枝，其他 1 条侧生脉不分枝）。雄蕊 35-48；花丝长 5-6 mm，上部条形，宽 0.3-0.4 mm，下部丝形，粗 0.1 mm；花药长圆形，0.7-1×0.25-0.35 mm，顶端尖或具小短尖头。心皮 4-9；子房斜倒卵形，长 1-1.8 mm，基部的柄长 1.8-2.2 mm；花柱长约 1 mm，顶端稍向弯曲；柱头长约 0.25 mm。瘦果扁，斜倒卵形，4-5.8×0.15-0.25 mm，每侧有 3 条细纵肋，基部的柄长 3-5 mm；宿存花柱长 1-1.4 mm。6-8 月开花。

Perennial herbs. Stems 50−120 cm tall, glabrescent, branched. Lower cauline leaves 4-ternate, up to 25 cm long; blades 7−25 cm; leaflets herbaceous, obovate, rhomic−elliptic, ovate or suborbicular, 0.6−3×0.5−2.5 cm, at apex obtuse or rounded, at base rounded, subcordate or broadly cuneate, 3-lobed or 3-fid, lobes obtusely 1−2-dentate or entire, nerves abaxially prominent and with short hairs 0.1−0.4 mm long, often glabrescent; petiolules 0.5−9 mm long; petioles ca. 6 cm long, at base vaginate. Thyrses terminal, 20−30 cm long, many-flowered, glabrous; pedicels 3−5 mm long. Flower glabrous: Sepals 4, caducous, white or tinged with yellowish-greenish, obovate or nearly oblong, 2.5−5×1.5-2.2 mm, apex rounded, each sepal with 5 basal nerves (midrib and 3 lateral nerves dichotomizing, and the another lateral nerve simple). Stamens 35−48; filaments 5−6 mm long, above linear, 0.3−0.4 mm broad, below filiform, 0.1 mm across; anthers oblong. 0.7−1×0.25−0.35 mm, apex acute or minutely mucronate. Carpels 4−9; ovary obliquely obovate, 1−1.8 mm long, base with a stipe 1.8-2.2 mm long; style ca. 1 mm long, apex slightly recurved; stigma ca. 0.25 mm long. Achene compressed, obliquely obovate, 4−5.8×0.15−0.25 mm, on each side with 3 thin longitudinal ribs, base with a stipe 3−5 mm long; persistent style 1−1.4 mm long.

分布于西藏东南部、四川、重庆、湖北西部、青海东部、甘肃和陕西的南部、河南西部、山西、河北西部和北部、内蒙古南部。生山地灌丛边、林下或草坡上；海拔高度，在陕西、甘肃以南为 2000-3500 m，在河北为 750-1700 m。

根有祛风之效，用于风湿病。花、果实用于肝炎等症。本种的根和根状茎在青海等地作"青海马尾莲"药材使用。根含长柄唐松草酮（thalprzewalskinone）和小檗碱

（berberine）等生物碱。（《中国药用植物志》，2016）

标本登录：西藏：江达，张永田，郎楷永 163，青藏队 76-12457，D. E. Boufford, B. Bartholomew et al. 41974。

四川：白玉，D. E. Boufford, B. Bartholomew et al. 37147, 37157；德格，崔友文 4406，贾慎修 2343，四川植物队 7487，D. E. Boufford, B. Bartholomew et al. 36521, 36540, 36801；色达，秦白生 6531，D. E. Boufford et al. 34637；垠塘，D. E. Boufford et al. 39068；甘孜，汪可宁 1718，D. E. Boufford et al. 33995；炉霍，张永田，郎楷永 103；道孚，姜恕 9666；新都桥，关克俭，王文采 839；小金，四川植物队 9774；绰斯甲，李谱雄 10424；马尔康，李谱雄 10765，李馨 71545, 74480，吴中伦 32049, 32507；刷经寺，张泽荣，周洪富 22365, 22506，李馨 74649, 74698；黑水，1885-07-27，G. N. Potanin s. n.；黑水，1957-1958，李馨 73229，张秀实 6139；理县，何铸，周子林 13643，李馨 74160；松潘，王作宾 7795，H. Smith 2692，D. E. Bouffford et al. 40435；九寨沟，李金喜 3113；若尔盖，郎楷永，李良千，费勇 1977，朱大海等 4577；红原，赵清盛 254；阿坝，郎楷永，李良千，费勇 2046。

重庆：巫溪，杨光辉 58947。

湖北：神农架，鄂植考队 25128，刘彬彬 2280；兴山，陈之端，马欣堂 96-1273。

青海：玉树，何廷农等 2013；同德，吴玉虎 6840；贵德，何廷农 1171，刘尚武 3278；民和，刘尚武 2731；乐都，吴玉虎 3662；湟源，刘更喜 s. n.；西宁，郝景盛 790，张志和 4282；大通，刘继孟 5994，张志和 4354；海晏，郭本兆 12278；门源，青甘队 60-2551，郭本兆 12123。

甘肃：舟曲，王作宾 14414，杨亲二，孔宏智 98-38；岷县，郝景盛 651，王作宾 4782；Radja, 1926-06, J. F. Rock 14125；卓尼，王作宾 5340；夏河，王作宾 7328, 6892, 6897，傅坤俊 1228；莲花山，孙学刚 2096, 2474；渭源，黄河队甘一分队 56-1607；天水，夏纬瑛 5715；泾川，王作宾 17073；兴隆山，崔友文 10301，黄河队 56-3353，何业祺 6950；永登，孙学刚 6012；天祝，何业祺 4512, 4821；Alpes Mudshik, 1880-06, N. M. Przewalski s. n.

陕西：凤县，刘继孟 10703；太白山，刘慎谔，钟补求 938，吴振海 20, 64，Chen, Xu. Wang 2547；佛坪，郭本兆 1727；宁陕，孔宪武 3012；雒南，王作宾 15829。

河南：卢氏，刘继孟 5002, 5053；嵩县，关克俭，戴天伦 2132。

山西：中条山，南天门，alt. 3100 m，冷杉林中，应俊生等 546（GH）；垣曲，H. Smith 6648；介休，刘继孟 1242；关帝山，刘心源 21284；离石，黄河队 55-2117, 55-2518；兴县，黄河队 55-2559；五寨，黄河队 55-2911；管涔山，张晓台 4025；宁武，刘继孟 1737；五台山，关克俭，陈艺林 1549；繁峙，段长虹 178；高帝山，夏纬瑛 1126，1225。

河北：阜平，阜平队 98；涞源，河北农大队Ⅳ 2306；涞水，刘继孟 2282，2533；小五台山，李建藩 11927，孔宪武 4769，王启无 61512，61692，段俊喜 421，黄秀兰等 1387，1463，1652；雾灵山，刘慎谔等 4769；遵化，东陵，刘继孟 412；青龙，王忠涛等 14-258；承德，承德队 1547；涿鹿，杨朝广 471。

北京：西灵山，杨朝广 76，303，1172；东灵山，杨朝广 1385，1687；平谷，镇罗营，平谷队 72。

47. 散花唐松草（《东北植物检索表》）　图版 47

Thalictrum sparsiflorum Turcz. ex Fisch. & Mey. Ind. Sem. Hort. Petrop. 1: 40. 1835; Lecoy. in Bull. Soc. Bot. Belg. 24: 155. 1885; Finet & Gagnep. in Bull. Soc. Bot. France 50: 609. 1903; Kom. & K.-Alisova, Key Pl. Far East USSR 1: 554. 1931; Nevski in Fl. URSS 7: 513. 1937; Kitag. Lineam. Pl. Mansh. 228. 1939; Boivin in Rhodora 46: 369. 1944; S. H. Li in Fl. Pl. Herb. Chinae Bor.-Or. 3: 204, pl. 88: 4-5. 1975; W. T. Wang & S. H. Wang in Fl. Reip. Pop. Sin. 27: 548, pl. 135: 5-6. 1979; S. Y. He, Fl. Beijing, rev. ed., 1: 239, fig. 297-2. 1984; S. H. Li in Clav. Pl. Chinae Bor.-Or., ed. 2: 207, pl. 96: 7. 1995;Park & Festerling Jr. in Fl. N. Amer. 3: 264. 1997; W. T. Wang in High. Pl. China 3: 470. 2000; D. Z. Fu & G. Zhu in Fl. China 6: 298. 2001. Described from Dauria.

多年生草本植物，全部无毛。茎高约 90 cm，上部分枝。中部茎生叶为 3-4 回三出复叶，具短柄；叶片长 6-9 cm；小叶膜质，倒卵形、卵形或近圆形，0.5-2×0.4-1.8 cm，顶端圆形，有短尖头，基部楔形或宽楔形，有时浅心形，3 浅裂或不分裂，裂片有 1-2 钝齿或全缘，脉平，脉网不明显；小叶柄长 0.5-7 mm；叶柄长约 3 cm。复单歧聚花序顶生，伞房状，宽 3-8 cm，有少数或多数花；花梗长 6-15 mm。果期增长至 1-3 cm，顶端钩状反曲。花：萼片 4，早落，白色，近长圆形，2.5-4×1-1.5 mm，顶端圆形，每一萼片有 5 条基生脉（中脉和 1 侧生脉二歧状分枝，其他 3 侧生脉不分枝）。雄蕊 10-30；花丝狭条形，长 5-6 mm，上部宽 0.2-0.25 mm，下部宽 0.1 mm；花药长圆形，约 1×0.4 mm，顶端钝或有长 0.05 mm 的小短尖头。心皮 4-7；子房近圆形，约 1.4×1 mm，基部的柄长约 2 mm；花柱长约 1 mm。瘦果下垂，扁，斜倒卵形，长 5-7 mm，每侧有 3 条弧状弯曲的细纵肋，基部的柄长 2-3.5 mm；宿存花柱 1-1.5 mm。6-7 月开花。

Perennial herbs, totally glabrous. Stems ca. 90 cm tall, above branched. Middle cauline leaves 3-4-ternate, shortly petiolate; blades 6-9 cm long; leaflets membranous, obovate,

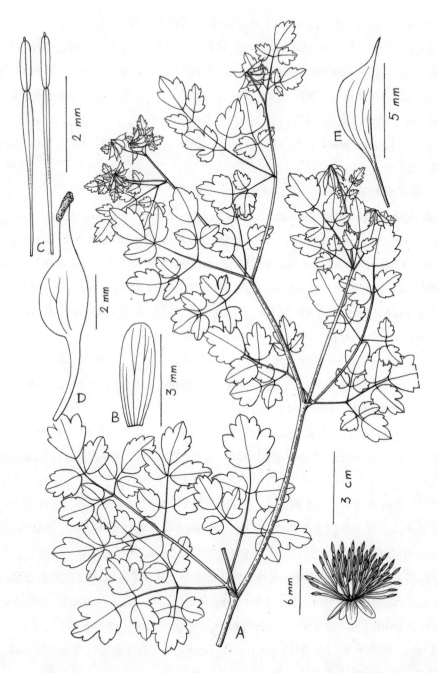

图版47. 散花唐松草　A. 开花茎上部，B. 萼片，C. 二雄蕊，D. 心皮（据密云队502），E. 瘦果（据傅沛云等1163）。

Plate 47. **Thalictrum sparsiflorum**　A. upper part of flowering stem, B. sepal, C. two stamens, D. carpel (from Miyun Exped. 502), E. achene (from P. Y. Fu et al. 1163).

ovate or suborbicular, 0.5–2×0.4–1.8 cm, at apex rounded, mucronate, at base cuneate or broadly cuneate, sometimes subcordate, 3-lobed or undivided, lobes obtusely 1–2-dentate or entire, nerves flat, nervous nets inconspicuous; petiolules 0.5–7 mm long; petioles ca. 3 cm long. Compound monochasia terminal, corymbiform, 3–8 cm broad, few- or many-flowered; pedicels 6–15 mm long, in fruiting time 10–30 mm long and apex hooked-reflexed. Flower: Sepals 4, caducous, white, nearly oblong, 2.5–4×1–1.5 mm, apex rounded, each sepal with 5 basal nerves (midrib and 1 lateral nerve dichotomizing, and the another 3 lateral nerves simple). Stamens 10–30; filaments narrow-linear, 5–6 mm long, above 0.2–0.25 mm broad, below 0.1 mm broad; anthers oblong, ca. 1×0.4 mm, apex obtuse or with small mucrones 0.05 mm long. Carpels 4–7; ovary suborbicular, ca. 1.4×1 mm, base with a stipe ca. 2 mm long; style ca. 1 mm long. Achene pendulous, compressed, obliquely obovate, 5–7 mm long, on each side with 3 arcuate thin longitudinal ribs, base with a stipe 2–3.5 mm long; persistent style 1–1.5 mm long.

分布于中国河北北部、吉林和黑龙江的东部；以及朝鲜、俄罗斯远东地区和北美洲。生山地草坡或落叶松林下。

标本登录：河北：兴隆，雾灵山，密云队 502。

北京：门头沟，齐家庄，中草药普查队 417。

吉林：安图，傅沛云等 1163，延边一组 149，洪德元等 33176，曹子余等 235；长白山，王薇等 2351。

黑龙江：带岭，中德一队 1443。

组 6. 尖叶唐松草组

Sect. **Physocarpum** DC. Syst. 1: 171. 1817; et Prodr. 1: 11. 1824; Boivin in Rhodora 46: 360. 1944; H. Eichler in Biblioth. Bot. 124: 5. 1958; Tamura in Sci. Rep. Osaka Univ. 17: 51. 1968; et in Hiepko, Nat. Pflanzenfam., ZWei. Aufl., 17a Ⅳ: 486. 1995; Emura in J. Fac. Sci. Univ. Tokyo, Sect. 3, Bot. 9: 101. 1972; Park & Ferterling Jr. in Fl. N. Amer. 3: 263. 1997. Type: *T. clavatum* DC.

Subsect. *Claviformes* Lecoy. in Bull. Soc. Bot. Belg. 24: 112. 1885, p.p.

花组成伞房状或近伞形状单歧聚伞花序。萼片具（1-2-）3 条通常不分枝的基生脉。

雄蕊花丝分化成上、下二部分。心皮具短，稀长，花柱，基部具细柄。瘦果具长柄和细纵肋，无翅。

Flowers in corymbiform or subumbelliform monochasia. Sepal with (1–2–)3 usually simple basal nerves. Stamen filament differentiated into upper and lower two parts. Carpel with a short, rarely long, style, and at base with a slender stipe. Achene long stipitate, thinly longitudinally ribbed, not winged.

28 种，中国有 9 种，隶属 2 系。

系 1. 尖叶唐松草系

Ser. **Clavata** W. T. Wang & S. H. Wang in Fl. Reip. Pop. Sin. 27: 550, 619. 1979. Type: *T. clavatum* DC.

心皮的花柱很短，比子房短甚多，在腹面具密集柱头组织。

Carpel's style very short, much shorter than ovary, ventrally with dense stigmatic tissue.

25 种分布于东亚，2 种分布于北美洲东部。中国有 8 种，广布秦岭以南地区和东北地区。

48. 阴地唐松草《中国植物志》 图版 48

Thalictrum umbricola Ulbr. in Notizbl. Bot. Gart. Berl. 9: 221. 1925; W. T. Wang & S. H. Wang in Fl. Reip Pop. Sin. 27: 551, pl. 143: 7–9. 1979; W. T. Wang in Bull. Bot. Lab. N.-E. Forest. Inst. 8: 29. 1980; in Iconogr. Corm. Sin. Suppl. 1: 425, fig. 8611. 1982; et in Fl. Guangxi 1: 276, pl. 118: 7–8. 1991; K. M. Liu, Fl. Hunan 2: 656, fig. 2–511. 2000; W. T. Wang in High. Pl. China 3: 470, fig. 747. 2000; D. Z. Fu & G. Zhu in Fl. China 6: 300. 2001; R. J. Wang in Fl. Guangdong 5: 20. 2003; W. T. Wang in Fl. Jiangxi 2: 777, fig. 184. 2004; Z. C. Luo & Y. Luo, Pl. Xinning142. 2008; W. T. Wang in High. Pl. China in Colour 3: 388. 2016. Holotype：江西：上犹，alt. 1300 m, 1921–05–14, 胡先骕 893（B, not seen）.

T. gueguenii B. Boivin in Rhodora 46: 366, fig. 17. 1944. Type：湖南：宜章, Pai Mu village, Ping-tou shan, 1934–04–05, 曾怀德 23486（holotype, GH; isotypes, IBSC, PE, US），23392（paratypes, GH, IBSC, PE, US）。江西：Hing San（与福建交界），1936–06–23, Gressit 1455（paratype, GH）.

T. acutifolium auct. non (Hand.-Mazz.) B. Boivin: Y. Yang et al. in Fl. Hong Kong 1: 73. 2007, p. p. quoad photogr. 148, quae *T. umbricola* est; W. B. Liao et al. Nat. Prot. Rare &

六、分类学处理 207

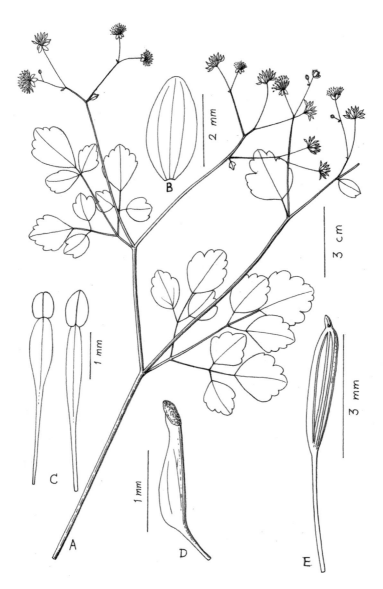

图版48. **阴地唐松草** A. 开花茎，B. 萼片，C. 二雄蕊，D. 心皮（据江西科考队2003026），E. 瘦果（据聂敏祥8281）。

Plate 48. **Thalictrum umbricola** A. flowering stem, B. sepal, C. two stamens, D. carpel (from Jiangxi Sci. Exped. 2003026), E. achene (from M. X. Nie 8281).

Endang. Pl. Shenzhen Shi 55. 2018.

多年生草本植物，全部无毛。茎高15–50 cm，纤细，分枝。基生叶为2–3回三出复叶，长达30 cm；小叶膜质，近圆形、圆卵形或宽倒卵形，1–2.5×1–2.5 cm，基部圆形，浅心形或钝，3浅裂，裂片具1–2钝齿，脉平或在下面稍隆起；小叶柄长6–14 mm；叶柄长达12 cm。茎生叶较小，为1–2回三出复叶。复单歧聚伞花序顶生，伞房状，宽3–7 cm，有3–9花；花梗细，长1–24 cm。花：萼片4，早落，白色，卵形，2–2.6×1.4 mm，顶端钝，每一萼片具3条不分枝的基生脉。雄蕊11–16；花丝长2–3.8 mm，上部狭倒披针形，宽0.3–0.4 mm，下部丝形，粗0.1 mm；花药黄色，宽椭圆形，0.5–0.8×0.3–0.4 mm，顶端钝。心皮6–9；子房狭卵形，约1.2×0.4 mm，基部具长0.8 mm的柄；花柱长约0.3 mm；柱头长约0.4 mm。瘦果近纺锤形，2–3×0.4–0.8 mm，每侧有3条细纵肋，基部的柄长1–3 mm；宿存花柱长0.3 mm。3–4月开花。

Perennial herbs, totally glabrous. Stems 15–50 cm tall, slender, branched. Basal leaves 2–3-ternate, up to 30 cm long; leaflets membranous, suborbicular, broad-obovate or orbicular-ovate, 1–2.5×1–2.5 cm, at base rounded, subcordate or obtuse, 3-lobed, lobes obtusely 1–2-dentate, nerves flat or abaxially slightly prominent; petiolules 6–14 mm long; petioles up to 12 cm long. Cauline leaves smaller, 1–2-ternate. Compound monochasia terminal, corymbiform, 3–7 cm broad, 3–9-flowered; pedicels slender, 1–2.4 cm long. Flower: Sepals 4, caducous, white, ovate. 2–2.6×1.4 mm, apex obtuse, each sepal with 3 simple basal nerves. Stamens 11–16; filaments 2–3.8 mm long, above narrow-oblanceolate, 0.3–0.4 mm broad, below filiform, 0.1 mm across; anthers yellow, broad-elliptic, 0.5–0.8×0.3–0.4 mm, apex obtuse. Carpels 6–9; ovary narrow-ovate, ca. 1.2×0.4 mm, base with a stipe 0.8 mm long; style ca. 0.3 mm long; stigma ca. 0.4 mm long. Achene nearly fusiform, 2–3×0.4–0.8 mm, on each side with 3 thin longitudinal ribs, base with a stipe 1–3 mm long; persistent style 0.3 mm long.

分布于广西东部、香港、广东、湖南东南部、江西西南部和西部。生山地林中、沟边或陡崖阴湿处，海拔650–1300 m。

标本登录：广西：象州，黄志39463。

香港：Ma on Shan，胡秀英10514（GH，PE）。

广东：深圳，张寿洲，李良千1957，3137；阳山，邓良1305；乳源，刘媖光400。

湖南：新宁，罗毅波2075，罗林波492。

江西：大余，聂敏祥 9527；上犹，聂敏祥 8281，刘勇 2，庐山植物园队 2003-06；安福，武功山，岳俊三 3254。

49. 稀蕊唐松草（《秦岭植物志》）　图版 49

Thalictrum oligandrum Maxim. in Acta Hort. Petrop. 11: 16. 1890; Fl. Tsinling. 1(2): 243. 1974; W. T. Wang & S. H. Wang in Fl. Reip Pop. Sin. 27: 553, pl. 136: 3-6. 1979; W. T. Wang in Iconogr. Corm. Sin. Suppl. 1: 425, fig. 8612. 1982; et in Vasc. Pl. Hengduan Mount. 1: 501. 1993; L. H. Zhou in Fl. Qinghai. 1: 332. 1997; W. T. Wang in High. Pl. China 3: 471, fig. 148. 2000; D. Z. Fu & G. Zhu in Fl. China 6: 294. 2001; L. Q. Li et al. Pl. Baishuijiang St. Nat. Resserve Gansu Prov. 71. 2014. Type: 四川：黑水河河谷，1886-07-24, G. N. Potanin s. n.（holotype, LE; isotype, PE）.

多年生草本，全部无毛。茎高 20-60 cm；不分枝或上部分枝。基生叶为 3 回三出复叶，具长柄；叶片长约 6 cm；小叶膜质，宽倒卵形，楔状倒卵形或宽卵形，1-1.5×0.7-1.2 cm，顶端圆形，基部宽楔形或近圆形，3 浅裂，裂片有 1-2 牙齿，脉平，脉网不明显；小叶柄长 1.5-7 mm；叶柄长 3.8-5 cm，基部有狭鞘。聚伞圆锥花序顶生，有稀疏分枝；花梗细，长 2-15 mm。花：萼片 4，脱落，白色，宽椭圆形，1.2-1.5×0.8-1 mm，顶端圆形，每一萼片具 2-3 条不分枝的基生脉。雄蕊 3-7；花丝长 1.8-2.2 mm，上部条形，宽 0.25 mm，下部丝形，粗 0.1 mm；花药黄色，宽长圆形或长圆形，0.5-0.6×0.25-0.4 mm，顶端钝。心皮 6-10；子房卵形，约 1×0.5 mm，基部具长 0.5 mm 的柄；花柱长约 0.8 mm，顶端钩状弯曲；柱头长约 0.6 mm。瘦果长椭圆形，2-3×0.8-1 mm，每侧有 3 条细纵肋；基部的柄长 1-3.5 mm，宿存花柱长约 0.4 mm。7 月开花。

Perennial herbs, totally glabrous. Stems 20-60 cm tall, simple or broanched. Basal leaves 3-ternate, long petiolate; blades ca. 6 cm long; leaflets membranous, broad-obovate, cuneate-obovate or broad-ovate, 1-1.5×0.7-1.2 cm, at apex rounded, at base broadly cuneate or subrounded, 3-lobed, lobes 1-2-dentate, nerves flat, nervous nets inconspicuows; petiolules 1.5-7 mm long; petioles 3.8-5 cm long, at base narrowly vaginate. Thyrses terminal, sparsely branched; pedicels slender, 2-15 mm long. Flower: Sepals 4, deciduous, white, broad-elliptic, 1.2-1.5×0.8-1 mm, apex rounded, each sepal with 2-3 simple basal nerves. Stamens 3-7; filaments 1.8-2.2 mm long, above linear, 0.25 mm broad, below filiform, 0.1 mm across; anthers yellow, broad-oblong or oblong, 0.5-0.6×0.25-0.4 mm, apex obtuse. Carpels 6-10; ovary ovate, ca. 1×0.5 mm, base with a stipe 0.5 mm long; style ca. 0.8 mm long, apex

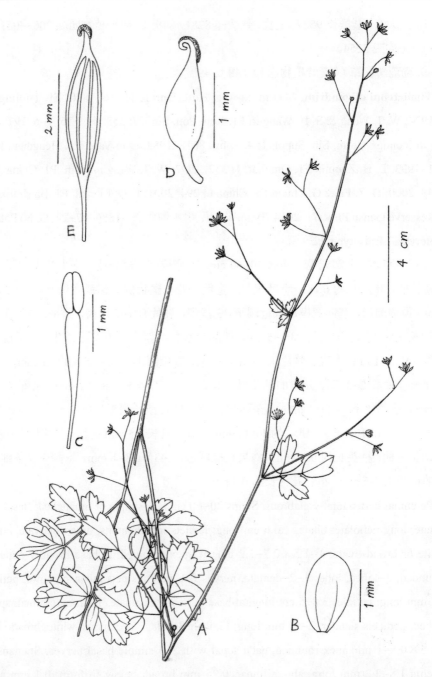

图版49. **稀蕊唐松草** A. 开花茎上部，B. 萼片，C. 雄蕊，D. 心皮，E. 瘦果。（据郎楷永，李良千，费勇 750）

Plate 49. **Thalictrum oligandrum** A. upper part of flowering stem, B. sepal, C. stamen, D. carpel, E. achene. (from K. Y. Lang, L. Q. Li & Y. Fei 750)

hooked; stigma ca. 0.6 mm long. Achene long elliptic, 2–3×0.8–1 mm, on each side with 3 thin longitudinal ribs, base with a stipe 1–3.5 mm long; persistent style ca. 0.4 mm long.

分布于四川西部、青海东部、甘肃南部和陕西南部（太白山）。生山地林中或草坡上，海拔 2300–3400 m。

标本登录：四川：康定，郎楷永，李良千，费勇 750；小金，两河口，张秀实，任有铣 6345。

青海：泽库，麦秀林区，周立华 1971。

甘肃：舟曲，杨亲二，孔宏智 98-41；榆中，兴隆山，何业祺 5472；永登，连城自然保护区，孙学刚 s.n.。

陕西：太白山，平安寺，刘慎谔，钟补求 2616。

50. **岳西唐松草**（《植物科学学报》） 图版 50

Thalictrum yuexiense W. T. Wang in Pl. Sci. J. 32(6): 567. fig. 1. 2014. Holotype：安徽：岳西，alt. 1670 m, 1997-08-13, 谢中稳，郑鹿邻 97-021（PE）．

多年生小草本植物，全部无毛。茎高约 20 cm，具 1–3 条分枝。基生叶约 2，为 3 回三出复叶，具长柄；叶片 2.5–5×4.5–7 cm；小叶纸质，倒卵形，稀宽倒卵形或卵形，0.6–1.3×0.4–0.9 cm，基部宽楔形或圆形，近顶端 3 微裂或具 3 齿，具三出脉，脉上面平，下面不明显隆起；叶柄长 3–3.8 cm。茎生叶约 3，较小，为 1–2 回三出复叶，近无柄。复单歧聚伞花序顶生并腋生，伞房状，有 5–8 花；下部苞片具 3 小叶；花梗丝形，长 4–7 mm。花：萼片 4，早落，紫红色，长圆状椭圆形，约 1.6×0.4 mm，顶端圆形，基部宽楔形，每一萼片具 3 条不分枝的基生脉。雄蕊 20–24；花丝长约 3 mm，上部狭倒披针形，与花药等宽，下部丝形；花药淡黄色，长圆形，长 0.6–1 mm，顶端钝。心皮 2–5；子房新月形，长 1.2–2 mm，基部具长 0.5–0.8 mm 的柄；花柱长 0.1–0.3 mm，顶端具扁头形柱头。瘦果扁，宽新月形，约 3×1 mm，每侧具 3 条细纵肋，基部具长 1 mm 的柄；宿存柱头长约 0.3 mm。7–8 月开花。

Small perennial herbs, totally glabrous. Stem ca. 20 cm tall, 1–3-branched. Basal leaves ca. 2, 3-ternate, long petiolate; blades 2.5–5×4.5–7 cm; leaflets papery obovate, rarely broad-obovate or ovate, 0.6–1.3×0.4–0.9 cm, at base broadly cuneate or rounded, near apex 3-lobulate or 3-dentate, 3-nerved, nerves adaxially flat, abaxially inconspicuously prominent; petioles 3–3.8 cm long. Cauline leaves ca. 3, smaller, 1–2-ternate, subsessile. Compound monochasia terminal and axillary, corymbiform, 5–8-flowered; lower bracts with 3 leaflets;

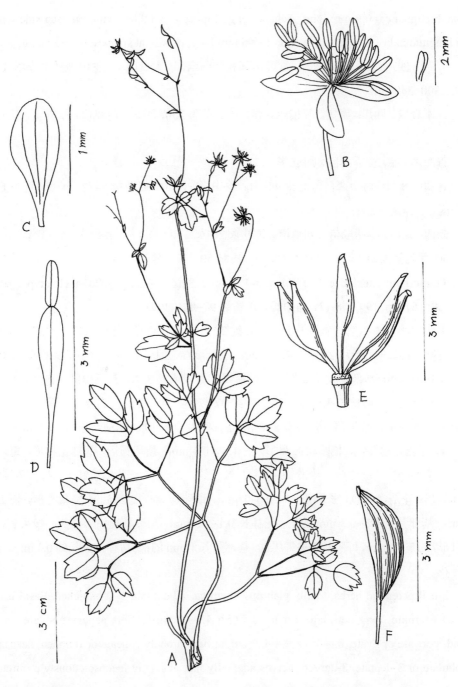

图版50. **岳西唐松草** A. 植株全形，B. 花，C. 萼片，D. 雄蕊，E. 雌蕊群，F. 瘦果。（据holotype）
Plate 50. **Thalictrum yuoxiense** A. habit, B. flower, C. sepal, D. stamen, E. gynoecium, F. achene. (from holotype)

pedicels filiform, 4–7 mm long. Flower: Sepals 4, caducous, purple-red, oblong-elliptic, ca. 1.6×0.4 mm, apex rounded, base broadly cuneate, each sepal with 3 simple basal nerves. Stamens 20–24; filaments ca. 3 mm long, above narrow-oblanceolate, as broad as anthers, below filiform; anthers yellowish, oblong, 0.6–1 mm long, apex obtuse. Carpels 2–5; ovary lunate, 1.2–2 mm long, base with a stipe 0.5–0.8 mm long; style 0.1–0.3 mm long, at apex with a depressed-capitate stigma. Acheme compressed, broad-lunate, ca. 3×1 mm, on each side with 3 thin longitudinal ribs, base with a stipe ca. 1 mm long; persistent stigma ca. 0.3 mm long.

特产安徽岳西县。生山地多石山坡石缝中，海拔 1670 m。

51. **尖叶唐松草**(《中国高等植物图鉴》) 图版 51

Thalictrum acutifolium (Hand.-Mazz.) B. Boivin in Rhodora 46: 364, fig. 14: a-c. 1944; Iconogr. Corm. Sin. 1: 677, fig. 1354. 1972; W. T. Wang & S. H. Wang in Fl. Reip. Pop. Sin. 27: 553, pl. 137: 5–8. 1979; W. T. Wang in Bull. Bot. Lab. N.-E. Forest. Inst. 8: 29. 1980; S. R. Lin & x. Z. Zhao in Fl. Fujien. 2: 18. 1985; X. W. Wang in Fl. Anhui 2: 324, fig. 627. 1987; Y. K. Li in Fl. Guizhou. 3: 73, pl. 29: 4. 1990; W. T. Wang in Fl. Guangxi 1: 276, pl. 118: 46. 1991; Z. H. Lin in Fl. Zhejiang 2: 375, fig. 2–364. 1992; W. T. Wang in Keys Vasc. Pl. Wuling Mount. 165. 1995; K. M. Liu, Fl. Hunan 2: 654, fig. 2–510. 2000; W. T. Wang in High. Pl. China 3: 471, fig. 740: 5–8. 2000; D. Z. Fu & G. Zhu in Fl. China 6: 285. 2001; R. J. Wang in Fl. Guangdong 5: 19, fig. 14. 2003; W. T. Wang in Fl. Jiangxi 2: 176, fig. 183. 2004; Z. C. Luo & Y. B. Luo, Pl. Xinning 142. 2008; C. X. Yang et al. Keys Vasc. Pl. Chongqing 216. 2009; L. Q. Li et al. Wild Pl. Huangjing For. Gulin Xian Sichuan Prov. 165. 2015; W. T. Wang in High. Pl. China in Colour 3: 389. 2016; L. Xie in Med. Pl. China 3: 255, with fig. & photo. 2016. —*T. clavatum* DC. var. *acutifolium* Hand.-Mazz. in Sitzungsb. Akad, Wiss. Math. Naturw. Wien 43: 1. 1926; et Symb. Sin. 7: 310. 1931, p.p. excl, syn. *T. tuberiferum* Maxim. Holotype: 湖南：武岗，云山，alt. 600–1300 m, H. Handel-Mazzetti 11173 (not seen; photo, PE).

T. clavatum DC. var. *cavaleriei* Lévl. In Repert. Sp. Nov. Reg. Veg. 7: 357. 1909. Holotype: 贵州：Ma-jo, 1908–05, Cavalerie 3003 (E, not seen).

T. declinatum B. Boivin in Rhodora 46: 364, fig. 13: a–b. 1944. Type: 贵州：都匀, alt. 880 m, 1930-07-07，蒋英 5662（holotype, NY）.

T. unguiculatum B. Boivin in l. c. 365, fig. 16: a–b. Type: 贵州：都匀, alt. 880 m, 1930-07-07，蒋英 5662（holotype, GH, not seen）.

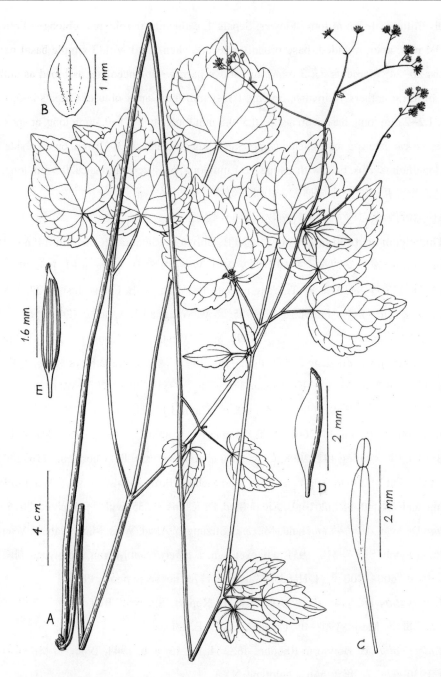

图版51. **尖叶唐松草** A. 植株全形，B. 萼片，C. 雄蕊，D. 心皮，E. 瘦果。（A、C、D 据李明红，旷月秋989，B据北京队648，E据黄向旭等10266）

Plate 51. **Thalictrum acutifolium** A. habit, B. sepal, C. stamen, D. carpel, E. achene. (A、C、D from M. H. Li & Y. Q. Kuang 989, B from Beijing Exped. 648, E from X. X. Huang et al. 10266).

T. chiaonis B. Boivin in 1. c. 368, fig 21: a–b. Type: 江西：庐山，牯岭，alt. 3400 m, 1928-07-27, 焦启源 18719（holotype, NY; isotype, NAS）；庐山，1922-07-17, A. N. Steward 2495（paratype, N, NY）。安徽：黄山，1925-07-15，秦仁昌 8593（paratype, US, not seen）。

多年生草本植物，全部无毛，有时小叶下面疏被短柔毛（重庆南部和贵州的一些居群）。茎高 25–65 cm，中部以上分枝。基生叶 2–3，为 2 回三出复叶，有长柄；叶片长 7–18 cm；小叶草质，卵形或宽卵形，1–5×0.7–3 cm，顶端急尖或钝，基部圆形、圆楔形或心形，不分裂或不明显 3 浅裂，边缘有疏牙齿，脉下面稍隆起；小叶柄长 0.6–2.5 cm；叶柄长 10–20 cm。茎生叶较小，有短柄。复单歧聚花序顶生，伞房状，宽 2–7 cm，有 5–20 花；花梗长 3–8 mm。花：萼片 4，早落，白色或带粉红色，卵形，约 1.8×1.2 mm，顶端微尖，每一萼片具 3 条不分枝的基生脉。雄蕊 15–20；花丝长 2–5.5 mm，上部倒披针形，宽 0.3–0.6 mm，下部丝形，粗 0.1 mm；花药长圆形或宽长圆形，0.8–1.3×0.3–0.6 mm，顶端钝。心皮 2–12；子房狭椭圆形，约 2×0.8 mm，基部具长约 0.5 mm 的柄；花柱长约 0.6 mm；柱头长约 0.5 mm，瘦果长椭圆形，有时稍镰状弯曲，3–3.8（–4.5）×0.6–0.8（–1.2）mm，每侧有 3 条细纵肋，基部具长 1–2.5 mm 的柄；宿存柱头长约 0.2 mm。4–7 月开花。

Perennial herbs, totally glabrous, in some populations found in southern Chongqing Shi and Guizhou Province, the leaflets abaxially sparsely puberulous. Stems 25–60 cm tall, above the middle branched. Basal leaves 2–3, 2-ternate, long petiolate; blades 7–18 cm long; leaflets herbaceous, ovate or broad-ovate, 1–5×0.7–3 cm, at apex acute or obtuse, at base rounded, rounded-cuneate or cordate, undivided or inconspicuously 3-lobed, margin sparsely dentate, nerves abaxially slightly prominent; petiolules 0.6–2.5 cm long; petioles 10–20 cm long. Cauline leaves smaller, shortly petiolate. Compound monochasia terminal, corymbiform, 2–7 cm broad, 5–20-flowered; pedicels 3–8 mm long. Flower: Sepals 4, caducous, white or tinged with pink, ovate, ca. 1.8×1.2 mm, apex slightly acute, each sepal with 3 simple basal nerves. Stamens 15–20; filaments 2–5.5 mm long, above oblanceolate, 0.3–0.6 mm broad, below filiform, 0.1 mm across; anthers oblong or broad-oblong, 0.8–1.3×0.3–0.6 mm, apex obtuse. Carpels 2–12; ovary narrow-elliptic, ca. 2×0.8 mm, base with a stipe ca. 0.5 mm long; style ca. 0.6 mm long; stigma ca. 0.5 mm long. Achene long elliptic, sometimes slightly falcate, 3–3.8(–4.5)×0.6–0.8(–1.2)mm, on each side with 3 thin longitudinal ribs, base with a stipe 1–3.5 mm long; persistent stigma ca. 0.2 mm long.

分布于云南东北部、贵州、广西、广东、福建、浙江、安徽南部、江西、湖南、湖北西部、重庆、四川和陕西东南部。生山谷坡地、沟边或林边湿润处，海拔 300–2800 m。

全草、根和根状茎均可供药用，用于痢疾、目赤肿痛、湿热黄疸等症。根含氧化小檗碱 (oxyberberine) 等生物碱。(《中国药用植物志》, 2016)

标本登录：云南：大关，1700 m，孙必兴 759。

贵州：安龙，1300 m，张志松，张永田 3020；兴仁，张志松，张永田 8515；雷山，黔南队 989，简焯坡等 50648，51085；凯里，1350 m，黔南队 1014，1666；施秉，黔南队 2655；印江，武陵山队 2034；梵净山 1100–2200 m，钟补勤 1781，张无休 20，武陵山队 1186；正安，1670 m，西部科学院 3272。

广西：融水，九万山，1000 m，陈少卿 14785，14985；融水，秋芽山，吕清华 2544；融水，元宝山，1200 m，吕清华 2846；三防，九峰山，秦仁昌 5850；临桂，陈照宙 50812；龙胜，广福林区队 286，778；兴安，猫儿山，广西队 53-419。

广东：乐昌，左景烈 20651，高锡朋 54584；乳源，高锡朋 52521；连南，350 m，王兆芬 120161，曾飞燕 475；饶平，陈念劬 42810；深圳：刘全儒 1155。

福建：上杭，林来官 3562；Yenping, 1925-06-08, H. H. Chung 3256；仙游，林英 454；南平，300–600 m，林镕 2246，何国生 370，640，李振宇 11187；顺昌，李明生，李振宇 5416，5584；泰宁，李明生 107；崇安，540–700 m，236 组 929，武夷山队 257；武夷山，1800 m，简焯坡等 400048，400619。

浙江：开化，植物资源队 59-26212，59-26641；遂昌，钟观光 s. n.；云和，钟观光 343；平阳，? 8361；瑞安，章绍尧 6626；乐清，章绍尧 5308；泰顺，章绍尧 5620；雁宕，钟观光 s.n.；天台山，? 39；昌化，贺贤育 22752。

安徽：金寨，覃海宁，于胜祥 20053；潜山，天柱山，沈保安 399；黄山，秦仁昌 4203，刘慎谔，钟补求 2642b，关克俭 75-276。

江西：龙南，236 组 1113；寻邬，江西师院队 1305；井冈山，江西药物所 86-4206；赣县，杨志斌，姚淦 1041；宜春，明月山，1530 m，岳俊三 3340；武功山，1800 m，江西队 54-1129，岳俊三 3694；靖安，谭策明 95-351；武宁，600 m，赖书绅 2740；庐山，五老峰，1400 m，王文采 s. n.，刘勇 1。

湖南：汝城，黄向旭 10266；宜章，莽山，肖吉中 3612；新宁，万峰，800 m，罗毅波 2335；武岗，云山，1020 m，邓云飞 10244；衡山，胡忠辉 501，旷月秋 989；浏阳，大围山实习队 09-6045；临湘，? 678；永顺，300 m，北京队 648；桑植，张兵，向新 425040。

湖北：咸丰，王映明 6622；宣恩，1270 m，王映明 4081，4269，鹤峰，王映明 5214。

重庆：南川，金佛山，1750 m，李国凤 62899，熊济华，周子林 91411。刘正宇 449，1680；武隆，易思荣 357；酉阳，西南师院队 2565；石柱，曹亚玲，溥发鼎 908，三峡队 1637。

四川：金阳，2800 m，植物资源队 59-3098；古蔺，2200 m，古蔺队 0723。

陕西：镇坪，陈彦生 369（WUK）。

52. **盾叶唐松草**（《东北植物检索表》） 图版 52

Thalictrum ichangense Lecoy. ex Oliv. in Hook. Ic. Pl. 18: pl. 1765. 1888; Finet & Gagnep. in Bull. Soc. Bot. France 50: 605. 1903; Finet in J. de Bot. 21: 19. 1908; Hand.-Mazz. Symb. Sin. 7: 310. 1931; Kitag. Lineam. Fl. Mansh. 227. 1939; Boivin in Rhodora 46: 368. 1944; C. Pei & T. Y. Chou, Icon. Med. Pl. China 7: fig. 325. 1964; Iconogr. Corm. Sin. 1: 678, fig. 1355. 1972; Fl. Tsinling. 1(2): 243. 1974; S. H. Li in Fl. Pl. Herb. Chinae Bor.-Or. 3: 209, pl. 91. 1975; S. H. Fu, Fl. Hupeh. 1: 353, fig. 492. 1976; W. T. Wang & S. H. Wang in Fl. Reip. Pop. Sin. 27: 555, pl. 138: 68. 1979; B. Z. Ding et al. Fl. Henan. 1: 467, fig. 546. 1981; W. T. Wang in Bull. Res. Bot. Harbin 6(1): 23. 1986; Y. K. Li in Fl. Guizhou. 3: 68, Pl. 28: 4-8. 1990; W. T. Wang in Fl Guangxi 1: 276. 1991; Z. H. Lin in Fl. Zhejiang 2: 374, fig. 2-363. 1992; W. T. Wang in Keys Vasc. Pl. Wuling Mount. 165. 1995; K. M. Liu, Fl. Hunan 2: 656, fig. 2-502. 2000; W. T. Wang in Fl. Yunnan. 11: 164. 2000; et in High. Pl. China 3: 472. 2000; D. Z. Fu & G. Zhu in Fl. China 6: 292. 2001; R. J. Wang in Fl. Guangdong 5: 19. 2003; Y. M. Shui & W. H. Chen, Seed Pl. Honghe Reg. S.-E. Yunnan 78. 2003; Z. Y. Li & M. Ogisu, Pl. Mount Emei 255. 2007; C. X. Yang et al. Keys Vasc. Pl. Chongqing 215. 2009; Q. L. Gan, Fl. Zhuxi Suppl. 242. 2011; L. Q. Li in High. Pl. Daba Mount. 117. 2014; L. Q. Li et al Pl. Baishuijiang st. Nat. Reserve Gansu Prov. 71. 2014; W. T. Wang in High. Pl. China in Colour 3: 389. 2016; L. Xie in Med. Fl. China 3: 256, with fig. & photo. 2016. Holotype：湖北：宜昌，Nanto, 1887-10，A. Henry 3932 (K, not seen; photo, PE).

T. tripeltatum Maxim. in Acta Hort. Petrop. 11: 13. 1890. Type：四川：Inter vicos Siao-pu et I-tang, 1885-08-18, G. N. Potanin s. n. (holotype, LE; isotype, PE).

T. coreanum Lévl. in Bull. Acad. Geogr. Bot. 11: 297. 1902; Finet & Gagnep. in Bull. Soc. Bot. France 50: 635. 1903; Yabe, Enum. Pl. S. Manch. 53. 1912; Boivin in Rhodora 46: 368. 1944; et in J. Arn. Arb. 26: 116. 1945; Kitag. in J. Jap. Bot. 36: 18. 1961; Lauener & Green in

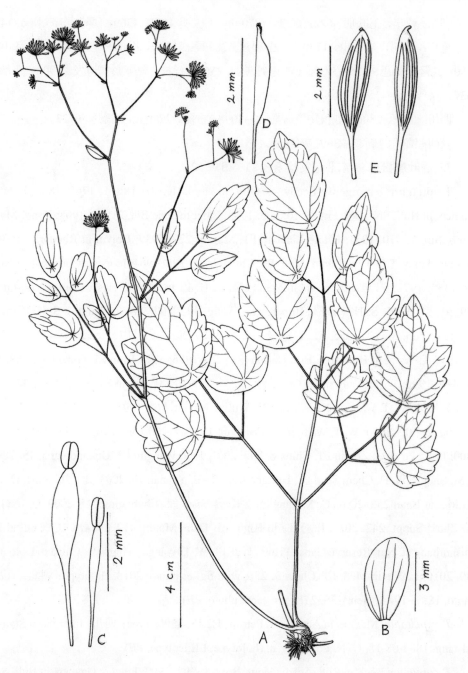

图版52. **盾叶唐松草**　A. 植株全形，B. 萼片，C. 二雄蕊，D. 心皮（据巴山队5112），E. 二瘦果（据巴山队1051）。

Plate 52. **Thalictrum ichangense**　A. habit, B. sepal, C. two stamens, D. carpel (from Bashan Exped. 5112), E. two achenes (from Bashan Exped. 1051).

Not. Bot. Gard. Edinburgh 23: 591. 1961. —*T. ichangense* race *coreanum* (Lévl.) Lévl. in Repert. Sp. Nov. Reg. Veg. 7: 100. 1909. —*T. ichangense* var. *coreanum* (Lévl.) Lévl. ex Tamura in Acta Phytotax. Geobot. 15: 83. 1953; W. T. Wang in Bull. Res. Bot. Harbin 6(1): 23. 1986; S. H. Li in Clav. Pl. Chinae Bor.-Or., ed. 2: 207, pl. 96: 2. 1995; Y. J. Zheng in Fl. Shandong 2: 30, fig. 22. 1997. Holotype: Korea: in montibus interioris, 1901–09, Faurie 23 (E, not seen).

Isopyrum multipeltatum Pamp. in Nour. Gior. Bot. Ital., n. ser., 18: 115. 1911.

Thalictrum multipeltatum Pamp. In l. c. 167.

多年生草本，全部无毛。根状茎有密集须根和纺锤形小块根。茎高 14-32 cm，不分枝或上部分枝。基生叶为 1-3 回三出复叶，长 8-25 cm；叶片长 4-14 cm；小叶草质或纸质，卵形、宽卵形、宽椭圆形或近圆形，2-4×1.5-4 cm，顶端钝或圆形，基部盾状，3 浅裂或不分裂，边缘有疏钝齿，脉平；小叶柄长 0.5-3 cm；叶柄长 5-12 cm。茎生叶 1-3，较小。复单歧聚伞花序顶生或腋生，伞房状，宽 3-9 cm，有 5 至多数花；花梗丝形，长 0.3-2 cm。花：萼片 4，早落，白色，椭圆状倒卵形，约 3.5×1 mm，顶端圆形，每一萼片具 3 条基生脉（中脉和 1 侧生脉不分枝，另一侧生脉二歧状分枝）。雄蕊 15-30；花丝长 1.8-4.8 mm，上部狭倒披针形，宽 0.3-0.6 mm，下部丝形，粗 0.1 mm；花药宽椭圆形或长圆形，0.5-0.6×0.25-0.4 mm，顶端钝。心皮 5-12（-16）；子房斜狭倒卵形，约 2×0.4 mm，基部具长 1 mm 的柄；花柱长约 0.4 mm；柱头狭倒卵形，长 0.4 mm。瘦果纺锤形，2.2-2.6×0.6-1 mm，每侧有 2-3 条细纵肋，基部具长 1.5-1.8 mm 的细柄；宿存柱头长 0.3 mm。5-8 月开花。

Perennial herbs, totally glabrous. Rhizome with dense fibrous roots and small fusiform root tubers. Stems 14–32 cm tall, simple or above branched. Basal leaves 1–3-ternate, 8–25 cm long; blades 4–14 cm long; leaflets herbaceous or papery, ovate, broad-ovate, broad-elliptic or suborbicular, 2–4×1.5–4 cm, at apex obtuse or rounded, at base peltate, 3-lobed or undivided, margin sparsely obtusely dentate, nerves flat; petiolules 0.5–3 cm long; petioles 5–12 cm long. Cauline leaves 1–3, smaller. Compound monochasia terminal or axillary, corymbiform, 3–9 cm broad, 5-many-flowered; pedicels filiform, 0.3–2 cm long. Flower: Sepals 4, caducous, white, elliptic-obovate, ca. 3.5×1 mm, apex rounded, each sepal with 3 basal nerves (midrib and 1 lateral nerve simple, the another lateral nerve dichotomizing). Stamens 15–30; filaments 1.8–4.8 mm long, above narrow-lanceolate, 0.3–0.6 mm broad, below filiform, 0.1 mm across; anthers broad-elliptic or oblong, 0.5–0.6×0.25–0.4 mm, apex obtuse. Carpels 5–12(–16);

ovary obliquely narrow-obovate, ca. 2×0.4 mm, base with a stipe 1 mm long; style ca. 0.4 mm long; stigma narrow-obovate, 0.4 mm long. Achene fusiform, 2.2–2.6×0.6–1 mm, on each side with 2–3 thin longitudinal ribs, base with a slender stipe 1.5–1.8 mm long; persistent stigma 0.3 mm long.

分布于中国的云南中部和东部、贵州、广西、广东北部、湖南、浙江南部、湖北西部、重庆、四川、陕西东南部、山东东部和辽宁东南部；以及朝鲜和越南北部。生山地沟边、灌丛中或林中，在四川、湖北一带分布于海拔 1300–1900 m 间山区，在辽宁分布于海拔 600 m 低山区。

全草或根供药用，用于丹毒、跌打损伤等症。全草含去氢海罂粟碱（dehydroglaucine）等生物碱。(《中国药用植物志》，2016)

标本登录：云南：石林，税玉民等 64781，65266；砚山，王启无 83622；马关，税玉民等 32200；屏边，王启无 82423；河口，税玉民 3548。

贵州：兴义，张永田，张志松 7040；安龙，张永田，张志松 4905；贵阳，何顺志 s.n.；德江，安明志 3322；务川，钟补勤 659; Ma-jo, Ling-ly. J. Cavalerie 2431。

广西：马山，钟树权 462227；环江，P. J. Cribb 等 ASBK417；融水，王文采 s. n.；融水，元宝山，侯满福 1；融安，P. J. Cribb 等 ASBK523；桂林，李光照 13789；临桂，李光照 14901；阳朔，单人骅 780。

广东：阳山，邓良 1501。

湖南：沅陵，武陵山队 213；桑植，天平山，北京队 4533；石门，壶瓶山队 219，432，806，1099。

浙江：景宁，章绍尧 5263。

湖北：宣恩，李洪钧 3110，3170；建始，王荣 11；巴东，周鹤昌 562，傅国勋 1183。

重庆：南川，植物资源队 59-122；巫山，杨光辉 58726，三峡队 1413；巫溪，杨光辉 59402；城口，巴山队 1051。

四川：雷波，管中天 8484；峨眉山，彭新伯 6048，王蜀秀 463；天全，俞德浚 1826；宝兴，张秀实，任有铣 4782，5061；成都，汪发缵 22134；旺苍，巴山队 4909，5112，5420；万源，李培元 6125。

陕西：山阳，杨金祥 2593。

山东：海阳，山东中医研究所 10。

辽宁：千山，刘慎谔等 604；岫岩，朱有昌等 859；草河口，野光田藏等 609。

53. 深山唐松草（《东北植物检索表》） 图版 53

Thalictrum tuberiferum Maxim. in Bull. Acad. Sci. St.-Petersb. 22: 227. 1876; Lecoy. in Bull. Soc. Bot. Belg. 24: 161. 1885; Kom. & K.-Alisova, Key Pl. Far. East USSR 1: 554. 1931; Nevski in Fl. URSS 7: 515. 1937; Kitag. Lineam. Fl. Mansh. 228. 1939; Boivin in Rhodora 46: 367, fig. 19. 1944; et in J. Arn. Arb. 26: 117. 1945; S. H. Li in Fl. Pl. Herb. Chinae Bor.-Or. 3: 206, pl. 90: 1–4. 1975; W. T. Wang & S. H. Wang in Fl. Reip. Pop. Sin. 27: 556, pl. 138: 1–5. 1979; S. H. Li in Clav. Pl. Chinae Bor.-Or., ed. 2: 205, pl. 96: 4. 1995; W. T. Wang in High. Pl. China 3: 472, fig. 75. 2000; D. Z. Fu & G. Zhu in Fl. China 6: 300. 2001; Y. Kadota in I watsuki et al. Fl. Japan 2a: 335. 2006; L. Xie in Med. Fl. China 3: 257, with fig. & photo. 2016. Described from the Ol'ga Gulf area.

多年生草本植物，全部无毛。须根有纺锤形小块根。茎高 50–70 cm，上部分枝。基生叶 1，通常为 3 回三出复叶，长 25–30 cm；叶片长 13–23 cm；小叶草质，卵形或菱状椭圆形，3.5–4.5×2.5–3.5 cm，基部圆形或浅心形，3 浅裂或不分裂，边缘有少数钝牙齿，脉下面稍隆起；小叶柄长 1–2.8 cm；叶柄长 11–19 cm。茎生叶小，2，对生，有时为 1 或 2 回三出复叶。复单歧聚伞花序顶生，伞房状，宽 4–7 cm；下部苞片三出；花梗长 5–9 mm。花：萼片早落，椭圆形，长约 2 mm。雄蕊约 35；花丝长 4–6 mm，上部倒披针形，宽 0.5–1 mm，下部丝形，粗 0.1 mm；花药长圆形，0.5×0.2–0.4 mm。心皮 3–5；子房椭圆形，约 2×0.8 mm，基部具长 0.8 mm 的柄；花柱长约 0.2 mm；柱头狭倒卵形，长约 0.3 mm。瘦果斜狭椭圆形，3.5–4×1.5 mm，每侧有 3 条细纵肋，基部具长 1.6 mm 的柄；宿存柱头长约 0.3 mm。6–7 月开花。

Perennial herbs, totally glabrous. Fibrous roots with small fusiform root tubers. Stems 50–70 cm, above branched. Basal leaf 1, usually 3-ternate, 25–30 cm long; blades 13–23 cm long; leaflets herbaceous, ovate or rhombic-elliptic, 3.5–4.5×2.5–3.5 cm, at base rounded or subcordate, 3-lobed or undivided, margin obtusely few-dentate, nerves abaxially slightly prominent; petiolules 1–2.8 cm long; petioles 11–19 cm long. Compound monochasia terminal, corymbiform, 4–7 cm broad; pedicels 5–9 mm. Flower: Sepals caducous, elliptis ca. 2 mm long. Stamens ca. 35; filaments 4–6 mm long, above oblanceolate, 0.5–1 mm broad, below filiform, 0.1 mm across; anthers oblong, 0.5×0.2–0.4 mm. Carpels 3–5; ovary elliptic, ca. 2×0.8 mm, base with a stipe 0.8 mm long; style ca. 0.2 mm long; stigma narrow-obovate, ca. 0.3 mm long. Achene obliquely narrow-elliptic, 3.5–4×1.5 mm, on each side with 3 thin

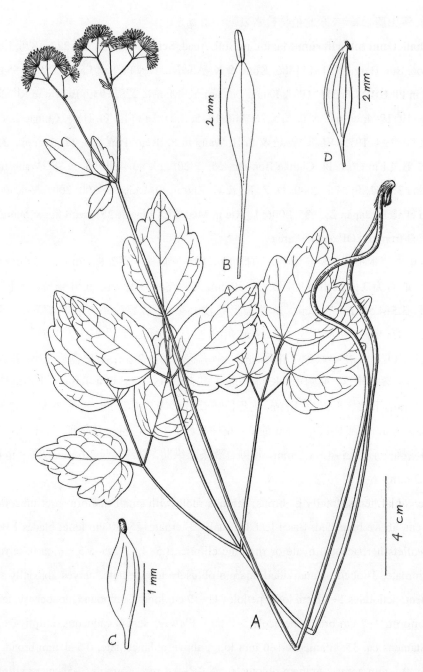

图版53. **深山唐松草** A. 开花植株，B. 雄蕊，C. 心皮（据温带森林组98），D. 瘦果（据曹子余等240）。

Plate 53. **Thalictrum tuberiferum** A. flowering plant, B. stamen, C. carpel (from Temperate zone forest group 98), D. achene (from Z. Y. Cao et al. 240).

longitudinal ribs, base with a stipe 1.6 mm long; persistent stigma ca. 0.3 mm long.

分布于辽宁、吉林、黑龙江；以及俄国远东地区、朝鲜、日本。生山地草坡或灌丛边，海拔 780–1100 m。

全草和根可供药用，用于水肿、肾结石等症。全草含小檗碱（berberine）等生物碱。（《中国药用植物志》，2016）

标本登录：吉林：安图，长白山，温带森林组 98，261，王薇等 2010，俞德浚，陆玲娣 75-186，孔宏智 96-026，曹子余等 240；抚松，周以良等 305，洪德元等 33337，33932。

黑龙江：1860-06，C. J. Maximowicz. s. n.

54. 花唐松草（《东北植物检索表》） 图版 54

Thalictrum filamentosum Maxim. Prim. Fl. Amur. 13. 1859; Lecoy. in Bull. Soc. Bot. Belg. 24: 159. 1885; Kom. & K.-Alisova, Key Pl. Far East USSR 1: 553. 1931; Nevski in Fl. URSS 7: 516. 1937; Kitag. Lineam. Fl. Mansh. 227. 1939; Boivin in Rhodora 46: 363, fig. 11. 1944; S. H. Li in Fl. Pl. Herb. Chinae Bor.-Or. 3: 209, pl. 90: 5–7. 1975; W. T. Wang & S. H. Wang in Fl. Reip Pop. Sin. 27: 556, pl. 136: 7–8. 1979; S. H. Li in Clav. Pl. Chinae Bor.-Or., ed. 2: 205, pl. 96: 5. 1995; W. T. Wang in High. Pl. China 3: 472. 2000; D. Z. Fu & G. Zhu in Fl. China 6: 289. 2001. — *T. clavatum* DC. var. *filamentosum* (Maxim) Finet & Gagnep. in Bull. Soc. Bot. France 50: 605. 1903. Type: Amur, 1855-05-24, C. J. Maximowicz s. n. (holotype, LE, not seen; isotype, PE).

多年生草本植物。须根有球形或椭圆球形小块根。茎高 15–36 cm，无毛。基生叶 1，为 2 回三出复叶，长 20–25 cm；叶片长 6.5–9 cm；小叶草质，宽卵形或圆卵形，2–5×1.7–4.5 cm，基部心形或圆形，不明显 3 浅裂，边缘有少数钝齿，无毛或下面有短柔毛，脉近平；小叶柄长 1.5–2.8 cm；叶柄长 7–13 cm。茎生叶为单叶，2 枚对生，无柄，宽卵形或卵形，长 3.2–3.6 cm，边缘每侧有 2 钝齿。复单歧聚伞花序顶生，伞房状，长 1.5–5.5 cm；花梗长 5–15 mm。花：萼片 4，早落，狭倒卵形，约 2×1 mm，顶端圆形，每一萼片只有 1 条中脉，无侧生脉。雄蕊约 20；花丝长 2–4 mm，上部倒披针形，宽 0.25–0.8 mm，下部狭条形，宽约 0.15 mm；花药长圆形或宽长圆形，0.5–0.6×0.2–0.4 mm。心皮 2–5；子房新月形，约 3.6×1.2 mm，基部有长约 1.5 mm 的柄；花柱长约 0.2 mm；柱头长 0.25–0.5 mm。5 月开花。

Perennial herbs. Fibrous roots with small globose or ellipsoid root tubers. Stems 15–36 cm tall, glabrous. Basal leaf 1, 2-ternate, 20–25 cm long; blades 6.5–9 cm long; leaflets

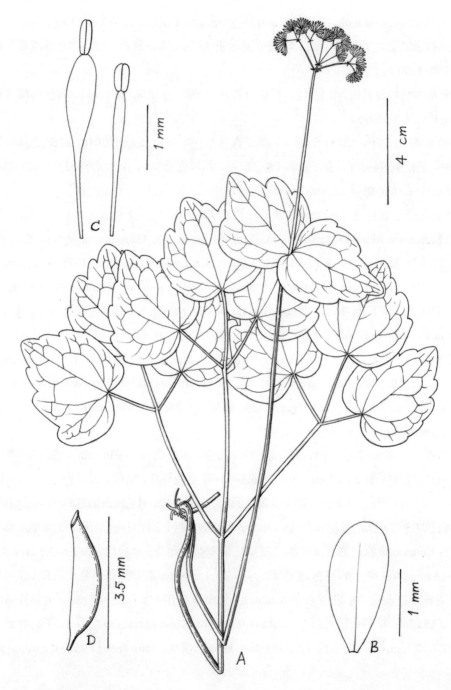

图版54. **花唐松草** A. 开花植株，B. 萼片，C. 二雄蕊，D. 心皮（据C. J. Maximowicz s. n.）。
Plate 54. **Thalictrum filamentosum** A. flowering plant, B. sepal, C. two stamens, D. carpel (from C. J. Maximowicz s, n.).

herbaceous, broad-ovate or orbicular-ovate, 2–5×1.7–4.5 cm, at base cordate or rounded, inconspicuously 3-lobed, at margin obtusely few-dentate, glabrous or abaxially puberulous, nerves nearly flat; petiolules 1.5–2.8 cm long; petioles 7–13 cm long. Cauline leaves simple, 2 opposite, sessile, broad-ovate or ovate, 3.2–3.6 cm long, margin at each side obtusely 2-dentate. Compound monochasia terminal, corymbiform, 1.5–5.5 cm long; pedicels 5–15 mm long. Flower: Sepals 4, cadacous, narrow-obovate, ca. 2×1 mm, apex rounded, each sepal with only 1 midrib, without lateral nerves. Stamens ca. 20; filaments 2–4 mm long, above oblanceolate, 0.25–0.8 mm, below narrow-linear, ca. 0.15 mm broad; anthers oblong or broad-oblong, 0.5–0.6×0.2–0.4 mm. Carpels 2–5; ovary lunate, ca. 3.6×1.2 mm, base with a stipe 0.4 mm long; style ca. 0.2 mm long; stigma 0.25–0.5 mm long.

分布于中国吉林、黑龙江以及俄国远东地区。生山地林中或灌丛中。

55. 小果唐松草（《药学学报》） 图版 55

Thalictrum microgynum Lecoy. ex Oliv. in Hook. Ic. Pl. 18: pl. 1766. 1886; Boivin in Rhodora 46: 365, fig. 15 1944; S. H. Fu, Fl. Hupeh. 1: 358, fig. 501.1976; W. T. Wang & S. H. Wang in Fl. Reip. Pop. Sin. 27: 558, pl. 138: 9–12. 1979; W. T. Wang in Bull. Bot. Lab. N.-E. Forest. Inst. 8: 30. 1980; in Iconogr. Corm. Sin. Suppl. 1: 426, fig. 8615. 1982; in Vasc. Pl. Hengduan Mount. 1: 501. 1993; in Keys Vasc. Pl. Wuling Mount. 165. 1995; in Fl. Yunnan. 11: 166. 2000; et in High. Pl. China 3: 472, fig. 751. 2000; K. M. Liu, Fl. Hunan 2: 654, fig. 2–509. 2000; D. Z. Fu & G. Zhu in Fl. China 6: 293. 2001; Z. Y. Li & M. Ogisu, Pl. Mount Emei 255. 2007; C. X. Yang et al. Keys Vasc. Pl. Chongqing 216. 2009; Q. L. Gan, Fl. Zhuxi Suppl. 245, fig. 294. 2011; L. Q. Li in High. Pl. Data Mount. 118. 2014; W. T. Wang in High. Pl. China in Colour 3: 389. 2016; L. Xie in Med. Fl. China 3: 258, with fig. & photo. 2016. Holotype: 湖北：宜昌，Nanto, 1885–1888, A. Henry3992（K, not seen; photo, PE）.

T. scaposum W. E. Evans in Not. Bot. Gard. Edinburgh 13: 187. 1921. Holotype：云南：怒江－澜沧江分水岭，1919–07，G. Forrest 18194 (E, not seen; photo, PE).

多年生草本植物，全部无毛。须根有倒圆锥形的小块根。茎高 20–42 cm，上部分枝。基生叶 1，为 2–3 回三出复叶；叶片长 10–15 cm；小叶膜质，楔状倒卵形、菱形或卵形，2–6.4（–9.5）×1.5–3.8（–4.8）mm，3 浅裂或不分裂，边缘有粗齿，脉平，不明显；小叶柄长 0.5–3.5 cm；叶柄长 8–15 cm。茎生叶 1–2，似基生叶，但较小。复单歧聚伞花序顶生，似伞形花序，具密集的花；苞片椭圆形或长椭圆形，长 1.5–16 mm；花梗丝形，长达 1.5 cm。

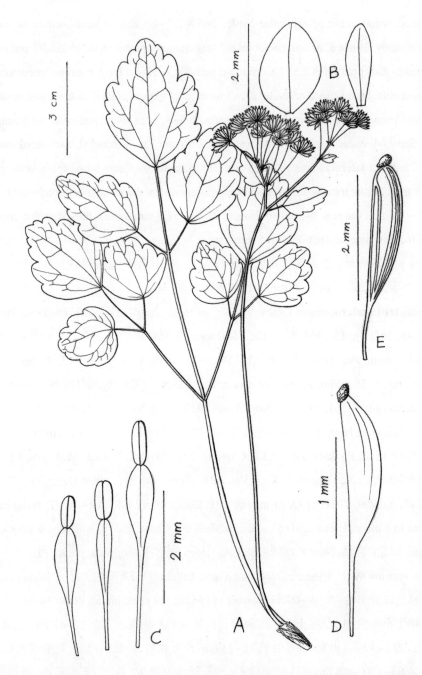

图版55. **小果唐松草** A. 开花植株,B. 二萼片,C. 三雄蕊,D. 心皮(据刘瑛411),E. 瘦果(据罗毅波,向巧萍198)。

Plate 55. **Thalictrum microgynum** A. flowering plant, B. two sepals, C. three stamens, D. carpel (from Y. Liou 411), E. achene (from Y. B. Luo & Q. P. Xiang 198).

花：萼片 4，早落，白色，宽椭圆形或长圆形，1.6-2×0.6-1 mm，顶端钝，每一萼片只具 1 条中脉，无侧生脉。雄蕊约 8；花丝长 2-3 mm，上部倒披针形，宽 0.3-0.9 mm，下部丝形，粗 0.1 mm；花药长圆形或宽长圆形，0.6-0.8×0.3-0.5 mm，顶端钝。心皮 4-16；子房半倒卵形，1.5-1.8×0.4 mm，基部具长 0.8-1 mm 的柄；花柱长约 0.1 mm；柱头椭圆形，长约 0.15 mm。瘦果下垂，倒卵形，约 2×0.6 mm，每侧有 2-3 条细纵肋，基部具长 0.8-1.2 mm 的柄；宿存柱头长约 0.2 mm。4-7 月开花。

Perennial herbs, totally glabrous. Fibrous roots with small obconical root tubers. Stems 20-42 cm tall, above branched. Basal leaf 1, 2-3-ternate; blades 10-15 cm long; leaflets membranous, cuneate-obovate, rhombic or obvate, 2-6.4(-9.5) ×1.5-3.8(-4.8)cm, 3-lobed or undivided, margin coarsely dentate, nerves flat, inconspicuous; petiolules 0.5-3.5 cm long; petioles 8-15 cm long. Cauline leaves 1-2, similar to basal leaf, but smaller. Compound monochasia terminal, umbelliform, with dense flowers; bracts elliptic or long elliptic, 1.5-16 mm; pedicels filiform, up to 1.5 cm long. Flower: Sepals 4, caducous, white, broad-elliptic or oblong, 1.6-2×0.6-1 mm, apex obtuse, each sepal with only 1 midrib, without lateral nerves. Stamens ca. 8; filaments 2-3 mm long, above oblanceolate, 0.3-0.9 mm broad, below filiform, 0.1 mm across; anthers oblong or broad-oblong, 0.6-0.8×0.3-0.5 mm, apex obtuse. Carpels 4-16; ovary semi-obovate, 1.5-1.8×0.4 mm, base with a stipe 0.8-1 mm long; style ca. 0.1 mm long; stigma elliptic, ca. 0.15 mm long. Achene pendulous, obovate, ca. 2×0.6 mm, on each side with 2-3 thin longitudinal ribs, base with a stipe 0.8-1.2 mm long; persistent stigma ca. 0.2 mm long.

分布于云南西部、贵州、四川、重庆，湖北西北部、湖北西部和陕西南部。生山地林下，草坡，石崖上或石边较阴湿处，海拔 700-2800 m。

全草有清热解毒之效，用于黄疸等症。全草含箭头唐松草米定碱（thalicsimidine）等生物碱。(《中国药用植物志》，2016)

标本登录：云南：腾冲，1820 m，Gaoligong Shan Biodiv. Survey 29570 (GH,PE)。

贵州：梵净山，2300 m，何顺志 94076；Lou-Kouang-kiao, 1907. J. Cavalerie 3003。

四川：峨眉山，杜大华 109；灌县，汪发缵 20559；南充，巴中队 5613；金阳，经济植物队 59-3098。

重庆：南川：曲桂龄 1214，李国凤 61763，62620，熊济华，周子林 90994，92319，刘正宇 1768，陈心启，郎楷永 2389；武隆，易思荣 252；酉阳，1500 m，西南师院队 2565；开县，川大生物系 100381；奉节，张泽荣 25182；周洪富，粟和毅 107806，108182；巫山，杨光辉 57986；巫溪，孔宏智 96-575，96-607；城口，戴天伦 105653。

湖南：龙山，刘林翰 1643，10159；桑植，刘林翰 9116，李丙贵 75-221；石门，壶瓶山队 806。

湖北：利川，罗毅波 198；宣恩，李洪钧 5261；巴东，鄂植考队 24111；兴山，老君山，刘瑛 411，656；神农架，236 部队六所 2499。

陕西：山阳，杨金祥 2579；佛坪，孔宏智 1013。

系 2. 拟盾叶唐松草系

Ser. **Pseudoichangensia** W. T. Wang, ser. nov. Type: *T. pseudoichangense* Q. E. Yang & G. Zhu

Series nova haec. a Ser. *Clavatis* W. T. Wang & S. H. Wang differt carpelli stylo elongato aciculari ovario aequilongo ventro superne textura stigmatica inconspicua praedito.

本系与尖叶唐松草系 Ser. *Clavata* 的区别在于本系心皮的花柱伸长，呈针形，与子房等长，腹面上部有不明显的柱头组织。

1 种，特产贵州。

56. 拟盾叶唐松草　图版 56

Thalictrum pseudoichangense Q. E. Yang & G. Zhu in Novon 14(4): 510, fig. 1. 2004. Holotype：贵州：无野外记录，产地及采集日期均不明，H. J. Esquirol s. n.（K，not seen）。

多年生草本植物，全部无毛。茎高 15-25 cm，上部有少数分枝。基生叶 2，为 2-3 回三出复叶，具长柄；叶片近三角形，6-12×5 cm；小叶薄草质，倒卵形、宽椭圆形或宽卵形，0.8-2×0.7-1.5 cm，顶端微尖，基部圆盾形，3 浅裂（裂片全缘，或中央裂片具 1-2 小齿），具三出脉，脉两面平；小叶柄长 1.7-3 cm；叶柄长 4-5 cm。茎生叶 3，似基生叶，但较小，1-2 回三出复叶。复单歧聚伞花序顶生并腋生，伞房状，宽约 4 cm，具少数花；花梗细，长 5-12 mm。花：萼片早落，未见。雄蕊长约 6 mm；花丝长约 4.5 mm，上部倒披形，宽约 0.5 mm，下部狭条形，宽约 0.15 mm；花药长圆形，约 0.5×0.2 mm，顶端钝。心皮 5-8；子房长约 3 mm，具 1 细柄；花柱针形，与子房等长，腹面上部有不明显柱头组织。瘦果狭纺锤形，约 3×0.5 mm，每侧有 3 条细纵肋，基部具长约 2 mm 的细柄；宿存花柱针形，长约 3 mm。

Perennial herbs, totally glabrous. Stems 15-25 cm tall, above few-branched. Basal leaves 2, 2-3-ternate, long petiolate; blades subtriangular, 6-12×5 cm; leaflets thin-herbaceous,

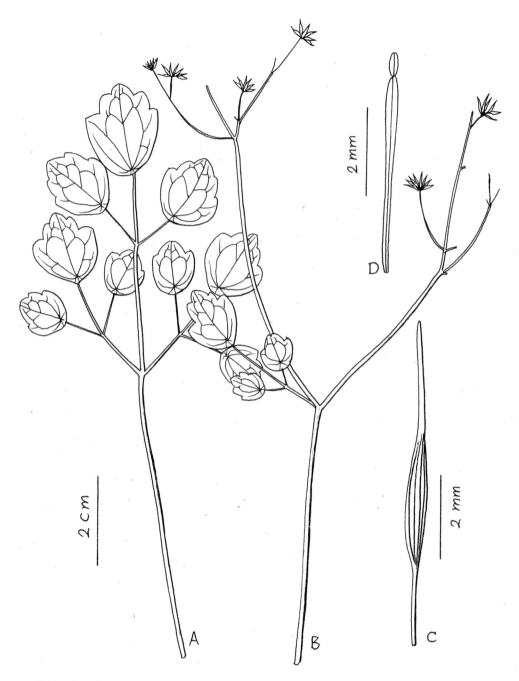

图版56. **拟盾叶唐松草** A.基生叶，B.结果植株上部，C.瘦果，D.雄蕊。（仿自原始文献插图，2014）
Plate 56. **Thalictrum pseudoichangense** A. basal leaf, B. upper part of fruiting plant, C. achene, D. stamen. (redrawn from the original figure, 2014)

obovate, broad-elliptic, or broad-ovate, 0.8–2×0.7–1.5 cm, at apex slightly acute, at base rounded-peltate, 3-lobed (lobes entire. or central lobe 1–2-denticulate), tri-nerved, nerves on both surfaces flat; petiolules 1.7–3 cm long; petioles 4–5 cm long. Cauline leaves 3, similar to basal leaves, smaller, 1–2-ternate. Compound monochasia terminal and axillary, corymbiform, ca. 4 cm broad, few-flowered; pedicels slender, 5–12 mm long. Flower: Sepals caducous, not seen. Stamens ca. 6 mm long; filaments ca. 4.5 mm long, above oblanceolate, ca. 0.5 mm broad, below nerrow-linear, ca. 0.15 mm broad; anthers oblong, ca. 0.5×0.2 mm, apex obtuse. Carpels 5–8; ovary ca. 3 mm long, base with a slender stipe; style acicular, as long as ovarg, ventrally above with inconspicuous stigmatic tissue. Achene narrow-fusiform, ca. 3×0.5 mm, on each side with 3 thin longitudinal ribs, base with a slender stipe ca. 2 mm long; persistent style acicular, 3 mm long.

特产贵州。

组 7. 唐松草组

Sect. **Tripterium** DC. Syst. 1: 169. 1817; et Prodr. 1: 11. 1824; Ledeb. Fl. Ross. 1: 5. 1842; Boivin in Rhodora 46: 375. 1944; Tamura in Sci. Rep. Osaka Univ. 17: 52. 1968; et in Hiepko, Nat. Pflanzenfam. 17a Ⅳ: 487. 1995; Emura in J. Fac. Sci. Univ. Tokyo, Sect. 3, Bot. 9: 128. 1972.—Subgen. *Tripterium* (DC.) Reichb. Consp. Reg. Veg. 192. 1828. —Ser. *Tripterea* W. T. Wang & S. H. Wang in Fl. Reip. Pop. Sin. 27: 547, 619. 1979. Type: *T. aquilegifolium* L.

花组成伞房状复单歧聚伞花序。萼片具 3 条基生脉。雄蕊花丝分化成上、下两部分。心皮具细柄，短花柱和小柱头。瘦果具细柄和 4 条纵翅。

Flowers in corymbiform compound monochasia. Sepal with 3 basal nerves. Stamen filament differentiated into upper and lower two parts. Carpel slenderly stipitate, with a short style and a small stigma. Achene slenderly stipitate, with 4 longitudinal wings.

1 种，广布欧洲和亚洲。中国有 1 变种。

57. **唐松草**（《中国植物图鉴》） 图版 57

Thalictrum aquilegifolium L. var. **sibiricum** Regel & Tiling, Fl. Ajan. 23. 1858; Hara in J. Fac. Sci. Univ. Tokyo, Sect. 3, Bot. 6: 55. 1952; Lauener & Green in Not. Bot. Gard. Einburgh 23(4):591. 1961; Iconogr. Corm. Sin. 1: 679, fig. 1357. 1972; S. H. Li in Fl. Pl.

Herb. Chinae Bor.-Or. 3: 202, pl. 87: 1–4. 1975; W. T. Wang & S. H. Wang in Fl. Reip. Pop. Sin. 27: 57, pl. 134. 1979; Tamura & Shimizu in Satake, Wild Flow. Pl. Japan. 2: 85, pl. 84: 2. 1984; W. T. Wang in Acta Bot. Yunnan. 4(2):377. 1984; J. W. Wang in Fl. Hebei 1: 441, fig. 437. 1986; D. Z. Ma, Fl. Ningxia. 1: 173. 1986; X. W. Wang in Fl. Anhui 2: 429, fig. 628. 1987; W. T. Wang in Bull. Bot. Res. Harbin 9(2): 6. 1989; Y. Z. Zhao in Fl. Intramongol., ed. 2: 451, pl. 182; 1990; X. Y. Yu et al. in Fl. Shanxi. 1: 596, fig. 378. 1992; Z. H. Lin in Fl. Zhejiang 2: 377, fig. 2–367. 1992; S. H. Li in Clav. Pl. Chinae Bor.-Or., ed. 2: 205. 1995; L. H. Zhou in Fl. Qinghai. 1: 330. 1997; Y. J. Zheng in Fl. Shandong 2: 31, fig. 24. 1997; K. M. Liu, Fl. Hunan 2: 652, fig. 2–508. 2000; W. T. Wang in High. Pl. China 3: 469, fig. 744. 2000; D. Z. Fu & G. Zhu in Fl. China 6: 286. 2001; Y. Kadota in Iwatsuki et al. Fl. Japan 2a:374. 2006; Y. Zhou, Pl. Resouc. Changbai Mount 225, pl. 116: 8, 11. 2010; Q. L. Gan, Fl. Zhuxi Suppl. 242. 2011; M. B. Deng & K. Ye in Fl. Jiangsu 2: 84, fig. 2–135. 2013; F. H. Li et al. Pl. Yanqing 144. 2015; W. T. Wang in High. Pl. China in Colour 3: 388. 2016; L. Xie in Med. Fl. China 3: 252, with fig. & photo. 2016.

T. contortum L. Sp. Pl. 547. 1753; DC. Syst. 1: 170. 1817; Nevski in Fl. URSS 7: 513, t. 5: 4. 1937; C. Pei & T. Y. Chou, Icon. Chin. Med. Pl. 7: fig. 334. 1964.

T. aquilegifolium L. var. *asiaticum* Nakai in J. Jap. Bot. 13: 173. 1937. —*T. aquilegiafolium* L. ssp. *asiaticum* (Nakai) Kitag. Lineam. Fl. Mansh. 226. 1939.

T. aquilegifolium auct. non L.: Finet & Gagnep. in Bull. Soc. Bot. France 50: 604. 1903.

多年生草本植物，全部无毛。茎粗壮，高 60-150 cm，分枝，茎生叶为 3-4 回三出复叶；叶片长 10-30 cm；小叶草质，倒卵形或宽长圆形，1.5-2.5×1.2-3 cm；顶端圆形或微钝，基部圆楔形或浅心形，3 浅裂，裂片全缘或有 1-2 齿，脉平或在下面稍隆起；小叶柄长 0.5-2 cm；叶柄长 4.5-8 cm，基部具鞘。复单歧聚伞花序顶生，伞房状，宽约 20 cm，有多数花，花梗长 4-7 mm。花：萼片 4，早落，白色或带紫色，宽椭圆形，3-3.5×1.8-2 mm，顶端圆形，每一萼片具 3 条基生脉（中脉和 1 侧生脉二歧状分枝，另一侧生脉不分枝）。雄蕊 28-40；花丝长 3-5.8 mm，上部狭倒披针形，宽 0.3-0.5 mm，下部狭条形，宽约 0.2 mm，有时整个花丝呈狭条形，上部宽 0.25 mm，下部宽 0.2 mm；花药长圆形，1.2×0.2-0.4 mm，顶端钝。心皮 4-8；子房椭圆体形，约 1×0.4 mm，基部具长 1.5 mm 的细柄，花柱长约 0.2 mm；柱头长约 0.2 mm。瘦果倒卵形，4-6×1.6-1.8 mm，有 4 条纵翅（翅狭新月形，4-5×1 mm），基部具长 3-5 mm 的细柄；宿存柱头长 0.3-0.5 mm。7

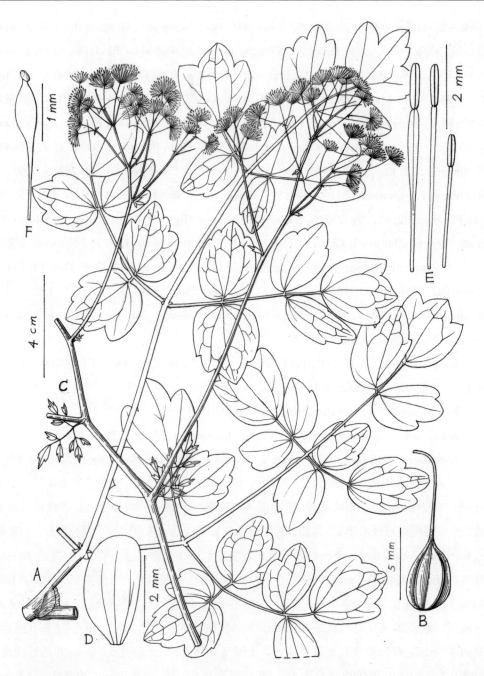

图版57. **唐松草** A. 上部茎生叶，B. 瘦果（据钱家驹等389），C. 聚伞圆锥花序，D. 萼片，E. 三雄蕊，F. 心皮（据郝景盛16024）。

Plate 57. **Thalictrum aquilegifolium** var. **sibiricum** A. upper cauline leaf, B. achene (from J. J. Qian et al. 389), C. thyrse, D. sepal, E. three stamens, F. carpel (from J. S. Hao 16024).

月开花。

Perennial herbs, totally glabrous. Stems robust, 60−150 cm tall, branched. Cauline leaves 3−4-ternate; blades 10−30 cm long; leaflets herbaceous, obovate or broad−oblong, 1.5−2.5×1.2−3 cm, at apex rounded or slightly obtuse, at base rounded-cuneate or subcordate, 3-lobed, lobes entire or 1−2-dentate, nerves flat or abaxially prominent; petiolules 0.5−2 cm; petioles 4.5−8 cm long, at base vaginate. Compound monochasia terminal, corymbiform, ca. 20 cm broad, many-flowered; pedicels 4−7 mm long. Flower: Sepals 4, caducous, white or tinged with purple, broad-elliptic, 3−3.5×1.8−2 mm, apex rounded, each sepal with 3 basal nerves (midrib and 1 lateral nerve dichotomizing, and the another lateral nerve simple). Stamens 28−40; filaments 3−5.8 mm long, above narrow-oblanceolate, 0.3−0.5 mm broad, below narrow-linear, ca. 0.2 mm broad, sometimes total-filaments narrow-linear, above 0.25 mm below 0.2 mm broad; anthers oblong, 1.2×0.2−0.4 mm, apex obtuse. Carpels 4−8; ovary ellipsoid, ca. 1×0.4 mm, base with a slender stipe 1.5 mm long; style ca. 0.2 mm long, stigma ca. 0.3 mm long. Achene obovate, 4−6×1.6−1.8 mm, longitudinally 4-winged (wings narrow-lunate, 4−5×1 mm), base with a slender stipe 3−5 mm long; persistent stigma 0.3−0.5 mm long.

分布于中国湖南、湖北西北部、浙江西北部、安徽西南部、江苏东北部、山东东部、河北、山西、内蒙古、辽宁、吉林、黑龙江；以及俄罗斯西伯利亚地区、朝鲜、日本。生草原，山地草坡或林中，海拔350−1800 m。

全草、根和根状茎供药用，用于肺热咳嗽，黄疸性肝炎等症。地上部分含箭头唐松草碱（thalsimine）等生物碱。(《中国药用植物志》，2016)

标本登录：浙江：昌化，1090 m，贺贤育26058；天目山，贺贤育24887。

安徽：金寨，沈显生897，1342；霍山，刘淼A50081。

山东：崂山，焦启源2685，周太炎等1100；昆嵛山，刘慎谔，刘继孟1502，刘继孟511。

河北：小五台山，孔宪武334，杨承元，李祖桂6458；围场，1600 m，傅国勋33，1081。

山西：五台山，关克俭，陈艺林1833。

内蒙古：巴林右旗，赤峰队2-023；乌兰浩特，乌兰浩特队9062449；凉城，马毓泉65-60；颇尔古纳旗，王战1502；阿尔山，牙资南队769；牙克石，牙资南队234；呼伦贝尔盟，姜恕134。

辽宁: 本溪, 王薇等 395; 抚顺, 张玉良等 384; 清源, 曹伟 193; 西丰, 李文俊 s. n.。

吉林: 通化, 曹子余等 2; 安图, 洪德元 32-719, 32-986, 曹子余等 140; 长白山, 刘慎谔等 1654, 钱家驹 389, 607, 洪德元等 86-085; 珲春, 傅沛云等 675; 临江, 刘慎谔 986; 额穆, 孔宪武 1718, 1885; 威虎岭, 孔宪武 1551。

黑龙江: 尚志, 中德队 7262; 集贤, 张玉良 1680; 富锦, 张玉良 1577; 鹤岗, 中德队 31, 东北植物队 95-41183; 带岭, 刘慎谔 3309; 北安, 东北队 249; 塔河, 350 m, 张治明 126; 三股流, 1951-07-21, 郝景盛 16024。

组 8. 芸香叶唐松草组

Sect. **Rutifolia** (Prantl) W. T. Wang, st. nov. —Sect. *Comptonota* Prantl subsect. *Rutifolia* Prantl in Bot. Jahrb. 9: 271. 1888; Tamura in Hiepko, Nat. Pflanzenfam., Zwei. Aufl., 17a IV: 479. 1995. —Sect. *Leptostigma* ser. *Rutipolia* W. T. Wang & S. H. Wang in Fl. Reip. Pop. Sin. 27: 559. 1979. Type: *T. rutifolium* Hook. f. & Thoms.

Sect. *Cincinneria* Boivin in J. Arn. Arb. 26: 117. 1945, non Boivin 1944. Type: *T. mairei* Lévl.

复单歧聚伞花序狭窄, 似总状花序。花具 4(-5) 萼片, 每一萼片具 3(-6) 条基生脉。花丝狭条形或丝形。花柱直, 稀稍向后弯曲; 柱头明显, 无翅或有 1-2 条狭翅。

Compound monochasia narrow, raceme-like. Sepals 4, rarely 5 per flower, each sepal with 3, rarely 6 basal nerves. Style straight, rarely slightly recurved; stigma not winged or with 1 or 2 narrow wings.

6 种, 隶属 4 系, 分布于中国西南部, 其中 2 种分布到喜马拉雅山区。

系 1. 小金唐松草系

Ser. **Xiaojinensia** W. T. Wang, ser. nov. Type: *T. xiaojinense* W. T. Wang

Leaflets abaxially hirtellous. Sepals 4 per flower; each sepal with 3 or 6 basal nerves. Stamens 12-32 per flower; filaments narrow-linear. Style straight, nearly as long as ovary; stigma not winged.

小叶下面被微硬毛。花具 4 枚萼片, 每一萼片有 3 或 6 条基生脉。花具 12-32 枚雄蕊; 花丝狭条形。花柱直, 约与子房等长; 柱头无翅。

2 种，特产四川。

58. 小金唐松草（《广西植物》） 图版 58

Thalictrum xiaojinense W. T. Wang in Guihaia 37(4): 415, fig. 5. 2017. Holotype：四川：小金县，自县城往夹金山路上，alt. 3350–3400 m，2007-07-28，D. E. Boufford et al. 38515(PE).

多年生草本。茎高 80–100 cm，无毛，中部之上分枝。基生叶和下部茎生叶未见，中部和上部茎生叶为 2–3 回三出羽状复叶，具短柄；叶片三角形，4–13×4–10 cm，在叶轴近小叶柄处疏被短柔毛（毛长 0.1–0.2 mm）；小叶薄纸质，宽倒卵形、宽卵形、倒卵形或椭圆形，0.4–1.4×0.3–1.3 cm，顶端圆形或钝，常具短尖头，基部圆形、宽楔形或近心形，3 浅裂（裂片全缘，或具 1–2 齿），上面无毛，下面在脉上被微硬毛（毛长 0.1–0.2 mm），具 3 或 5 出脉，基生脉在两面均隆起；小叶柄长 0.5–7 mm；叶柄长 1–10 cm，无毛，基部具鞘。复单歧聚伞花序顶生，生茎顶端者长 15 cm，似总状花序，疏具 6 花，生于分枝顶端者长 4–5 cm，约有 2 花；苞片叶状；花梗细，长 0.5–2 cm。花无毛：萼片 4，白色，椭圆状卵形，3–3.8×1.8 mm，在顶端圆形，每一萼片具 3 条不分枝的基生脉。雄蕊 16–24；花丝狭条形，长约 4 mm，上部宽 0.15 mm，下部宽 0.1 mm；花药黄色，狭长圆形，约 1.2×0.3 mm，顶端钝或具极小的短尖头。心皮 8–10；子房扁压，狭倒卵形，1.2×0.6–0.8 mm，基部具长 0.2 mm 的短柄；花柱长 1–1.2 mm；柱头狭长圆形，0.8–1×0.2–0.25 mm。7–8 月开花。

Perennial herbs. Stems 80–100 cm tall, glabrous, above the middle branched. Basal and lower cauline leaves not seen, middle and upper cauline leaves 2–3-ternate-pinnate, shortly petiolate; blades triangular, 4–13×4–10 cm, on rachis near petiolules sparsely puberulous (hairs 0.1–0.2 mm long); leaflets thinly papery, broadly obovate, broadly ovate, obovate or elliptic, 0.4–1.4×0.3–1.3 cm, at apex rounded or obtuse, often mucronate, at base rounded, broadly cuneate or subcordate, 3-lobed (lobes entire or 1–2-denticulate), adaxially glabrous, abaxially on nerves hirtellous (hairs 0.1–0.2 mm long), 3–5-nerved, basal nerves on both surfaces prominent; petiolules 0.5–7 mm long; petioles 1–10 cm long, glabrous, at base vaginate. Compound monochasia terminal, those borne on stem apex ca. 15 cm long, raceme-like, sparsely ca. 6-flowered, and those borne on branch apex 4–5 cm long. ca. 2-flowered; bracts foliaceous; pedicels slender, 0.5–2 cm long. Flower glabrous: Sepals 4, white, elliptic-ovate, 3–3.8×1.8 mm, at apex rounded, each sepal with 3 simple basal nerves. Stamens 16–24;

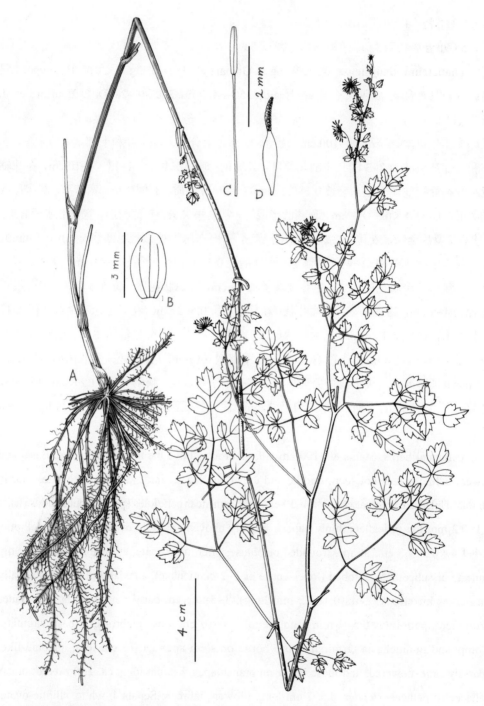

图版58. 小金唐松草　A.植株全形，B.萼片，C.雄蕊，D.心皮。（据holotype）
Plate 58. **Thalictrum xiaojinense**　A. habit, B. sepal, C. stamen, D. carpel. (from holotype)

filaments narrow-linear, ca. 4 mm long, above 0.15 mm broad, below 0.1 mm broad; anthers yellow, narrow-oblong, ca. 1.2×0.3 mm, apex obtuse or minutely mucronate. Carpels 8–10; ovary compressed, narrow-obovate, 1.2×0.6–0.8 mm, at base with a short stipe 0.2 mm long; style 1–1.2 mm long; stigma narrow-oblong. 0.8–1×0.2–0.25 mm.

特产四川小金。生高山山坡冷杉林边阴处，海拔 3350–3400 m。

59. 六脉萼唐松草（《广西植物》） 图版 59

Thalictrum sexnervisepalum W. T. Wang in Guihaia 37（6）：679, fig. 3. 2017. Holotype：四川：崇州市，鞍子河自然保护区，鸡冠山, alt. 3100 m, 2007-07-29, 冯正波, 朱大海, 李小杰 LD4214（PE）.

多年生草本。茎高约 1 米，近节处疏被长 0.1 mm 的短毛，其他部分无毛，上部分枝。上部茎生叶为 3 回三出羽状复叶，具短柄；叶片 7–15×7–15 cm；小叶纸质，宽卵形、卵形、倒卵形或宽椭圆形，0.6–1.6×0.5–1.4 cm，在顶端微尖或圆形，在基部圆形或近心形，3 浅裂（裂片具 1–2 齿或全缘），上面无毛，下面具直径约 0.1 mm 的白色小颗粒，并在脉上被微硬毛（毛长 0.2–0.4 mm），具 3–5 出脉，基生脉在叶上面平或稍凹陷，在下面隆起；小叶柄丝形，长 0.05–1.5 cm，无毛；叶柄长 0.5–1 cm，无毛，基部具狭鞘。复单歧聚伞花序顶生，生茎顶端者长约 15 cm，似总状花序，疏具 10 花，生分枝顶端者长 2.5–5 cm，具 3–4 花；花序轴和花梗无毛；苞片叶状；花梗丝形，长 6–10 mm。花：萼片 4（？），白色，宽椭圆形，约 3×2 mm，无毛，每一萼片具 6 条基生脉（中脉和 3 条侧生脉不分枝，近萼片边缘的 2 条侧生脉 1 或 2 回二歧状分枝）。雄蕊 12–32，无毛，其中一些为退化雄蕊，具败育花药；花丝狭条形，长 2.8–4.2 mm，上部宽 0.1–0.2 mm，下部宽 0.05–0.1 mm；能育雄蕊的花药近条形或狭长圆形，0.8–1×0.2 mm，退化雄蕊的花药近条形或狭新月形，0.4–1×0.1–0.15 mm。心皮 8–10，长约 3 mm；子房狭倒卵形，约 1.5×0.5 mm，具长 0.05–0.1 mm 的短毛，基部具长 0.5 mm 的短柄，花柱长约 1.5 mm；柱头狭长圆形，约长 0.8 mm。7–8 月开花。

Perennial herbs. Stem ca. 1 m tall, near nodes with sparse short hairs ca. 0.1 mm long, elsewhere glabrous, above branched. Upper cauline leaves 3-ternate-pinnate, shortly petiolate; blades 7–15×7–15 cm; leaflets papery, broadly ovate, ovate, obovate or elliptic, 0.6–1.6×0.5–1.4 cm, at apex slightly acute or rounded, at base rounded or subcordate, 3-lobed (lobes 1–2-denticulate or entire), adaxially glabrous, abaxially on nerves hirtellous (hairs 0.2–0.4 mm long), 3–5-nerved, basal nerves adaxially flat or slightly impressed, abaxially

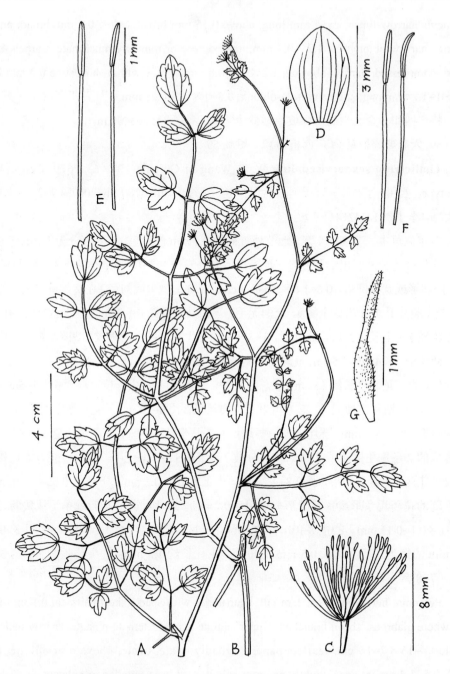

图版59. **六脉萼唐松草** A. 开花茎上部，B. 中部茎生叶，C. 花，D. 萼片，E. 二能育雄蕊，F. 二不育雄蕊，G. 心皮。（据holotype）

Plate 59. **Thalictrum sexnervisepalum** A. upper part of flowering stem, B. middle cauline leaf, C. flower, D. sepal, E. two fertile stamens, F. two sterile stamens, G. carpel. (from holotype)

prominent; petiolules filiform, 0.05–1.5 mm long, glabrous; petioles 0.5–1 cm long, glabrous, base narrowly vaginate. Compound monochasia terminal, those borne on stem apex ca. 15 cm long, raceme-like, sparsely 10-flowered, those borne on branch apex 2.5–5 cm long, 3–4-flowered; rachis with pedicels glabrous; bracts foliaceous; pedicels filiform, 6–10 mm long. Flower: Sepals 4(?), white, broadly elliptic, ca. 3×2 mm, glabrous, each sepal with 6 basal nerves (midrib and 3 lateral nerves simple, and the other 2 lateral nerves near sepal margins 1 or twice dichotomizing). Stamens 12–32, glabrous, of them some being staminods bearing abortive anthers; filaments narrow-linear, 2.8–4.2 mm long, above 0.1–0.2 mm broad, below 0.05–0.1 mm broad; anthers of fertile stamens sublinear of narrow-oblong, 0.8–1×0.2 mm, at apex obtuse, those of staminodia sublinear or narrowly lunulate, 0.4–1×0.1–0.15 mm. Carpels 8–10, ca. 3 mm long; ovary narrow-obovate, ca. 1.5×0.5 mm, with short hairs 0.05–0.1 mm long, base with a short stipe 0.5 mm long; style ca 1.5 mm long; stigma narrow-oblong, ca. 0.8 mm long.

特产四川崇州市鸡冠山，生山坡草丛，海拔 3100 m。

系 2. 芸香叶唐松草系

Ser. **Rutifolia** W. T. Wang & S. H. Wang in Fl. Reip. Pop. Sin. 27: 559. 1979, p.p. excl. *T. leuconoto* Franch. Type: *T. rutifolium* Hook. f. & Thoms.

植株全部无毛。花具 4 枚萼片，每一萼片具 3 条基生脉。花具 5–24 枚雄蕊；花丝丝形。花柱直或顶端稍内弯曲，约与子房等长；柱头无翅。

Plants totally glabrous. Sepals 4 per flower; each sepal with 3 basal nerves. Stamens 5–24 per flower; filaments filiform. Style straight or at apex slightly recurved; stigma not winged.

1 种，分布于中国西南和西北南部，以及喜马拉雅山区。

60. 芸香叶唐松草（《中国高等植物图鉴》） 图版 60

Thalictrum rutifolium Hook. f. & Thoms. Fl. Ind. 14. 1855; et in Hook. f. Fl. Brit. Ind. 1: 12. 1872; Lecoy. in Bull. Soc. Bot. Belg. 24: 195. 1885; Maxim. Fl. Tangut. 5. 1889; Finet & Gagnep. in Bull. Soc. Bot. France 50: 606. 1903; Iconogr. Corm. Sin. 1: 679, fig. 1358. 1972; W. T. Wang & S. H. Wang in Fl. Reip. Pop. Sin. 27: 559, pl. 140: 1–3. 1979; Grierson, Fl. Bhutan 1(2): 276. 1984; W. T. Wang in Fl. Xizang. 2: 69, fig. 20: 4–6. 1985; et in Vasc. Pl. Hengduan Mount. 1: 501. 1993; Rau in Sharma et al. Fl. India 1: 141. 1993; L. H. Zhou in Fl. Qinghai. 1: 332, pl. 72: 7–11. 1997; W. T. Wang in Fl Yunnan. 11: 166. 2000; et in High Pl. China 3: 473,

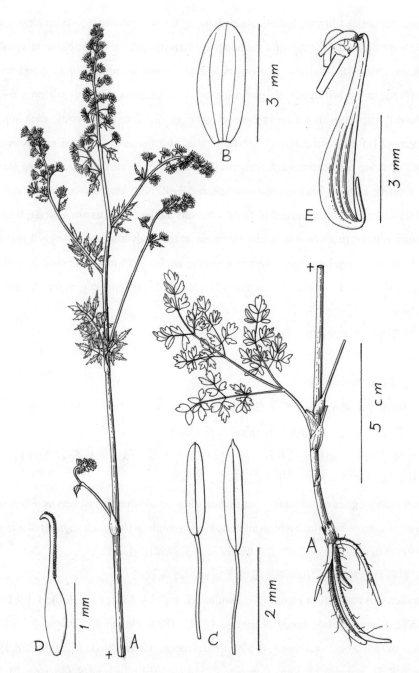

图版60. **芸香叶唐松草** A. 植株全形，B. 萼片，C. 二雄蕊，D. 心皮（据青藏队73-55），E. 瘦果（据陶德定11073）。

Plate 60. **Thalictrum rutifolium** A. habit, B. sepal. C. two stamens, D. carpel, (from Qinghai-Xizang Exped. 73-55), E. achene (from D. D. Tao 11073).

fig. 752. 2000; D. Z. Fu & G. Zhu in Fl. China 6: 297. 2001; W. T. Wang in High. Pl. China in Colour 3: 390. 2016; L. Xie in Med. Pl. China 3: 259, with fig. & photo. 2016. Syntypes: Sikkim, alt. 12 000 ft., J. D. Hooker s. n. (K, not seen). 中国：西藏西部，alt. 10–14 000 ft., T. Thomson s. n. (K. not seen).

多年生草本，全部无毛。茎高 11–50 cm，上部分枝。基生叶和下部茎生叶有长柄，为 3–4 回羽状复叶；叶片长 3.2–11 cm；小叶草质，楔状倒卵形、菱形、椭圆形或近圆形，3–8×2–7 mm，在顶端圆形，在基部楔形或圆形，3 裂或不分裂，全缘，脉两面平；小叶柄长 0–3.5 mm；叶柄长达 6 cm，基部有短鞘；托叶膜质，分裂。复单歧聚伞花序顶生和腋生，似总状花序，长 2–15 cm，分枝 4–20 条，密集，长 0.5–1 cm，有（1–）2 花；下部苞片叶状，上部苞片为三出复叶或不分裂，呈长椭圆形或披针形，长 4–7 mm；花梗长 2–7 mm。花：萼片 4，早落，带淡紫色，近长圆形或卵形，1.5–30×8–1.8 mm，在顶端钝，每一萼片具 3 条不分枝基生脉。雄蕊 5–24；花丝丝形，长 1–2 mm，花药长圆形或狭长圆形，0.5–2.2×0.3–0.5 mm，顶端有长 0.05–0.1 mm 的短尖头。心皮 3–5；子房狭倒卵形，0.8–1.2×0.3–0.5 mm，基部渐狭；花柱长 0.6–1.2 mm，直或顶端稍向内弯曲，腹面密生柱头组织。瘦果倒垂，稍扁，狭倒卵形，约 3×1.2 mm，每侧具 3 条纵肋，基部具长 1–2 mm 的柄；宿存花柱长约 1 mm。6 月开花。

Perennial herbs, totally glabrous. Stems 11–50 cm tall, above branched. Basal and lower cauline leaves long petiolate, 3–4-pinnate; blades 3.2–11 cm long; leaflets herbaceous, cuneate-obovate, rhombic, elliptic or suborbicular, 3–8×2–7 mm, at apex rounded, at base cuneate or rounded, 3-lobed or undivided, entire, nerves on both surfaces flat; petiolules 0–3.5 mm long; petioles up to 6 cm long, base shortly vaginate; stipules membranous, divided. Compound monochasia terminal and lateral, raceme-like, 2–15 cm long; branches 4–20, dense, 0.5–1 cm long, (1–)2-flowered; lower bracts foliaceous, upper bracts ternate or simple, long elliptic or lanceolate, 4–7 mm long; pedicels 2–7 mm long. Flower: Sepals 4, caducous, purplish-tinged, suboblong or ovate, 1.5–3×0.8–1.8 mm, at apex obtuse, each sepal with 3 simple basal nerves. Stamens 5–24; filaments filiform, 1–2 mm long; anthers oblong or narrow-oblong, 0.5–2.2×0.3–0.5 mm, apex with mucro 0.05–0.1 mm long. Carpels 3–5; ovary narrow-obovate, 0.8–1.2×0.3–0.5 mm, base attenuate; style 0.6–1.2 mm long, straight or at apex slightly recurved, adaxially with dense stigmatic tissue. Achenes pendulous, slightly compressed, narrow-obovate, ca. 3×1.2 mm, on each side longitudinally 3-ribbed, base with a stipe 1–

2 mm long; persistent style ca. 1 mm long.

分布于中国西藏、云南西北部、四川西部、青海和甘肃中部以南；以及不丹和印度北部。生高原草坡、河滩上或山谷中，海拔 2280–4300 m。

全草和根有清热解毒之效，用于炭疽病、疮疡不愈等症。(《中国药用植物志》, 2016)

标本登录：西藏：日达，高原生物所西藏队 3492a；革吉，青藏队 76-215；普兰，青藏队 76-8529；聂雄拉山口，张永田，郎楷永 4189；聂拉木，张永田，郎楷永 3857, 3958, 4103, 4624；日喀则，傅国勋 374；萨迦，西藏队 1678；南木林，西藏中草药队 827；拉萨，张永田，郎楷永 1669；当雄，综考会队 85-288；班戈，郎楷永 7498；墨竹工卡，杨竞生 90-219；嘉黎，陶德定 10453；那曲，陶德定 10533, 10867；巴青，陶德定 11073；工布江达，杨竞生 90-154；林芝，张永田，郎楷永 1166；芒康，青藏队 76-11995；察雅，青藏队 76-12277；类乌齐，综考会队 76-246；昌都，青藏队 73-55，D. E. Boufford, B. Bartholomew et al. 41338；江达，张永田，郎楷永 205，青藏队 76-12458，杨竞生 91-560。

四川：木里，俞德浚 6236, 6548；乡城，青藏队 81-4096；稻城，姜恕 5349, D. E. Boufford, B. Bartholomew et al. 28236；康定，关克俭，王文采等 836, 1175；雅江，姜恕 5212；刷经寺，张泽云，周洪富 22263；埌塘，姜恕 9002；甘孜，姜恕 2525；德格，四川植物队 7050；石渠，姜恕 9254。

青海：囊谦，杨永昌 1040，青藏队 72-1042；杂多，刘尚武 49；玉树，? 166；玛多，吴玉虎 1604；达日，何廷农等 1094；玛沁，何廷农等 496；同德，吴玉虎 4985；兴海，何廷农等 97, 176, 1404；门源，刘继孟 7152；刚察，郭本兆 11350；天峻，青甘队 1078；德令哈，郭本兆 11623；祁连，青甘队 2766。

甘肃：夏河，王作宾 5524, 5675，傅坤俊 1394；连城保护区，孙学刚等 6131；天祝，? 235；安元镇，何业祺 4610。

系 3. 长柱唐松草系

Ser. **Sinomacrostigmata** W. T. Wang, ser. nov. Type: *T. sinomacrostigma* W. T. Wang

Plants totally glabrous. Sepals 5 per flower; each sepal with 3 basal nerves. Stamens ca. 50 per flower; filaments filiform. Style elongately subulate, twice longer than ovary; stigma with a narrow-linear long wing.

植株全部无毛。花具 5 枚萼片，每一萼片具 3 条基生脉。花具约 50 枚雄蕊；花丝丝形。花柱长钻形，比子房长 2 倍；柱头具 1 条狭条形长翅。

1 种，特产四川西部。

61. 长柱唐松草《植物分类学报》 图版 61

Thalictrum sinomacrostigma W. T. Wang in Acta Phytotax. Sin. 31(3): 213. 1993. – *T. macrostigma* Finet & Gagnep. in Bull. Soc. Bot. France 53: 125. 1906, non Edgew. 1851. Holotype: 四川西部, 1904–06, E. H. Wilson 3085a (P).

T. leuconotum auct. non Franch. : W. T. Wang & S. H. Wang in Fl. Reip. Pop. Sin. 27: 559. 1979, p.p. quoad syn. *T. macrostigma* Finet & Gagnep.; W. T. Wang in Bull. Bot. Lab. N.-E. Forest. Inst. 8: 31. 1980, p. p. quoad syn.; D. Z. Fu & G. Zhu in Fl. China 6: 293. 2001, p.p. quoad syn. *T. macrostigma* Finet & Gagnep.

多年生草本，全部无毛。茎高约 60 cm，不分枝，约具 4 叶。基生叶未见。茎生叶为 2–3 回三出羽状复叶或 2–3 回三出复叶，具长柄；叶片 3–12×3–8 cm；小叶 18–27，纸质，宽倒卵形，宽菱形或宽卵形，1–2.2×0.8–2 cm，在基部圆形或近截形，具 5 出脉，3 中裂，中央裂片楔形，顶端具 3 钝齿或全缘，侧裂片较小，三角状卵形，具 1 齿或全缘，所有裂片在顶端钝或圆钝，基生脉在叶两面均平；小叶柄丝形，长 0.5–12 mm；小托叶宽卵形，长约 1 mm；叶柄细，长 1–10 cm，基部具狭鞘。复单歧聚伞花序顶生，狭窄，似总状花序，长约 4 cm，具少数花，基部具 1 条长 2 cm 的分枝；苞片小，叶状；花梗长 4–5 mm。花：萼片 5，白色，卵状披针形或狭卵形，5.2–6×2.8–3.5 mm，在顶端急尖，每一萼片具 3 条二歧状分枝的基生脉。雄蕊约 50；花丝丝形，长 0.6–1 mm；花药长圆形或狭长圆形，1.5–2×0.3–0.5 mm，顶端尖。心皮 5–6；子房近长圆形，约 2×0.5 mm；花柱长钻形，长约 4.2 mm，基部直径约 0.2 mm；柱头不明显，具 1 条狭长形，长约 3 mm 的白色膜质翅。6 月开花。

Perennial herbs, totally glabrous. Stem ca. 60 cm tall, simple, ca. 4-leaved. Basal leaves not seen. Caulin leaves 2–3-ternate-pinnate or 2–3-ternate, long petiolate; blades 3–12×3–8 cm; leaflets 18–27, papery, broadly obovate, broadly rhombic or broadly ovate, 1–2.2×0.8–2 cm, at base rounded or subtruncate, 5-nerved, 3-fid, central lobe cuneate, at apex obtusely 3-denticulate or entire, lateral lobes smaller, triangular-ovate,1-denticulate or entire, all lobes at apex obtuse or rounded obtuse, nerves on both surfaces flat; stipels broadly ovate, ca. 1 mm long; petiolules filiform, 0.5–12 mm long; petioles slender, 1–10 cm long, base narrowly vaginate. Compound monochasium terminal, narrow, raceme-like, ca. 4 cm long, few-flowered, at base with a branch 2 cm long; bracts small, foliaceous; pedicels 4–5 mm long. Flower: Sepals 5, whitish, ovate-lanceolate or narrow-ovate, 5.2–6×2.8–3.5 mm, at apex acute, each

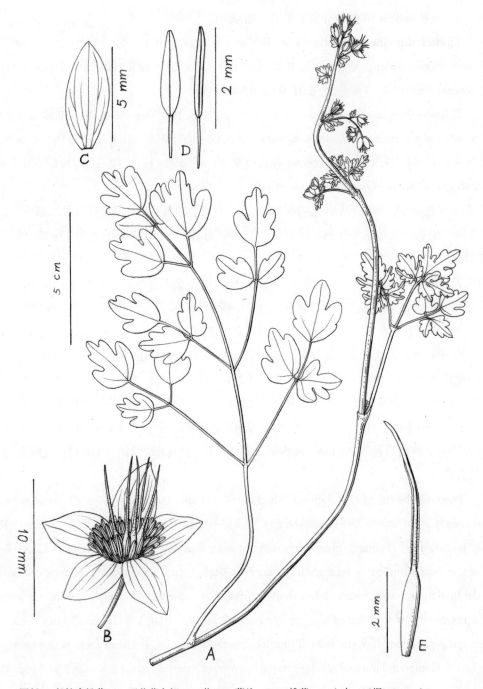

图版61. **长柱唐松草** A. 开花茎上部，B. 花，C. 萼片，D. 二雄蕊，E. 心皮。（据holotype）
Plate 61. **Thalictrum sinomacrostigma** A. upper part of flowering stem, B. flower, C. sepal, D. two stamens, E. carpel. (from holotype).

sepal with 3 dichotomizing basal nerves. Stamens ca. 50; filaments filiform, 0.6–1 mm long; anthers oblong or narrow-oblong, 1.5–2×0.3–0.5 mm, at apex acute. Carpels 5–6; ovary suboblong, ca. 2×0.5 mm; style elongate-subulate, ca. 4.2 mm long, at base ca. 0.2 mm in diam.; stigma inconspicuous, with a white, membranous, narrow-linear wing ca. 3 mm long.

特产四川西部。

本种的雄蕊数目多，相当短，花柱伸长，比子房长2倍，柱头具1条狭长的白色膜质翅，这些形态特征说明本种很独特，与近缘种容易区别。

此独特种的模式标本由著名植物采集家E. H. Wilson于1904年6月在四川西部发现，但在野外记录中缺少具体的地点和生长环境、海拔高度等内容，这对有关原产地模式标本（topotype）的采集造成困难。自E. H. Wilson于1904年发现本种后至今已过去了113年，但一直未再能采到此种的标本，这种情况说明此种是一个有极小分布区的狭域分布种。

系 4. 白茎唐松草系

Ser. **Leuconota** W. T. Wang, ser, nov. Type: *T. leuconotum* Franch.

Plants totally glabrous. Sepals 4 per flower; each sepal with 3 simple basal nerves. Stamens 7–16 per flower; filaments filiform. Style straight, shorter than ovary; stigma with 2 narrow lateral wings.

植株全部无毛。花有4枚萼片；每一枝萼片具3条不分枝的基生脉。花有7–16枚雄蕊；花丝丝形。花柱直，比子房短；柱头具2条狭侧生翅。

2种，分布于中国西南部和不丹。

62. 吉隆唐松草（《广西植物》） 图版 62

Thalictrum jilongense W. T. Wang in Guihaia 37(6): 679, fig. 4. 2017. Holotype：西藏：吉隆县，汝戈喇嘛庙后山，alt. 3600 m，林下，1972-06-17，西藏中草药队309（PE）。

多年生草本，全部无毛。茎高约60 cm，不分枝，约有5叶。中部及上部茎生叶为3回三出羽状复叶，具柄；叶片宽三角形，长达5.8 cm，宽达14 cm；小叶纸质，宽倒卵形或近圆形，0.4–1.3×0.3–0.9 cm，在顶端圆形，在基部圆形、宽楔形或近心形，2-3-浅裂或2-3-中裂（裂片在顶端圆形或钝，全缘，稀具1钝齿），具三出脉，基生脉在叶两面平，脉网明显；小叶柄长0–3 mm；小托叶白色，卵形，长2–3 mm，顶端急尖；叶柄长0.5–6.5 cm。复单歧聚伞花序顶生，长约4 cm，似总状花序，有密集的10花，侧生复单歧聚伞花序不存在；苞片叶状，长1–3 cm；花梗丝形，长3–7 mm。花：萼片4，白色，宽椭圆形，约2.8×2 mm，在顶端圆形，每一萼片具3条不分枝的基生脉。雄蕊7–10；花丝丝

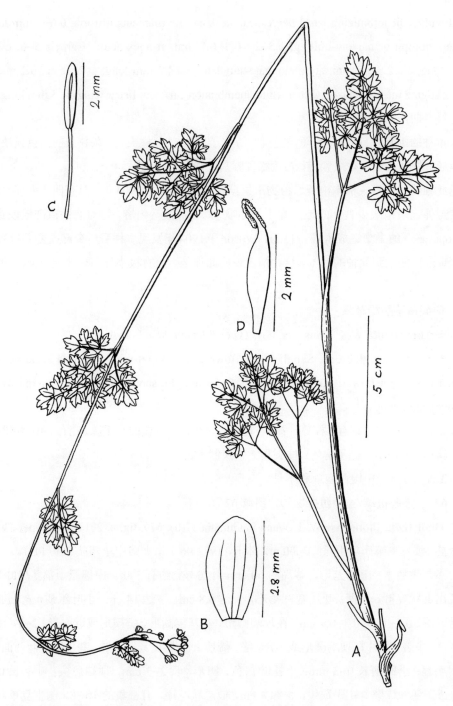

图版62. 吉隆唐松草　A. 植株全形，B. 萼片，C. 雄蕊，D. 心皮。（据holotype）
Plate 62. **Thalictrum jilongense**　A. habit, B. sepal, C. stamen, D. carpel. (from holotype)

形，长约 3 mm；花药黄色，狭长圆形，约 2×0.4 mm，顶端钝。心皮 3-4；子房狭倒卵形，2×0.5-0.6 mm；花柱长 1 mm；柱头狭长圆形，约 0.8×0.35 mm，在每侧均具 1 条狭翅。6 月开花。

Perennial herbs. Stem ca. 60 cm tall, simple, ca. 5-leaved. Middle and upper cauline leaves 3-ternate-pinnate, petiolate; blades broadly triangular, up to 5.8×14 cm; leaflets papery, broadly obovate or suborbicular, 0.4-1.3×0.3-0.9 cm, at apex rounded, at base rounded, broadly cuneate or subcordate, 2-3-lobed or 2-3-fid (lobes at apex rounded, entire, rarely obtusely 1-denticulate), 3-nerved, with basal nerves on both surfaces flat and conspiculous nervous nets; stipels white, ovate, 2-3 mm long, apex acute; petiolules 0-3 mm long; petioles 0.5-6.5 cm long. Compound monochasium terminal, ca. 2 cm long, raceme-like, densely 10-flowered, lateral compound monochasia wanting; bracts foliaceous, 1-3 cm long; pedicels filiform, 3-7 mm long. Flower: Sepals 4, whitish, broad-elliptic, ca. 2.8×2 mm, at apex rounded, each sepal with 3 simple basal nerves, Stamens 7-10; filaments filiform; anthers yellow, narrow-oblong, ca. 2×0.4 mm, apex obtuse. Carpels 3-4; ovary narrow-obovate, 2 × 0.5-0.6 mm; style 1 mm long; stigma narrow-oblong, ca. 0.8 × 0.35 mm, on each side with a narrow wing.

特产西藏吉隆县。生山地林下，海拔 3600 m。

63. 白茎唐松草（《中国高等植物图鉴》）　　图版 63

Thalictrum leuconotum Franch. Pl. Delav. 15. 1889; Finet & Gagnep. in Bull. Soc. Bot. France 50: 607. 1903; Marq. in J. Linn. Soc. Bot. 48: 154. 1929; Hand.-Mazz. Symb. Sin. 7: 311. 1931; Iconogr. Corm. Sin. 1: 675. 1972; W. T. Wang & S. H. Wang in Fl. Reip. Pop. Sin. 27: 559, pl. 139: 1-4. 1979; W. T. Wang in Bull. Bot. Lab. N.-E. Forest. Inst. 8: 31. 1980; et in Iconogr. Corm. Sin. Suppl. 1: 427, fig. 8614. 1982; Grierson, Fl. Bhutan 1(2): 298. 1984; W. T. Wang in Vasc. Pl. Hengduan Mount. 1: 502. 1993; L. H. Zhou in Fl. Qinghai. 1: 332. 1997; W. T. Wang in Fl. Yunnan. 11: 167. 2000; et in High Pl. China 3: 473, fig. 753. 2000; D. Z. Fu & G. Zhu in Fl. China 6: 293. 2001; L. Xie in Med. Fl. China 3: 260, with fig. 2016. Type: 云南：洱源，罗平山 (Lo-pin-chan), alt. 3200 m, 1886-05-25, Delavay 2854 (holotype, P, not seen; photo, PE; isotype, NAS).

T. mairei Lévl. in Repert. Sp. Nov. Reg. Veg. 7: 339. 1909; Boivin in J. Arn. Arb. 26: 117. 1945, p. p. excl. specim. T. T. Yu 11686 et 14355; Lauener & Green in Not. Bot. Gard.

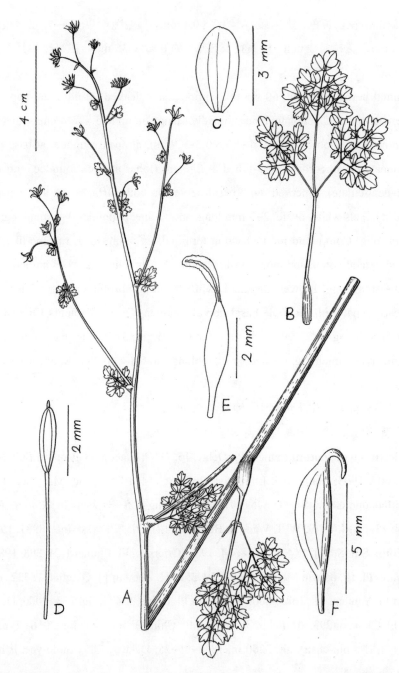

图版63. **白茎唐松草** A. 开花茎，B. 下部茎叶生，C. 萼片，D. 雄蕊，E. 心皮（据俞德浚11806），F. 瘦果（据吴中伦32507）。

Plate 63. **Thalictrum leuconutum** A. flowering stem, B. lower cauline leaf, C. sepal, D. stamen, E. carpel, (from T. T. Yü 11806) F. achene (from Z. L. Wu 32507).

Edinberg 23(4): 593. 1961. Holotype: 云南：无详细地点，1905–09，Maire 388 (not seen).

多年生草本，全部无毛。茎高 50–150 cm，分枝。基生叶和下部茎生叶有长柄，为 3 回三出或近羽状复叶；叶片长 6–14 cm；小叶草质，宽菱形，楔状倒卵形或近圆形，7–9×4.5–12 mm，在顶端圆形，在基部圆形，宽楔形或浅心形，3 浅裂，偶尔不分裂，全缘，脉在叶两面平，脉网不明显；小托叶圆卵形，长约 2 mm，顶端圆形；小叶柄长 0.8–6 mm；叶柄长达 10 cm，基部具鞘；托叶膜质，全缘。顶生复单歧聚伞花序长达 8 cm，似总状花序，约具 5 花，侧生复单歧聚伞花序约 5，长 2–7 cm，具少数花；苞片叶状；花梗长 3–10 mm。花：萼片 4，脱落，淡黄绿色，或带紫色，宽椭圆形或卵形，3–4×1.5–1.8 mm，在顶端圆钝，每一萼片具 3 条不分枝的基生脉。雄蕊 8–16；花丝丝形，长约 4.5 mm；花药长圆形，1.8–2×0.4 mm，顶端具长 0.2 mm 的短尖头。心皮 2–5；子房狭椭圆形，约 2.5×1 mm，基部具长 1 mm 的柄；花柱长 2 mm；柱头狭长圆形，1.8×0.5 mm，在每侧均有 1 条狭翅。瘦果扁，斜狭倒卵形，4.5–7×1.1–2 mm，每侧具 3 条纵肋；基部具长 0.5–2 mm 的柄；宿存花柱长 1.2–2 mm。5 月开花。

Perennial herbs, totally glabrous. Stems 50–150 cm tall, branched. Basal and lower cauline leaves long; petiolate, 3-ternate or 3-pinnate; blades 6–14 cm long; leaflets herbaceous, broadly rhombic, cuneate-obovate or suborbicular, 7–9×4.5–12 mm, at apex rounded, at base rounded, broadly cuneate or subcordate, 3-lobed, occasionally undivided, entire, nerves on both surfaces flat and nervous nets inconspicaous; stipels orbicular-ovate, ca. 2 mm long, apex rounded; petiolules 0.8–6 mm long; petioles up to 10 cm long, at base vaginate; stipules membranous, undivided. Terminal compound monochasium up to 8 cm long, raceme-like, ca. 5-flowered, lateral compound monochasia ca. 5, 2–7 cm long, also raceme-like, few-flowered; bracts foliaceous; pedicels 3–10 mm long. Flower: Sepals 4, deciduous, yellowish-greenish, or purple-tinged, broad-elliptic, or ovate, 3–4×1.5–1.8 mm, at apex rounded-obtuse, each sepal with 3 simple basal nerves. Stamens 8–16; filaments filiform, ca. 4.5 mm long; anthers oblong, 1.8–2×0.4 mm, apex with mucros ca. 0.2 mm long. Carpels 2–5; ovary narrow-elliptic, ca. 2.5×1 mm, base with a stipe 1 mm long; style 2 mm long; stigma narrow-oblong, ca. 1.8×0.5 mm, at each side with a narrow wing. Achenes compressed, obliquely narrow-obovate, 4.6–7×1.1–2 mm, on each side longitudinally 3-ribbed, base with a stipe 0.5–2 mm long; persistent style 1.2–2 mm long.

分布于中国西藏东南部、云南北部（洱源以北），四川西部和青海南部；以及不丹。

生山地草坡或灌丛草地，海拔 2500–3800 m。

根有清热解毒之效。(《中国药用植物志》，2016)

标本登录：西藏东南部：Rong-chu (Tumbatse), 1924-06-15, J. Kingdon Ward, 5788 (fide C. Y. B. Marquand)。

云南：洱源，杨竞生 809；丽江，秦仁昌 20652, 30228；维西，冯国楣 4283，植物所横断山队 1551；中甸，俞德浚 11806，杨亲二，袁琼 1985, D. E. Boufford et al. 42102；贡山，青藏队 80-8640；无准确地址，Bonati 7443 (Maire 采)(NY)。

四川：盐源，姜恕 5912；木里，俞德浚 5706, 6705；雅江，应俊生 3118；康定，姜恕 2874，王德泉 4820，汤宗孝，田效文 783，陈伟烈等 6322；泸定，郎楷永，李良千，费勇 244；巴郎山，汪发缵 21213；理县，何铸，周子林 12790；马尔康，吴中伦 32266, 32507；若尔盖，郎楷永，李良千，费勇 1993。

组 9. 帚枝唐松草组

Sect. **Platystachys** W. T. Wang, sect. nov. Type: *T. virgatum* Hook. f. & Thoms.

Sectio nova haec est affinis Sect. *Rutifoliis* (Prantl) W. T. Wang, a qua differt monochasiis compositis corymbiformibus vel paniculiformibus.

花组成伞房状复单歧聚伞花序或圆锥状聚伞圆锥花序。萼片 4(-5-6)；每一萼片具(3-4-)5(-6-7)条基生脉。花丝狭条形或丝形。心皮具 1 直花柱和 1 明显柱头。瘦果在每侧具(2-)3(-4)条纵肋，无翅，稀沿腹缝线和背缝线具狭翅(偏翅唐松草)。

Flowers in corymbiform compound monochasia or paniculiform thyrses. Sepals 4(-5-6); each sepal with (3-4-)5(-6-7) basal nerves. Filaments narrow-linear or filiform. Carpel with a straight style and a conspicuous stigma. Achene on each side longitudinally (2-)3-ribbed, not winged, rarely along ventral suture and dorsal suture narrowly winged (*T. delavayi*).

13 种，隶属 3 系，其中 12 种分布于中国西南部和喜马拉雅山区，1 种分布于华北。

系 1. 帚枝唐松草系

Ser. **Virgata** W. T. Wang & S. H. Wang in Fl. Reip. Pop. Sin. 27: 562, 619. 1979. Type: *T. virgatum* Hook. f. & Thoms.

叶为复叶。花组成少花的复单歧聚花序。萼片 4(-5)，长 3-14 mm；每一萼片具 3-6

条多为 1–3 回二歧状分枝的基生脉。花丝狭条形，稀上部条形，下部丝形（金丝马尾连）。柱头无翅，或具 2 侧生狭翅（攀枝花唐松草，金丝马尾连）。瘦果无翅。

Leaves compound. Flowers in few-flowered corymbiform compound monochasia. Sepals 4(–5), 3–14 mm long; each sepal with 3–6 mostly 1–thrice dichotomizing basal nerves. Filaments narrow-linear, rarely above linear, below filiform (*T. glandulosissimum*). Stigma not winged, or with 2 lateral narrow wings (*T. panzhihuaense* and *T. glandulosissimum*). Achenes not winged.

5 种，其中 4 种分布于中国西南部，1 种分布于华北。

64. 帚枝唐松草（《药学学报》）阴阳和（云南）　图版 64

Thalictrum virgatum Hook. f. & Thoms. Fl. Ind. 14. 1855; et in Hook. f. Fl. Brit. Ind. 1: 12. 1872; Lecoy. in Bull. Soc. Bot. Belg. 24: 189. 1885; Franch. Pl. Delav. 13. 1886; Finet & Gagnep. in Bull. Soc. Bot. France 50: 615. 1903; Lévl. in Repert. Sp.Nov. Reg. Veg. 7: 909. 1909; et Cat. Pl. Yunnan 227. 1917; Hand.–Mazz. Symb. Sin. 7: 311. 1931; Boivin in J. Arn. Arb. 26: 115. 1945; Lauener & Green in Not. Bot. Gard. Edinburgh 23(4):594. 1961; W. T. Wang & S. H. Wang in Fl. Reip. Pop. Sin. 27: 562, pl. 141: 1–5. 1979; W. T. Wang in Bull. Bot. Lab. N.-E. Forest. Inst. 8: 31. 1980; et in Iconogr. Corm. Sin. Suppl. 1: 427, fig. 8615. 1982; Grierson, Fl. Bhutan 1(2):296. 1984; W. T. Wang in Fl. Xizang. 2: 69. 1985; in Vasc. Pl. Hengduan Mount. 1: 502. 1993; in Fl. Yunnan. 11: 167, pl. 50: 1–5. 2000; et in High. Pl. China 3: 474, fig. 754. 2000; D. Z. Fu & G. Zhu in Fl. China 6: 301. 2001; W. T. Wang in High. Pl. China in Colour 3: 390. 2016; L. Xie in Med. Fl. China 3: 261, with fig. & photo. 2016; X. T. Ma et al. Wild Flow. Xizang 28, photo 55. 2016. Syntypes: Sikkim, alt. 6–10000 ft., J. D. Hooker s. n. (K, not seen). Bhutan, W. Griffith s. n. (K, not seen).

T. virgaturm var. *stipitatum* Franch. in Bull. Soc. Bot. France 33: 368. 1886. Holotype：云南：鹤庆之北，Hee-gui-chao, Delavay 103 (P, not seen).

T. verticillatum Lévl. in Repert. Sp. Nov. Reg. Veg. 7: 97. 1909. Holotype：云南：昆明。Tchong-Chan，1904-08-10，Ducloux 602 (E, not seen).

T. englerianum Ulbr. in Bot. Jahrb. 48: 623. 1913. Type：云南：Hochweiden des Pe-long-tsin, alt. 3300 m, 1910-08, E. E. Maire 2661 (holotype, B, not seen; isotype, NAS).

多年生草本，全部无毛。茎高 16–65 cm，分枝或不分枝。叶均茎生，7–10，为三出复叶，有短柄或无柄；小叶纸质，菱状三角形或宽菱形，0.9–2.5×0.6–2.4 cm，在顶端微尖或钝，

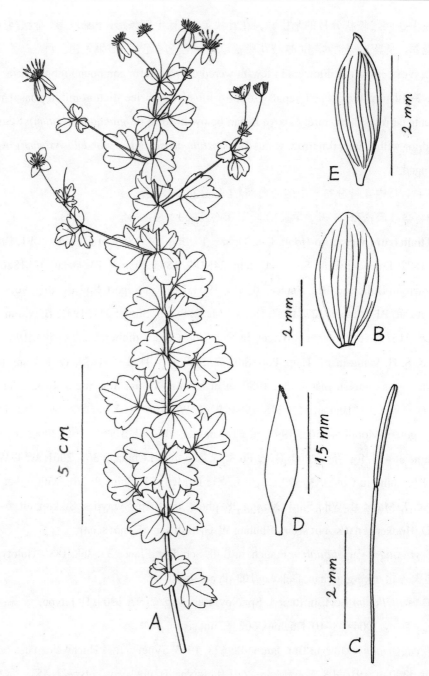

图版64. **帚枝唐松草** A. 植株全形，B. 萼片，C. 雄蕊，D. 心皮，（据秦仁昌 30875）E. 瘦果（据谢磊，毛建丰 29）。

Plate 64. **Thalictrum virgatum** A. habit, B. sepal, C. stamen, D. carpel, (from R. C. Ching 30875) E. achene. (from L. Xie & J. F. Mao 29)

在基部宽楔形或近截形,3 浅裂,边缘有少数圆齿,脉两面隆起,脉网明显;小叶柄长 0.8-13 mm。复单歧聚伞花序顶生,伞房状,宽 2.5-3.5 cm,有 3-5 花,有时,分枝的单歧聚伞花序简单,有(1-)2 花;花梗细,长 0.8-1.8 cm。花:萼片 4(-5),脱落,白色或带粉红色,卵形或宽椭圆形,4-8×2.5-4 mm,在顶端微钝,每一萼片具 3 条 3 回二歧状分枝的基生脉。雄蕊约 18;花丝狭条形,长 2-3.5 mm,上部宽 0.2 mm,下部宽 0.1 mm;花药条形或狭长圆形,1.8-2×0.2 mm,顶端钝。心皮 10-25;子房狭卵形,约 1.5×0.6 mm,基部的柄长 0.4 mm;花柱长 0.4 mm;柱头长 0.3 mm。瘦果扁,椭圆形,2-2.5×0.6-1 mm,每侧有 3-4 纵肋,基部的柄长约 0.4 mm;宿存花柱长约 0.2 mm。6-8 月开花。

Perennial herbs, totally glabrous. Stems 16–65 cm tall, branched or simple. Leaves all cauline, ternate, shortly petiolate or sessile; leaflets papery, rhombic-triangular or broadly rhombic, 0.9–2.5×0.6–2.4 cm, at apex slightly acute or obtuse, at base broadly cuneate or subtruncate, 3-lobed, at margin sparsely rotund-dentate, nerves on both surfaces prominent and nervous nets conspicuous; petiolules 0.8–13 mm long. Compound monochasia terminal, corymbiform, 2.5–3.5 cm broad, 3–5-flowered, sometimes monochasia borne on branch apexes simple, (1–)2-flowered; pedicels slender, 0.8–1.8 cm long. Flower: Sepals 4(–5), deciduous, white or pink-tinged, ovate or broadly elliptic, 4–8×2.5–4 mm, at apex slightly obtuse, each sepal with 3 often thrice dichotomizing basal nerves, Stamens ca. 18; filaments narrow-linear, 2–3.5 mm long, above 0.2 mm broad, below 0.1 mm broad,; anthers linear or narrow-oblong, 1.8–2×0.2 mm, apex obtuse. Carpels 10–25; ovary narrow-ovate, ca. 1.5×0.6 mm, base with a stipe 0.4 mm long; style 0.4 mm long; stigma 0.3 mm long. Achene compressed, elliptic, 2–2.5×0.6–1 mm, on each side longitudinally 3–4-ribbed, base with a stipe ca. 0.4 mm long; persistent style ca. 0.2 mm long.

分布于中国西藏南部、云南西部和北部和四川西部;以及不丹,尼泊尔,印度北部。生山地林下或林边岩石上,海拔 2300-3500 m。

根用于胃病。本种的根在四川、云南、西藏作"阴阳和"药材使用。(《中国药用植物志》,2016)

标本登录:西藏:吉隆,西藏中草药队 329,712;聂拉木,张永田,郎楷永 4529,西藏中草药队 1404;樟木,综考会队 75-578;自曲水至聂拉木,张永田,郎楷永 4588;卡马河下游,1959 年登山队 375;错那,PLPH Tibet Exped. 12-0977。

云南:镇康,俞德浚 16945,17189;宾川,谢磊,毛建丰 29;嵩明,D. E. Boufford et

al. 34981; 洱源, 秦仁昌 23078, 23278, 滇西北金沙江队 63-6198; 丽江, 蔡希陶 52966, 王启无 71534, 71595, 俞德浚 15313, 秦仁昌 30313, 30875, 冯国楣 8948, 青藏队 81-339, 植物所横断山队 81-2509, 杨亲二 9602, 杨亲二、孔宏智 98-501, 杨亲二、袁琼 143, 谢磊、毛建丰 40; 宁蒗, 韩裕丰 81-1073, 谢磊、毛建丰 61; 永宁, 姜恕 6094; 鹤庆, 秦仁昌 23933; 兰坪, 蔡希陶 53782; 中甸, 俞德浚 12047, 12389, 冯国楣 1385, 中甸队 62-1281, 汤宗孝 576, 杨亲二、孔宏智 98-112, 杨亲二、袁琼 2112, 谢磊、毛建丰 104。

四川: 盐源, 青藏 82-12114; 木里, 俞德浚 6595, 6771, 7283, 7690, 14544, 青藏队 83-13039, 83-14531, 陈又生 4324; 越西, 四川经济植物队 3837, 四川植物队 13953; 九龙, 应俊生 4013, D. E. Boufford et al. 33164; 稻城, 俞德浚 12962, 四川植被队 2057, 青藏队 81-4220, 郎楷永、李良千、费勇 2546; 雅江, 姜恕 5209, 郎楷永、李良千、费勇 2989; 康定, 姜恕 5023, 郎楷永、李良千、费勇 974。

65. 希陶唐松草　图版 65

Thalictrum tsaii W. T. Wang, sp. nov. Holotype: 云南: 无野外记录, 蔡希陶 (H. T. Tsai) 57289(PE, *T. smithii* Boivin 的 paratype).

T. smithii Boivin in J. Arn. Arb. 26: 114. 1945, p.p. quoad paratypum H. T. Tsai 57289.

Species nova haec affinis *T. virgato* Hook. f. & Thoms. est, a quo foliis caulinis 2–3-ternato-pinnatis vel 5-foliolatim pinnatis, staminum filamentis filiformibus brevioribus 1–2 mm longis, ovariis sessilibus, stigmatibus 0.6–0.8 mm longis differt. In *T. virgato*, folia caulina omnia ternata, filamenta anguste lineria 2–3.5 mm longa, ovaria stipitata, et stigmata breviora 0.3 mm longa sunt.

Perennial herbs, totally glabrous. Stems 40–45 cm tall, just supra base long 2–4-branched, branches below 1 mm in diam., with a few shorter slender secondary branches. Basal leaves absent. Cauline leaves 2–3-ternate-pinnate, and the upper-most one 5-foliolately pinnate; blades 3–11×2–6 cm; leaflets papery, broad-obovate, obovate or broad-elliptic, 0.5–1.6×0.3–1.4 cm, at base rounded, 3-lobed or 3-lobulate, rarely 3-fid (lobes at apex rounded or truncate, entire, rarely 1-denticulate), 3–5-nerved, nerves thin, flat; patiolules slender, 0.8–10 mm long; petioles 1–5 cm long, at base slightly dilated. Compound monochasia terminal and axillary, corymbiform, sparsely 5–10-flowered, long pedunculate; peduncles slender, 5–10 cm long; bracts ternate, or simple, leaflet-like; pedicels filiform, 0.2–2.2 cm long. Flower: Sepals 4, obovate, ca. 3×1.8 mm, at apex rounded, each sepal with 3 basal twice dichotomizing nerves.

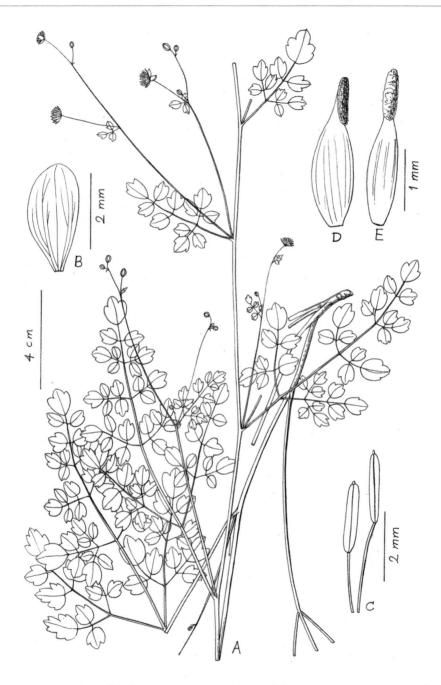

图版65. **希陶唐松草** A. 植株全形，B. 萼片，C. 二雄蕊，D. 心皮（侧面观），E. 心皮（腹面观）。（据蔡希陶57289）

Plate 65. **Thalictrum tsaii** A. habit, B. sepal, C. two stamens, D. carpel (lateral view), E. carpel (ventral view). (from H. T. Tsai 57289)

Stamens 16–18; filaments filiform, 1–2 mm long; anthers narrow-oblong, 2×0.3–0.5 mm, at apex with mucrones 0.1 mm long. Carpels ca. 9; ovary sessile, compressed, broad-elliptic, ca. 1.5×0.6 mm, on each side inconspicuously longitudinally 3–ribbed; style 0.7 mm long; stigma oblong or narrow-triangular, 0.6–0.8×0.4 mm.

多年生草本植物，全部无毛。茎高 40–45 cm，自基部之上具 2–4 条长分枝，枝下部直径 1 mm，具少数较短、二回细分枝。基生叶不存在。茎生叶为 2–3 回三出羽状复叶，最上部茎生叶为具 5 小叶的羽状复叶；叶片 3–11×2–6 cm；小叶纸质，宽倒卵形。倒卵形或宽椭圆形，0.5–1.6×0.3–1.4 cm，基部圆形，3 浅裂或 3 微裂，稀 3 中裂（裂片顶端圆形，全缘，稀具 1 小齿），具三出或五出脉，脉细平；小叶柄细，长 0.8–10 mm；叶柄长 1–5 cm，基部稍变宽。复单歧聚伞花序顶生并腋生，伞房状，有稀疏的 5–10 花，具长梗；花序梗细，长 5–10 cm；苞片为三出复叶，或为单叶，呈小叶状，花梗丝形，长 0.3–2.2 cm。花：萼片 4，倒卵形，约 3×1.8 mm，顶端圆形，每一萼片具 3 条二回二歧状分枝的基生脉。雄蕊 16–18；花丝丝形，长 1–2 mm；花药狭长圆形，2×0.3–0.5 mm，顶端具长 0.1 mm 的短尖头。心皮约 9；子房无柄，扁，宽椭圆形，约 1.5×0.6 mm，每侧具 3 条不明显纵肋；花柱长 0.7 mm；柱头长圆形或狭三角形，0.6–0.8×0.4 mm。

分布于云南西部。

本种与帚枝唐松草 *T. virgatum* 近缘，区别特征为本种的茎生叶为 2–3 回三出羽状复叶或为具 5 小叶的羽状复叶，花丝丝形，较短，长 1–2 mm，子房无柄，柱头长 0.6–0.8 mm。

中国著名植物学家蔡希陶教授（1911–1981）于 1931–1934 年在云南全省采集，采到大量植物标本，为《中国植物志》的编著做出重大贡献。本种模式标本 H. T. Tsai 57289，但无记载此号标本的野外记录本，因此其采集地点不明，经查存 PE 的蔡教授的其他野外记录本，此号标本之前 57000 号于 1934 年 2 月 20 日采自云南芒市（潞西），其之后的 57831 号于同年 9 月 9 日采自云南维西，据此推测 57289 号可能采自云南西部维西以南某地。

66. 丝叶唐松草（《东北植物检索表》） 图版 66

Thalictrum foeniculaceum Bunge in Mém. Sav. Etrang. Acad. Sci. St.-Petersb. 2: 76. 1833; Lecoy. in Bull. Soc. Bot. Belg. 24: 22. 1885; Forbes & Hemsl. in J. Linn. Soc. Bot. 23: 8. 1886; Maxim. in Acta Hort. Petrop. 11: 11. 1890; Finet & Gagnep. in Bull Soc. Bot. France 50: 623. 1903; Drumm. & Hutch. in Kew Bull. 1920: 169. 1920; Kitag. Lineam. Fl. Mansh.

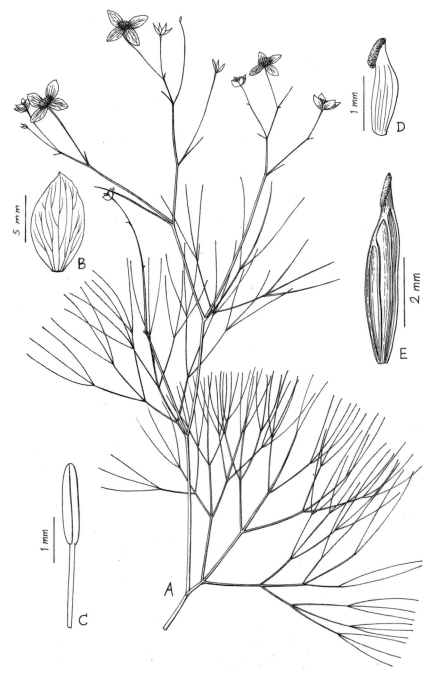

图版66. **丝叶唐松草** A. 开花茎上部，B. 萼片，C. 雄蕊，D. 心皮，E. 瘦果。（据迁西队579）
Plate 66. **Thalictrum foeniculaceum** A. upper part of flowering stem, B. sepal, C. stamen, D. carpel, D. achene. (from Qianxi Exped. 579)

227. 1939; Sealy in Hook. Icon. Pl. 35: pl. 3428. 1943; Lauener & Green in Not. Bot. Gard Edinburgh 23(4):590. 1961; Iconogr. Corm. Sin. 1: 680, fig. 1359. 1972; Fl. Tsinling. 1(2):244, fig. 208. 1974; S. H. Li in Fl. Pl. Herb. Chinae Bor.-Or. 3: 218, pl. 96. 1975; W. T. Wang & S. H. Wang in Fl. Reip. Pop. Sin. 27: 564, pl. 142: 58. 1979; S. Y. He, Fl. Beijing, rev. ed., 1: 242, fig. 303. 1984; J. W. Wang in Fl. Hebei. 1: 441, fig. 436. 1986; D. Z. Ma, Fl. Ningxia. 1: 172, fig. 149. 1986; X. Y. Yu et al. in Fl. Shanxi. 1: 600, fig. 380. 1992; S. H. Li in P. Y. Fu, Clav. Pl. Chinae Bor.-Or., ed. 2: 205, pl. 97: 5. 1995; L. H. Zhou in Fl. Qinghai. 1: 333. 1997; W. T. Wang in High. Pl. China 3: 474, fig. 755. 2000; D. Z. Fu & G. Zhu in Fl. China 6: 290. 2001; F. H. Li et al. Pl. Yanqing 141. 2015; W. T. Wang in High. Pl. China in Colour 3: 390. 2016; Y. Zhou, Handb. Endang. Pl. N.-E. China 1: 404. 2016. Type：北京（？），1831，A. Bunge s. n. (holotype, LE, not seen; isotype, PE).

Isopyrum trichophyllum Lévl. in Repert. Sp. Nov. Reg. Veg. 9: 224. 1911. Holotype: Pe-Tche-Ly, Kiu-Yang, 1910–08, L. Chanet 547 (not seen).

多年生草本，全部无毛。茎高 11–78 cm，不分枝或上部分枝。基生叶 2–6，长 5–18 cm，为 2–4 回三出复叶；叶片长 2.5–10 cm；小叶薄革质，狭条形或钻状狭条形，0.6–3×0.03–0.15 cm，在顶端尖锐，边缘常外卷；叶柄长 1.5–9 cm，基部有短鞘。茎生叶 3–4，似基生叶，但较小。复单歧聚伞花序伞房状，宽约 7 cm，约有稀疏 10 花；花梗纤细，长 2–4.5 cm。花：萼片 4，粉红色或白色，椭圆形或狭倒卵形，6–11×3–6 mm，在顶端急尖，每一萼片具 5 条 1–4 回二歧状分枝的基生脉。雄蕊 25–28，长达 3 mm；花丝狭条形，长 1–1.6 mm；花药长圆形，1.4–1.8×0.25–0.5 mm，顶端钝或有小短尖头。心皮 3–11，无柄；子房斜狭倒卵形，约 1×0.5 mm；花柱长约 0.5 mm；柱头狭长圆形，长约 0.6 mm。瘦果纺锤形，3.5–4.5×1.2 mm，每侧有 2–3 条纵肋；宿存花柱长约 1.2 mm。6–7 月开花。

Perennial herbs, totally glabrous. Stems 11–78 cm tall, simple or above branched. Basal leaves 2–6, 5–18 cm long. 2–4-ternate; blades 2.5–10 cm long; leaflets subcoriaceous, narrow-linear or narrow-subulate-linear, 0.6–3×0.03–0.15 cm, at apex pungent, margin often revolute; petioles 1.5–9 cm long, base shortly vaginate. Cauline leaves 3–4, similar to basal leaves, but smaller. Compound monochasia terminal, corymbiform, ca. 7 cm broad, ca. 10-flowered; pedicels slender, 2–4.5 cm long, Flower: Sepals 4, pink or white, elliptic or narrow-obovate, 6–11×3–6 mm, at apex acute, each sepal with 5 1–4 times dichotomizing basal nerves. Stamens 25–28, up to 3 mm long; filaments narrow-linear, 1–1.6 mm long; anthers oblong,

1.4–1.8×0.25–0.5 mm, apex obtuse or minutely mucronate. Carpels 3–11, sessile; ovary obliquely narrow-obovate, ca. 1×0.5 mm; style ca. 0.5 mm long; stigma narrow-oblong, ca. 0.6 mm long. Achene fusiform, 3.5–4.5×1.2 mm, on each side longitudinally 2–3-ribbed; persistent style ca. 1.2 mm long.

分布于甘肃中部、陕西北部、山西、山东中部、河北北部和辽宁西部。生山地干燥草坡、山脚沙地或多石砾处或平原草地，海拔 590–1000 m。

标本登录：甘肃：兰州，郝景盛 715；临洮，王作宾 14002；陇西，黄河队 4989；会宁，黄河队 5131；通渭，黄河队 5060；镇原，仲世奇 4。

陕西：商洛，崔友文 10332；延川，傅坤俊 7846；清涧，傅坤俊 7868。

山西：解州，P. Licent 7725；新绛，P. Licent 1969；太原，H. Smith 6834，夏纬瑛 1014，王作宾 2689，张耀 1012；临县，张德民 25。

山东：沂源，王德汇 s. n.。

河北：小五台山，郝景盛 624，杨承元 36153；迁西，PE 迁西队 579；张家口，鸡鸣山，王启无 62117（PE），诚静容 s. n. (PEM)。

北京：房山，PE 房山队 628；昌平，PE 昌平队 642。

天津：蓟县，盘山，河北普查队 438。

67. 大花唐松草（《秦岭植物志》） 图版 67

Thalictrum grandiflorum Maxim. in Acta. Hort. Petrop. 11: 11. 1890; Finet & Gagnep. in Bull. Soc. Bot. France 50: 614. 1903; Fl. Tsinling. 1(2): 245, fig. 209. 1974; W. T. Wang & S. H. Wang in Fl. Reip. Pop. Sin. 27: 564, pl. 142: 8–10. 1979; W. T. Wang in Vasc. Pl. Hengduan Mount. 1: 502. 1993; et in High. Pl. China 3: 474, fig. 756. 2000; D. Z. Fu & G. Zhu in Fl. China 6: 291. 2001; L. Q. Li et al. Pl. Baishuijiang St. Nat. Reserve Gansu Prov. 71. 2014. Type：甘肃：inter vicos Dsao-li-ping et Shun-dan-sien, 1885–09–07, G. N. Potanin s. n. (holotype, LE, not seen; isotype, PE).

多年生草本，全部无毛。茎高 20–28 cm，中部以上分枝。基生叶约 2，与下部茎生叶为 3 回三出羽状复叶，有稍长柄；叶片长 4.5–9.5 cm；小叶纸质，卵形或圆卵形，5–23×2–14 mm，在顶端急尖或圆形，在基部圆形，宽楔形或浅心形，全缘，稀 3 浅裂或有 1 钝齿，脉不明显；小叶柄长 1–12 mm；叶柄长 3–6 cm。复单歧聚伞花序顶生，伞房状，宽 4–8 cm，有 3–7 花；花梗长 1.6–5 cm。花：萼片 4，粉红色，狭卵形或卵形，1–2×0.5–1 cm，在顶端急尖，每一萼片有 6 条基生脉（5 条基生脉 1–3 回二歧状分枝，其他 1 条不分枝）。

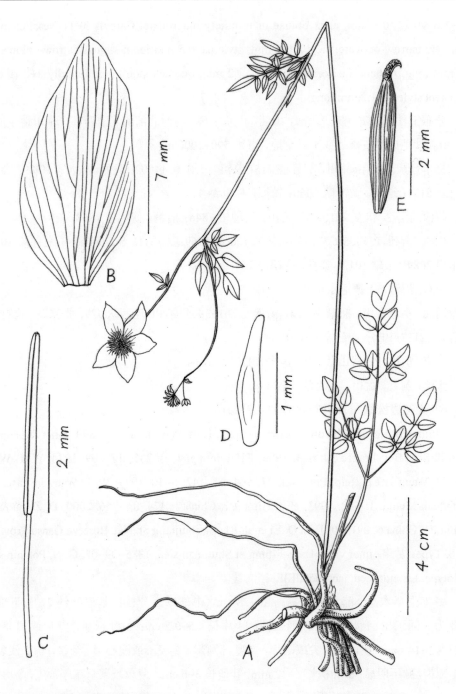

图版67. **大花唐松草**　A. 植株全形，B. 萼片，C. 雄蕊，D. 心皮，E. 瘦果。（据陈伟烈等8775）
Plate 67. **Thalictrum grandiflorum**　A. habit, B. sepal, C. stamen, D. carpel. E. achene. (from W. L. Chen et al.. 8775)

雄蕊为 40，长约 6 mm；花丝狭条形，2-2.8×0.25 mm；花药条形或狭长圆形，3-3.2×0.25 mm，顶端具长 0.2-0.25 mm 的短尖头。心皮 40-50；子房长圆形，长约 1-0.5 mm；花柱长约 0.7 mm；柱头长约 0.6 mm。瘦果纺锤形，约 3×0.8 mm，每侧有 2-3 条纵肋；宿存花柱长约 0.8 mm。8-10 月开花。

Perennial herbs, totally glabrous. Stems 20−28 cm tall, above the middle branched. Basal leaves ca. 2, with lower cauline leaves 3-ternate-pinnate, slightly long petiolate; blades 4.5−9.5 cm long; leaflets papery, ovate or rotund-ovate, 5−2.3×2−14 mm, at apex acute or rounded, at base rounded, broadly cuneate or subcordate, entire, rarely 3-lobulate or 1-crenate, nerves inconspicuous; petiolules 1−12 mm long; petioles 3−6 cm long. Compound monochasia terminal, corymbiform, 4−8 cm broad, 3−7-flowered; pedicels 1.6−5 cm long. Flower: Sepals 4, pink, narrow-ovate or ovate, 1−2×0.5−1 cm, at apex acute, each sepal with 6 basal nerves (5 nerves 1−thrice dichotomizing, and the another one simple). Stamens ca. 40, ca. 6 mm long; filaments narrow-linear, 2−2.8×0.25 mm; anthers linear or narrow-oblong, 3−3.2× 0.25 mm, apex with mucrones 0.2−0.25 mm long. Carpels 40−50; ovary oblong, ca. 1×0.5 mm; style ca. 0.7 mm long; stigma ca. 0.6 mm long. Achene fusiform, ca. 3×0.8 mm, on each side longitudinally 2−3-ribbed; persistent style ca. 0.8 mm long.

分布于四川北部（南坪）和甘肃南部（文县）。生海拔 1000 m 一带山地草坡上。

标本登录：四川：南坪，李全喜，赵兴存 2572（NWTC）。

甘肃：文县，陈伟烈、陈家瑞等 8775，杨亲二 92016。

68. 攀枝花唐松草（《植物研究》） 图版 68

Thalictrum panzhihuaense W. T. Wang in Bull. Bot. Res. Harbin 36(1): 3, fig. 1: E−F. 2016. Holotype: 四川：攀枝花市，苏铁山，alt. 1600−1700 m，2002−09−13，张钢民 638（PE）.

多年生草本。茎高约 30 cm，与叶柄、叶轴和花梗均密被短腺毛（毛长 0.1-0.2 mm），多分枝。茎生叶为具 5 小叶的羽状复叶或为三出复叶，具短柄；叶片狭卵形或正三角形，2.5-5×2.2-3.5 cm；小叶纸质，具柄，宽卵形或五角形。0.8-1.6×0.8-2.2 cm，在基部浅心形，3 微裂或不分裂，在边缘有稀疏浅齿或圆牙齿，上面密被短腺毛，下面脉上被短硬毛；基出脉 5 条，在上面平，在下面隆起；小叶柄纤细，长 0.4-1.5 cm；叶柄长 0.5-0.8 cm。单歧聚伞花序顶生，有 2-4 花；苞片为三出复叶或为单叶，似小叶；花梗纤细，长 0.5-1.7 cm。花：萼片和雄蕊均脱落，未见。心皮 2-4；子房长椭圆体形或狭披针形，1.5-1.8×0.4 mm，密被腺短毛；柱头与花柱贴生，狭三角形，约 1×0.3 mm，无毛，具 2

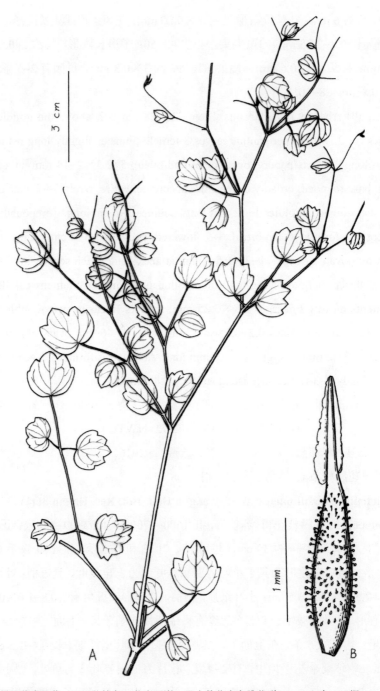

图版68. 攀枝花唐松草　A.开花枝条（花多凋谢，只少数花尚有雌蕊群），B.心皮。（据holotype）
Plate 68. **Thalictrum panzhihuaense**　A. flowering branch, with most flowers having wilted and a few of them still bearing gynoecium, B. carpel. (from holotype)

侧生膜质狭翅。8-9 月开花。

Perennial herbs. Stem ca. 30 cm tall, with petioles, leaf rachises and pedicels all densely glandular-puberulous (hairs 0.1–0.2 mm long), much branched. Basal leaves not seen. Cauline leaves 5–foliolately pinnate or ternate, shortly petiolate; blades narrowly ovate or deltoid in outline, 2.5–5×2.2–3.5 cm; leaflets papery, petiolulate, broadly ovate or pentagonal, 0.8–1.6×0.8–2.2 cm, at base subcordate, 3-lobulate or undivided, at margin sparsely crenate or rotund-dentate, adaxially densely glandular-puberulous, abaxially on nerves hirtellous; basal nerves 5, adaxially flat, abaxially prominent; petiolules slender, 0.4–1.5 cm long; petioles 0.5–0.8 cm long. Monochasia terminal, 2–4-flowered; bracts ternate or simple and similar to leaflets; pedicels slender, 0.5–1.7 cm long. Flower: Sepals and stamens dropped out. not seen. Carpels 2–4; ovary long ellipsoidal or narrow-lanceolate, 1.5–1.8×0.4 mm, densely glandular-puberulous; stigma entirely adnate with style, narrow-triangular, ca. 1×0.3 mm, glabrous, with 2 lateral narrow membranous wings.

特产四川攀枝花市。生山坡灌丛中，海拔 1600–1700 m。

69. 粘唐松草（《中国植物志》） 图版 69

Thalictrum viscosum C.Y. Wu ex W. T. Wang & S. H. Wang in Fl. Reip. Pop. Sin. 27: 567, 619, pl. 145: 1–3. 1979; W. T. Wang in Vasc. Pl. Hengduan Mount. 1: 502. 1993; et in Fl. Yunnan. 11: 169. 2000; D. Z. Fu & G. Zhu in Fl. China 6: 301. 2001. Type：云南：丽江，大具，虎跳峡，alt. 1810 m，1962-08-08，中甸队 62-1238（holotype, PE; isotype, KUN）.

多年生草本。茎高约 60 cm，上部和中轴均被小腺毛（毛长约 0.05 mm），多分枝。中部茎生叶为 3 回羽状复叶，有短柄，长约 14 cm；叶片长约 12 cm；小叶草质，宽卵形或宽菱状卵形，长和宽均为 0.6–1.1 cm，顶端圆形或近截形，基部心形，3 浅裂（裂片全缘或有 2–3 浅圆齿），两面密被小腺毛（毛长 0.05 mm），下面疏被短柔毛（毛长 0.1–0.3 mm）；小叶柄长 2.5–5.5 mm；叶柄长约 2 cm，有狭鞘。单歧聚伞花序顶生，长 2–3 cm，有 1–3 花，密被小腺毛；苞片为三出复叶；花梗长 0.3–1.4 cm。花：萼片 4，白色或淡紫色，椭圆形。3×1–1.5 mm，顶端钝，背面疏被短柔毛，每一萼片具 3 条基生脉（中脉不分枝，2 侧生脉 1–2 回二歧状分枝）。雄蕊 14–17，无毛；花丝狭条形，3–3.2×0.2 mm；花药长圆形，1×0.25–0.4 mm，顶端钝。心皮 1–3；子房狭卵形，约 1.5×0.5 mm，被小腺毛；花柱钻形，长 1.2–1.4 mm；柱头狭条形，长 0.8–1 mm。瘦果狭卵形或狭椭圆形，约 3.8–1.2 mm，密被小腺毛，每侧有 2–3 条纵肋；宿存花柱长 1.2–1.5 mm。7–8 月开花。

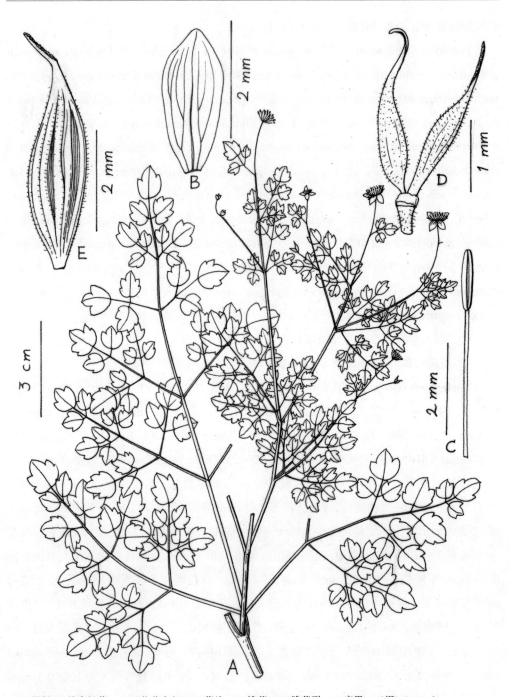

图版69. **粘唐松草**　A. 开花茎上部，B. 萼片，C. 雄蕊，D. 雌蕊群，E. 瘦果。（据holotype）
Plate 69. **Thalictrum viscosum**　A. upper part of flowering stem, B. sepal, C. stamen, D. gynoecium, E. achene. (from holotype)

Perennial herbs. Stems ca. 60 cm tall, above with leaf rachises minutely glandular-puberulous (hairs ca. 0.05 mm long), much branched. Middle cauline leaves 3-pinnate, shortly petiolate, ca. 14 cm long; blades ca. 12 cm long; leaflets herbaceous, broadly ovate or broadly rhombic-ovate, $0.6-1.1 \times 0.6-1.1$ cm, at apex rounded or subtruncate, at base subcordate, 3-lobed (lobes entire or 2–3-crenate), on both surfaces densely minutely glandular-puberulous (hairs 0.05 mm long), abaxially sparsely puberulous (hairs 0.1–0.3 mm long); petiolules 2.5–5.5 mm long; petioles ca. 2 cm long, base narrowly vaginate. Monochasia terminal, 2–3 cm long, 1–3-flowered, densely minutely glandular-puberulous; bracts ternate; pedicels 0.3–1.4 cm long. Flower: Sepals 4, white or purplish, elliptic, $3 \times 1-1.5$ mm, at apex obtuse, abaxially sparsely puberulous, each sepal with 3 basal nerves (midrib simple, 2 lateral nerves 1–twice dichotomizing). Stamens 14–17, glabrous, filaments narrow-linear, $3-3.2 \times 0.2$ mm; anthers oblong, $1 \times 0.25-0.4$ mm, apex obtuse. Carpels 1–3; ovary narrow-ovate, ca. 1.5×0.5 mm, minutely glandular-puberulous; style subulate, 1.2–1.4 mm long; stigma narrow-linear, 0.8–1 mm long. Achene narrow-ovate or narrow-elliptic, ca. 3.8×1.2 mm, densely minutely glandular-puberulous, on each side longitudinally 2–3-ribbed; persistent style 1.2–1.5 mm long.

特产云南丽江虎跳峡。生河谷山坡草丛中，海拔 1800 m。

70. 金丝马尾连　马尾连（云南宾川）　图版 70

Thalictrum glandulosissimum (Finet & Gagnep.) W. T. Wang & S. H. Wang in Fl. Reip Pop. Sin. 27: 567, pl. 145: 1–5. 1979; W. T. Wang in Iconogr. Corm. Sin. Suppl 1: 428, fig. 8616. 1982; et in Fl. Yunnan. 11: 169. 2000; D. Z. Fu & G. Zhu in Fl. China 6: 291. 2001; L. Xie in Med. Fl. China 3: 261 with fig. & photo. 2016. —*T. foetidum* L. var. *glandulosissimum* Finet & Gagnep in Bull. Soc. Bot. France 50: 619. 1903. Syntypes (not seen): 云南，近大理，Ma-eul-chan, 1877-08-16, Delavay 1840(P); 同地, 1888-09-27, Delavay 3203(P); Pi-jouse, 1887-07-16, Delavay 4813(P)

70a. var. glandulosissimum

多年生草木。根状茎有多数粗须根。茎高 60-85 cm，上部有腺毛（毛长 0.2 mm），分枝，约有 9 叶。叶为 3 回羽状复叶；叶片长 5-9.5 cm；小叶草质，宽倒卵形，椭圆形或近圆形，$0.7-1.6 \times 0.7-1.6$ cm，基部圆形或浅心形，3 浅裂（裂片全缘，或中央裂片有时具 2-3 浅圆齿），上面密被小腺毛，下面脉上密被短柔毛，脉平，不明显；小叶柄长 0.5–

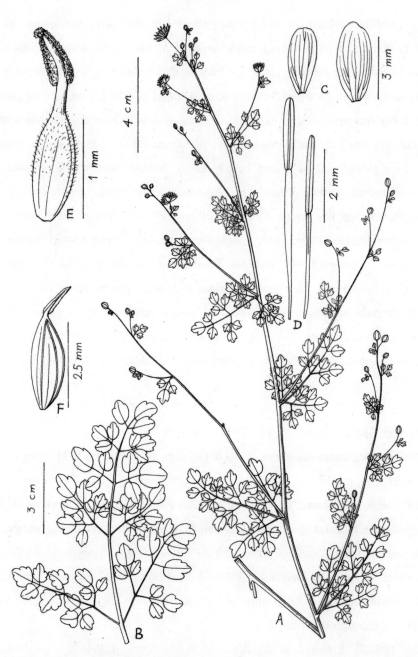

图版70. **金丝马尾连** A. 开花茎上部，B. 中部茎生叶，C. 二萼片，D. 二雄蕊，E. 心皮，（据秦仁昌 22925）F. 瘦果。（据吴征镒，刘德仪 20374）

Plate 70. **Thalictrum glandulosissimum** var. **glandulosissimum** A. upper part of flowering stem, B. middle cauline leaf, C. two sepals, D. two stamens, E. carpel, (from R. C. Ching 22925) F. achene. (from Z.Y. Wu & D. Y. Liu 20374)

1 cm；叶轴的毛长达 0.2 mm；叶柄长达 4 cm，基部有短鞘。复单歧聚伞花序顶生并腋生，多少伞房状，宽 1-5 cm，有 2-7 花，此类 4-5 个花序组成顶生长 12-25 cm 的聚伞圆锥花序；花梗细，长 0.3-2 cm。花：萼片 4，早落，黄白色，近长圆形或狭倒卵形，3-3.5×1.8-2 mm，顶端圆形，背面疏被短柔毛，每一萼片具 3-4 条基生脉（中脉不分枝或二回二歧状分枝，其他侧生脉 1-2 回二歧状分枝）。雄蕊 23-26，无毛；花丝长 2-3.2 mm，上部条形，宽 0.2 mm，下部狭条形，宽 0.1 mm；花药狭长圆形或条形，2×0.2-0.4 mm，顶端钝。心皮 4-5，无柄；子房椭圆体形，约 1×0.6 mm；花柱长 1.2 mm；柱头狭三角形，约 1×0.4 mm，有 2 侧生狭翅。瘦果纺锤形或椭圆体形，2.5-3×1.5 mm，密被短毛，每侧有 2-3 条纵肋；宿存柱头长约 1.2 mm。6-8 月开花。

Perennial herbs. Rhizome short, with many thick fibrous roots. Stems 60-85 cm tall, above with glandular hairs 0.2 mm long, branched, ca.9-leaved. Leaves 3-pinnate; blades 5-9.5 cm long; leaflets herbaceous, broadly oborate, elliptic or suborbicular, 0.7-1.6×0.7-1.6 cm, at base rounded or subcordate, 3-lobed (lobes entire, or central lobe sometimes 2-3-crenate), adaxially covered with dense minute glandular hairs; abaxially on nerves densely puberulous, nerves flat, inconspicuous; petiolules 0.5-1 cm long; rachis with hairs up to 0.2 mm long; petioles up to 4 cm long, at base shortly vaginate. Compound monochasia terminal and axillary, more or less corymbitorm, 1-5 cm broad, 2-7-flowered, these 4-5 monochasia forming terminal thgrse12-25 cm long; pedicels slender, 0.3-2 cm long. Flower: Sepals 4, caducous, yellowish-white, suboblong or narrowly obovate, 3-3.5×1.8-2 mm, at apex rounded, abaxially sparsely puberulous, each sepal with 3-4 basal nerves (rnidrib simple or twice dichotomizing, and the other lateral nerves once or twice dichotomizing). Stamens 23-26, glabrous; filaments 2-3.2 mm long, above linear, 0.2 mm broad, below narrow-linear, 0.1 mm broad; anthers narrow-oblong or linear, 2×0.2-0.4 mm, apex obtuse. Carpels 4-5; ovary ellipsoidal, ca. 1×0.6 mm; style 1.2 mm long; stigma narrow-triangular, ca. 1×0.4 mm. with 2 lateral narrow wings. Achene fusiform or ellipsoidal, 2.5-3×1.5 mm, densely puberalous, on each side longitudinally 2-3-ribbed; persistent stigma ca. 1.2 mm long.

特产云南（大理、宾川和东川一带）和四川西南部。生山坡草地，海拔 1800-2500 m。

须根深黄色，含小檗碱，在宾川一带当作马尾黄连收购，可治痢疾、肠类、急性结肠炎、急性咽喉炎等症（《云南省药品标准》）。

标本登录：云南：大理，秦仁昌 22925；同地，苍山，吴征镒，刘德仪 20374；洱源，王珂 92013；宾川，鸡足山，滇西北金沙江队 6410；东川，黄草岭，蓝顺彬 770。

四川：会理，四川中药材站 s.n.

70b. 昭通唐松草（《药学学报》，变种）

Var. **chaotungense** W. T. Wang & S. H. Wang in Fl Reip. Pop. Sin. 27: 569, 619, pl.145: 6.1979; W. T. Wang in Iconogr. Corm Sin. Suppl. 1: 429.1982; et in Fl. Yunnan. 11: 170.2000; D. Z. Fu& G. Zhu in Fl. China 6: 291. 2001; L. Xie in Med. Fl. China 3: 263. 2016. — *T. chaotungense* W. T. Wang & g. H. Wang in Acta Pharm. Sin. 12(11): 746, pl.2: c.1965, nom. nud. Holotype: 云南：昭通，炎山，alt. 1600 m, 1959-07-18, 胡月英等 6824（YIM）.

与金丝马尾连的区别：小叶较大，1.5–2.5×1.5–2.5 cm，叶轴的毛长达 0.4 mm；柱头较宽，三角形。

This variety differing from var. *glandulosissimum* in its larger leaflets 1.5–25 cm long and broad, in the leaf rachis with hairs up to 0.4 mm long, and in its broader triangular stigma.

特产云南昭通。生山地疏林中，海拔 1600 m。

根含小檗碱，可代替黄连药用，在昭通当作"马尾黄连"收购。

系 2. 圆叶唐松草系

Ser. **Indivisa** (DC.)W. T. Wang. st. nov — Subsect. *Indivisa* DC. Syst. 1: 185. 1817; et Prodr. 1: 15. 1824; Tamura in Hiepko, Nat. Pflanzenfam; Zwei. Aufl., 17a Ⅳ : 478. 1995. Lectotype: *T. rotundifclium* DC. (Tamara,1995)

Subsect. *Rotundifolia* Prantl in Bot. Jahrb. 9: 271. 1887

Piuttia Mattei in Malpighia 20: 332. 1906 — Sect. *Piuttia* (Mattei) Tamara in Sci. Rep. Osaka Univ. 17: 50.1968; et in Acta Phytotax. Geobot. 43: 55.1992. Lectotype: *T. rotundifolium* DC. (Tamura, 1968)

叶为单叶或为 1-3 回三出复叶；叶片圆形、近肾形或圆卵形，基部心形。花组成伞房状复单歧聚伞花序。萼片 4，每一萼片具 3 或 5 条基生脉。花丝狭条形或丝形。花柱直或稍向后弯曲，腹面具一狭长圆形或狭条形柱头。瘦果在每侧具 3 条纵肋，无翅。

Leaves simple or 1–3-ternate; blades orbicular, subreniform or orbicular-ovate, at base cordate. Flowers in corymbiform compound monochasia. Sepals 4; each sepal with 3 or 5 basal nerves. Filaments narrow-linear of filiform. Style straight or slightly recurved, ventrally with a narrow-oblong or narrow-linear stigma. Achenes on each side longitudinally 3-ribbed.

3 种，分布于中国西藏南部；以及尼泊尔和印度东北部。有 1 种，1 变种分布于西藏南部。

71. 圆叶唐松草（《植物分类学报》） 图版 71

Thalictrum rotundifolium DC. Syst. 1: 185. 1818; et Prodr. 1: 15. 1824; D. Don, Prodr. Fl. Nepal.193. 1825; Wall. Pl. As. Ray. 11: t.264.1832; Hook. f. & Thoms. Fl Ind.19. 1855; et in Hook. f. Fl. Brit. Ind. 1: 13. 1872; Lecoy. in Bull. Soc. Bot. Belg. 24: 214. 1855; Finet & Gagnep. in Bull. Soc. Bot France 50: 616. 1903; Tamura & Kitam. in Kihara, Fauna & Fl. Nepal Himal. 132.1955; Hara in Fl. East Himal. 2: 37: 1971; et in Hara & Williams, Enum. Flow. Pl. Nepal 2: 22. 1979; W. T. Wang in Acta Phytotax. Sin. 32(5): 470. 1994. Holotype: Nepal, F. Buchanan s.n.(not seen)

多年生草木。茎高 20-30 cm，疏被短柔毛，约有 3 叶。基生叶 1，为单叶，具长柄；叶片纸质，圆肾形，约 4.5×5.5 cm，在顶端圆形，在基部心形，5-7 微裂，边缘有小齿，两面脉网上被短腺毛，具 5-7 条脉，基生脉在上面稍隆起，在下面强烈隆起，与细脉形成明显脉网；叶柄长约 5 cm。茎生叶与基生叶相似，向上渐变小。单歧聚伞花序顶生并腋生，伞房状，长 2-5 cm，有 2-6 花；苞片长 0.3-1.2 cm，3-5 中裂或不分裂，狭卵形；花梗长 0.7-4 cm。花无毛：萼片 4，长椭圆形，约 6.8×3 mm，在顶端钝，每一萼片具 5 条基生脉。雄蕊约 20；花丝狭条形，1.5-5×0.2 mm；花药黄色，长圆形，1.2-1.5×0.25-0.4 mm，顶端钝或具小尖头。心皮约 30；子房斜狭椭圆形，约 1×0.25 mm；花柱长约 0.6 mm；柱头狭长圆形，长约 0.4 mm。瘦果斜狭倒卵形，1-1.6×0.4-0.5 mm，在每侧具 3 条纵肋；宿存花柱长约 0.8 mm。7-8 月开花。

Perennial herbs. Stems 20-30 cm tall, Sparsely puberulous, ca. 3-leaved. Basal leaf singular, simple, long petiolete; blades papery, orbicular-reniform, ca. 4.5×5.5 cm, at apex rounded, at base cordate, 5-7-lobulate, at margin denticulate, adaxially and abaxially on nerves with short glandular hairs, 5-7-nerved, basal nerves adaxially slightly prominent, abaxially strongly prominent, and with nervules forming conspicuous nervous nets; petioles ca. 5 cm long. Cauline leaves similar to basal leaf, but smaller. Monochasia terminal and axillary, corymbiform, 2-5 cm long, 2-6-flowered; bracts 0.3-1.2 cm long, 3-5-fid or undivided, narrow-ovate; pedicels 0.7-4 cm long. Flower glabrous: Sepals 4, long elliptic, ca.6.8×3 mm, at apex obtuse, each sepal with 5 basal nerves. Stamens ca.20; filaments narrow-linear, 1.5-5×0.2 mm; anthers yellow, oblong, 1.2-1.5×0.25-0.4 mm, apex obtuse or minutely

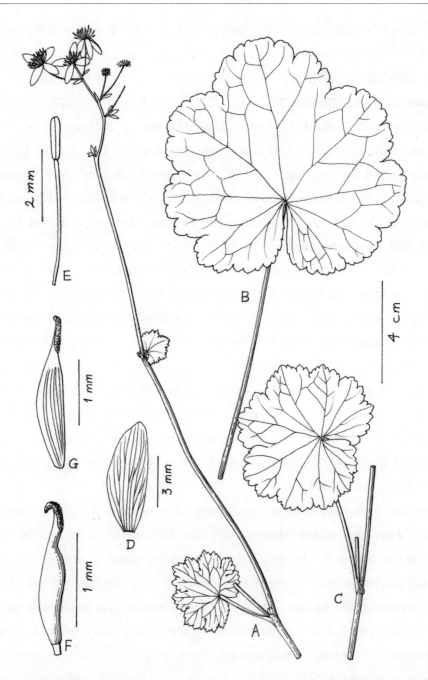

图版71. 圆叶唐松草　A. 开花茎，B.C. 茎生叶，D. 萼片，E. 雄蕊，F. 心皮，G. 瘦果。（据李渤生等 13321）

Plate 71. **Thalictrum rotundifoliam**　A. flowering stem, B. C. cauline leaves, D. sepal, E. stamen, F. carpel, G. achene. (from B. S. Li et al. 13321)

mucronate. Carpels ca.30; ovary obliquely narrow-elliptic, ca. 1×0.25 mm; style ca. 0.6 mm long; stigma narrow-oblong. ca. 0.4 mm long. Achene obliquely narrow-obovate, 1–1.6×0.4–0.5 mm, on each side longitudinally 3-ribbed; persistent style ca. 0.8 mm long.

分布于中国西藏南部以及尼泊尔。生山地乔松林下，海拔2800 m。

标本登录：西藏：吉隆，扎多，1990-08-31，李渤生，李辉，阳艺13321。

72. 心基唐松草（变种） 图版72

Thalictrum punduanum Wall. var. **hirtellum** W. T. Wang, var. nov. Type: 西藏(Xizang)：吉隆县，吉隆镇（Jilong Xian,Jilong Zhen），alt. 3100 m, 林边（at forest margin），2010-08-16, PE 西藏队（PE Xizang Exped）200（holotype, PE）；聂拉木县，樟木镇（Nielamu Xian, Zhangmu Zhen），alt. 2420 m, 2010-08-10，PE 西藏队1811（PE）；同地（same locality），alt. 3450 m, 林下（in forest），2010-08-17, PE西藏队1301（PE）；同地（same locality），alt. 3200 m, 河边灌丛（in bush on river bank），2013-07-29, PE 西藏队4387（PE）.

A var. *punduano* differt foliolis subtus ad nervos pilis brevibus crasis 0.1–0.5 mm longis dense hirtellis, acheniis puberulis. In var. *punduano*, foliola subtus glabra, achenia glabra sunt.

Perennial herbs. Stems ca. I m tall, glabrous, branched. Cauline leaves 2–3-ternate-pinnate; blades 8–30×7–28 cm; leaflets papery. orbicular-ovate, suborbicular or broad-ovate, 1–3 × 1.2–5.4 cm, at base cordate, at margin rotund-dentate, inconspicuously 3-lobed, adaxially glabrous, abaxially on nerves densely hirtellous (hairs short and thick, 0.1–0.5 mm long), 5-nerved, nerves adaxially flat, abaxially prominent; petiolules 0.3–3 cm long, glabrous; petioles short, at base narrowly vaginate. Thyrses terminal and axillary, 30–40 cm long, many-flowered, glabrous; lower bracts foliaceous, upper bracts simple, small, ovate; pedicels filiform, 0.7–3.2 cm long. Flower: Sepals 4. Yellow, elliptic, ca. 5×3 mm, glabrous, apex rounded, each sepal with 3 simple basal nerves. Stamens ca.20. glabrous; filaments filiform, 4–7 mm long; anthers narrow-oblong, 3–3.5×0.6–0.8 mm, at apex with mucrones 0.1 mm long. Carpels 5–9; ovary broad-elliptic, ca.2 mm long, at base abruptly narrowed into a short stipe; style ca. 2 mm long; stigma narrow-linear, 1.8 mm long. Young achene stipitate, compressed, obliquely obovate, ca. 3×1.4 mm, puberulous (hairs ca. 0.1 mm long), on each side with 3 longitudinal nerves, base with a glabrous stipe ca.1 mm long; persistent style 2 mm long; persistent stigma 1.8 mm long.

多年生草本植物。茎高约1 m，无毛，分枝。茎生叶为2-3回三出羽状复叶；叶片

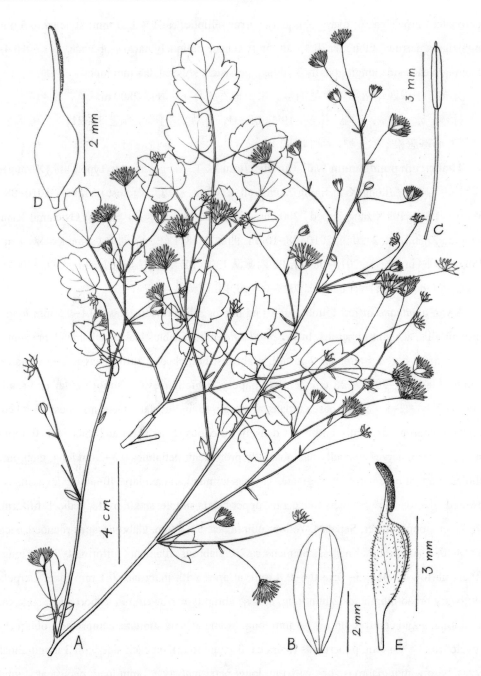

图版72. 心基唐松草（变种） A. 开花茎顶部，B. 萼片，C. 雄蕊，D. 心皮（据PE-西藏队200），E. 瘦果（据PE-西藏队1181）。

Plate 72. **Thalictrum punduanum** wall. var. **hirtellum** W. T. Wang A. apical part of flowering stem, B. sepal, C. stamen, D. carpel (from PE Xizang Exped. 200), E. achene (from PE Xiang Exped. 1181).

8–30×7–28 cm；小叶纸质，圆卵形，近圆形或宽倒卵形，1–3×1.2–5.4 cm，基部心形，边缘具圆牙齿，不明显 3 浅裂，上面无毛，下面在脉上密被短硬毛（毛短而粗，长 0.1–0.5 mm），具五出脉，脉上面平，下面隆起；小叶柄长 0.3–3 cm，无毛；叶柄短，基部具狭鞘。聚伞圆锥花序顶生并腋生，长 30–40 cm，具多数花，无毛；下部苞片叶状，上部苞片简单，小，卵形；花梗丝形，长 0.7–3.2 cm。花：萼片 4，黄色，椭圆形，约 5×3 mm，无毛，顶端圆形，每一萼片具 3 条不分枝的基生脉。雄蕊约 20，无毛；花丝丝形，长 4–7 mm；花药狭长圆形，3–3.5×0.6–0.8 mm，顶端具长 0.1 mm 的小短尖头。心皮 5–9；子房宽椭圆形，长约 2 mm，基部突变狭成短柄；花柱长约 2 mm；柱头狭条形，长 1.8 mm。幼瘦果具柄，扁，斜倒卵形，约 3×1.4 mm，被短柔毛（毛长约 0.1 mm），每侧 3 条纵脉，基部具一长约 1 mm 的无毛柄；宿存花柱长 2 mm；宿存柱头长 1.8 mm。7–8 月开花。

特产西藏南部吉隆至樟木一带。生山地林中或林边，或河边灌丛中，海拔 2420–3450 m。

本变种的小叶下面被短硬毛，瘦果被短柔毛而与模式变种 var. *punduanum*（分布于喜马拉雅山区及印度东北部）不同。在 var. *punduanum*，小叶和瘦果均无毛（Hook. f. & Thoms. in Hook. f. Fl. Brit. India 1: 13. 1872.）

系 3. 偏翅唐松草系

Ser. **Violacea** W. T. Wang & S. H. Wang in Fl. Reip. Pop. Sin. 27: 569, 620. 1979. — Subsect. *Violacea* (W. T. Wang & S. H. Wang) Tamura in Acta Phytotax. Geobot. 43: 56. 1992; et in Hiepko, Nat. Pflanzenfam., Zwei. Aufl., 17a Ⅳ : 481. 1995. Type: *T. delavayi* Franch.

叶为复叶。花组成聚伞圆锥花序，萼片 4(–5–6)，长 7.2–16 mm，紫色，淡紫或粉红色，稀白色（错那唐松草）；每一萼片有（3–）5(–7) 条常 1–5 回二歧状分枝的基生脉。花丝狭条形或丝形；花药顶端具短尖头。柱头无翅。瘦果通常强度扁压，无翅，稀沿腹、背缝线具狭翅（偏翅唐松草）。

Leaves compound. Flowers in thyrses. Sepals 4(–5–6), 7.2–16 mm long, purple, lilac or pink, rarely white (*T. cuonaense*); each sepal with(3–)5(–7) usually 1–5 times dichotomizing basal nerves. Filaments narrow-linear or filiform; anthers at apex mucronate. Stigma not winged. Achenes usually strongly compressed, not winged, rarely along ventral and dorsal sutures narrowly winged (*T. delavayi*).

6 种，分布于西南部山区。

73. 错那唐松草（《植物分类与资源学报》）　图版 73

Thalictrum cuonaense W. T. Wang in Pl. Divers. Resour. 36(6): 791, fig. 1. 2014. Holotype：西藏：错那县，那布茶场，alt. 2650 m, 1981-10-04，任再金 1364（PE）。

多年生草本。茎高 1-1.5 m，中部直径 9 mm，无毛，分枝。茎生叶为 2 回羽状复叶，长 6-25 cm，无毛；叶片三角形，长 5-20 cm；羽片 3 对，最下部羽片有 5-7 枚小叶，上部羽片有 3 枚小叶；小叶纸质，圆卵形或扁圆形，0.6-3.3×0.7-3.7 cm，在基部浅心形，3 浅裂（裂片顶端圆形，具小短尖头，边缘有 1-2 牙齿，或 1-2 浅圆齿或全缘），具 3 或 5 脉，叶脉在上面平，在下面不明显隆起；叶柄长 0.8-18 cm，基部具狭鞘。聚伞圆锥花序顶生，长约 16 cm，有多数花；花序轴和花梗无毛；苞片无毛，或为三出复叶，小叶狭椭圆形，长 1.5-5 mm，或为单叶，长椭圆形或披针形，长 1-6 mm；花梗长 0.9-1.5 cm。花：萼片 4，白色，卵形，狭卵形或椭圆形，10-12.8×3.5-7 mm，顶端稍钝，无毛，每一萼片具 3 条 1 或 3 回二歧分枝的基生脉。雄蕊 18-20，长约 9 mm；花丝狭条形，4-5.5×0.2 mm；花药狭条形，约 4×0.25 mm，顶端具长 0.15 mm 的短尖头。心皮 7-11；子房扁压，斜倒卵形，1.8-2.5×0.8 mm；密被小腺毛，基部具长 0.8 mm 的柄；花柱长 1.8-2.5 mm；柱头条形，长 1.2 mm。幼瘦果扁平，倒卵形，约 3×2 mm，每侧有 2 条纵肋，基部的柄长 1.2 mm；宿存花柱长 2 mm。9-10 月开花。

Perennial herbs. Stems 1-1.5 m tall, at the middle part ca.9 mm in diam., glabrous, branched. Cauline leaves 2-pinnate, 6-25 cm long, glabrous; blades triangular in ou tline, 5-20 cm long; pinnae 3-pairs, the lowest pinna 5-7-foliolate, and the upper pinnae 3-foliolate; leaflets papery orbicular-ovate or depressed-orbicular, 0.6-3.3×07-3.7 cm, at base subcordate, 3-lobed(lobes at apex rounded, and minutely mucronate, at margin 1-2-dentate, 1-2-crenate or entire), 3-5-nerved, nerves adaxially flat, abaxially inconspicuously prominent; petioles 0.8-1.8 cm long, at base narrow-vaginate. Thyrses teminul, ca.16 cm long, many-flowered; rachis and pedicels glabrous; bracts glabrous, ternate, with leaflets narrow-elliptic, 1.5-5 mm long, or simple, long elliptic or lanceolate, 1-6 mm long; pedicels 0.9-1.5 cm long. Flower: Sepals 4, white, ovate, narrow-ovate or elliptic, 10-12.8×3.5-7 mm, at apex slightly obtuse, glabrous, each sepal with 3 basal nerves once or thrice dichotomizing. Stamens 18-20, ca. 9 mm long; filaments narrow-linear, 4-5.5×0.2 mm; anthers narrow-linear, ca 4×0.25 mm, apex with mucrones 0.15 mm long. Carpels 7-11; ovary compressed, obliquely obovate, 1.5-2×0.8 mm, densely minutely glandular-puberulous, base with a stipe 0.8 mm long; style 1.8-2.5

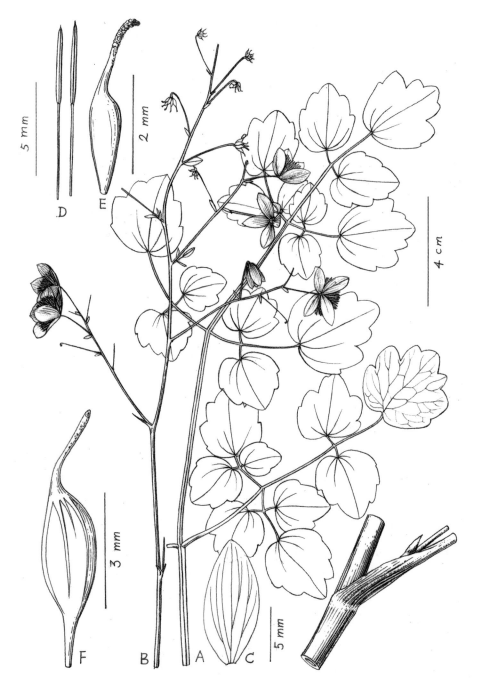

图版73. **错那唐松草** A. 茎生叶,B. 聚伞圆锥花序,C. 萼片,D. 二雄蕊,E. 心皮,F. 瘦果。(据 holotype)

Plate 73. **Thalictrum cuonaense** A. cauline leaf, B. thyrse, C. sepal, D. two stamens, E. carpel, F. achene. (from holotype)

mm long; stigma linear, 1.2 mm long. Young achene flattened, obovate, ca. 3×2 mm. on each side longitudinally 3-ribbed, base with a stipe 1.2 mm long; persistent style 2 mm long.

特产西藏错那县。生山坡灌丛草地，海拔 2650 m。

74. 珠芽唐松草（《中国植物志》）　　图版 74

Thalictrum chelidonii DC. Prodr. 1: 11. 1824; Hook f. & Thoms. in Hook. f. Fl. Brit. Ind. 1: 11. 1872; Lecoy. in Bull. Soc. Bot. Belg. 24: 185, t. 4: 3. 1885; W. T. Wang & S. H. Wang in Fl. Reip. Pop. Sin. 27: 574, pl.145: 12.1979; Grierson, Fl. Bhutan 1(2): 297. 1984, p. p.; W. T. Wang in Fl. Xizang. 2: 72. in 1985; Rau in Sharma et al. Fl. India 1: 134. 1993; D. Z. Fu & G. Zhu in Fl. China 6: 287. 2001; W. T. Wang in High. Pl. China in Colour 3: 392. 2016; L. Xie in Med. Fl. China 3: 267. 2016; X. T. Ma et al. Wild Flow. Xizang 26, Photo 49. 2016. Holotype: Nepal, Wallich s. n. (K, not seen).

T. cysticarpum Wall. [Cat. n. 3714. 1828] Pl. As. Rar. 2: 26, t.129.1831.

多年生草本植物，全部无毛。茎高约 1 m，分枝，在上部叶腋簇生少数珠芽。叶为 3 回羽状复叶；小叶草质，顶生小叶楔状长圆形、楔状倒卵形或圆卵形，侧生小叶卵形、圆卵形或心形，长 1–3 cm，不明显 3- 浅裂或不分裂，边缘有少数牙齿，叶脉在下面稍隆起，脉网明显；小叶柄长 0.1–1.8 cm；叶柄长 2–10 cm。聚伞圆锥花序长约 15 cm，有多数花；苞片叶状；花梗丝形，长 0.4–1 cm。花：萼片 4，淡粉红色，卵形，6–11×3–5.8 mm，在顶端钝，每一萼片具 5 条基生脉（中脉不分枝，2 条侧生脉 2 回二歧状分枝，另 2 条侧生脉 3 回二歧状分枝）。雄蕊 14–16；花丝狭条形，约 2.5×0.2 mm；花药黄色，狭长圆形，1.5–2×0.4–0.5 mm，顶端具长约 0.1 mm 的小尖头。心皮 5–8；子房椭圆形，约 0.8×0.5 mm，基部有长约 0.5 mm 的柄；花柱长约 1 mm；柱头狭条形，与花柱等长。瘦果扁平，半倒卵形，长约 4 mm，有细柄，8 月开花。

Perennial herbs, totally glabrous. Stems ca. 1 m tall, branched, above in leaf axils with a few bulbils. Leaves 3-pinnate; leaflets herbaceous, terminal leaflet cunaete-oblong, cuneate-obovate or orbicular-ovate, lateral leaflets ovate, orbicular-ovate or cordate, 1–3 cm long, inconspicuously 3-robed or undivided, at margin few-dentate, nerves abaxially slightly prominent, and nervous nets inconspicuous; petiolules 0.1–1.8 cm long; petioles 2–10 cm long. Thyrses ca.15 cm long, many-flowered; bracts foliaceous; pedicels filiform, 0.4–1 cm long. Flower: Sepals 4. pinkish, ovate, 6–11×3–5.8 mm, at apex obtuse, each Sepal with 5 basal nerves (midrib simple, 2 lateral nerves twice dichotomizing, and the another 2 lateral nerves

六、分类学处理　　　　　　　　　　　　　　　　　　　　　　277

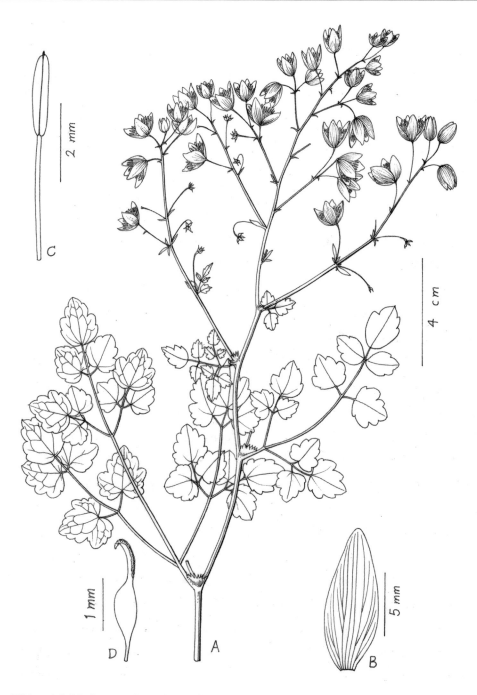

图版74. **珠芽唐松草**　A. 开花茎顶部，B. 萼片，C. 雄蕊，D. 心皮。（据倪志诚等1909）
Plate 74. **Thalictrum chelidonii**　A. apical part of flowering stem, B. sepal, C. stamen, D. carpel (from Z. C. Ni et al. 1909)

thrice dichotomizing). Stamens 14−16; filaments narrow-linear, ca.2.5×0.2 mm; anthers yellow, narrow-oblong,ca.0.8×0.5 mm, apex with mucrones ca. 0.1 mm long. Carpels 5−8; ovary elliptic, ca. 0.8×0.5 mm, base with a stipe ca. 0.5 mm long; style ca. 1 mm long; stigma as long as style, narrow-linear. Achene flattened, semi-obovate, ca. 4 mm long, slenderly stipitate.

分布于中国西藏南部, 以及尼泊尔、不丹。生山地林中或山坡灌丛草地, 海拔 2400-3400 m。

根和根状茎有解毒之效, 作黄连代用品。全草有舒筋之效。(《中国药用植物志》, 2016)

标本登录: 西藏: 聂拉木, 倪志诚等 1909; 吉隆, 倪志诚等 2267, 李渤生等 13366。

75. 美花唐松草(《广西植物》) 图版 75

Thalictrum callianthum W. T. Wang in Guihaia 33(5): 583, fig. 5: A−E. 2013; et in High. P1. China in Colour 3: 391. 2016. Holotype: 西藏: 米林, 派区, 格噶村, alt. 3400 m, 2012-08-01, 杨永、刘冰、林秦文等 171(PE)。

多年生草本植物, 全部无毛。茎高约 1.5 m, 分枝。中部茎生叶为 3-4 回三出复叶, 具短柄; 叶片正三角或三角形, 16−18×22−25 cm; 小叶薄纸质, 顶生小叶宽倒卵形, 侧生小叶斜卵形, 或斜狭倒卵形, 4−10×3−9 mm, 3− 浅裂或不分裂, 脉在两面平, 不明显; 小叶柄长 0.2−3 mm; 叶柄长约 5 cm, 基部具鞘。聚伞圆锥花序顶生, 长约 30 cm, 有多数花; 苞片叶状, 长 2.5−7 cm; 花梗丝形, 长 2−4.5 cm。花: 萼片 4(−5), 紫色, 长椭圆形或狭卵形, 11−16×5−10 mm, 顶端具短尖头, 每一萼片具 5 条基生脉(1 条脉 1 回二歧状分枝, 4 条脉 2 回二歧状分枝)。雄蕊 22−28, 长 5−8.5 mm; 花丝丝形, 长 4−5 mm, 狭条形或下部丝形, 上部宽 0.2 mm; 花药黄色, 近条形, 1.5−2×0.25−0.4 mm, 顶端具短尖头。心皮 10−15; 子房椭圆形, 约 1×0.4−0.6 mm, 基部具长 0.8−1 mm 的柄; 花柱暗蓝色, 长 1 mm; 柱头黄色, 狭条形, 长 0.8−1 mm。瘦果扁平, 新月形, 5×1.5−2 mm, 在每侧具 1−3 条细纵肋, 基部具长 1 mm 的柄; 宿存花柱长 2−2.2 mm, 顶端稍向后弯曲。7−8 月开花。

Perennial herbs, totally glabrous. Stems ca. 1.5 m tall, branched. Middle cauline leaves 3−4-ternate, shortly petiolate; blades deltoid or triangular, 16−18×22−25 cm; leaflets thinly papery, terminal leaflets broadly obovate, lateral leaflets obliquely ovate, or obliquely narrow-ovate, 4−10×3−9 mm, at apex acute, 3-lobed or undivided, nerves on both surfaces flat and nervous nets inconspicuous; petiolules 0.2−3 mm long; petioles ca. 5 cm long, at base vaginate.

六、分类学处理   279

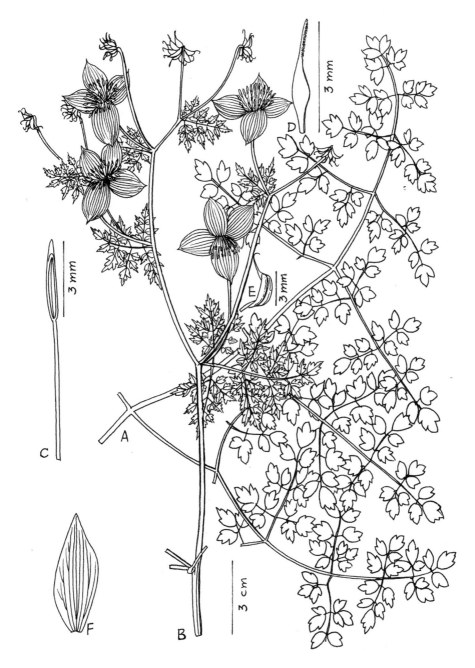

图版75. 美花唐松草 A. 中部茎生叶，B. 聚伞圆锥花序，C. 萼片，D. 雄蕊，E. 心皮，F. 瘦果。（据 holotype）

Plate 75. **Thalictrum callianthum** A. middle cauline leaf, B. thyrse, C. sepal, D. stamen, E. carpel, F. achene. (from holotype)

Thyrses terminal, ca. 30 cm long, many-flowered; bracts foliaceous, 2.5–7 cm long; pedicels filiform, 2–4.5 cm long. Flower: Sepals 4(–5), purple, long elliptic or narrow-ovate, 11–16×5–10 mm, at apex mucronate, each sepal with 5 basal nerves (1 nerve once dichotomizing, and 4 nerves twice dichotomizing). Stamens 22–28, 5–8.5 mm long; filaments narrow-linear, 4–5 mm long, above 0.2 mm broad, or below filiform; anthers yellow, sublinear, 1.5–2×0.25–0.4 mm, apex mucronate. Carpels 10–15; ovary elliptic, ca. 1×0.4–0.6 mm; base with a stipe 0.8–1 mm long; style dark-blue, 1 mm long; stigma yellow, narrow-linear, 0.8–1 mm long. Achene flattened, lunate, 5×1.5–2 mm, on each side longitudinally and thinly 1–3-ribbed, base with a stipe 1 mm long; persistent style 2–2.2 mm long, apex stightly recurved.

特产西藏东南部（米林至林芝一带）。生高山草甸灌丛地区，海拔 3400 m。

标本登录：西藏，林芝：鲁朗，2013-07-16，金孝锋 3143。

76. 美丽唐松草（《中国高等植物图鉴》）　鹅整（西藏药名）　图版 76

Thalictrum reniforme Wall. [Cat. n. 3716. 1828] Pl. As. Rar. 2: 26. 1831; Hook. f. & Thoms. in Hook. f. Fl. Brit. Ind. 1: 11. 1872; Lecoy. in Bull. Soc. Bot. Belg. 24: 187, t. 4: 4. 1885; A. C. Smith in Curtis's Bot. Mag., n. s.,178: t. 592. 1971; W. T. Wang & S. H. Wang in Fl. Reip. Pop. Sin. 27: 571, pl.145: 7–10. 1979; W. T. Wang in Fl. Xizang. 2: 70, fig. 20: 4–5.1985; et in High. Pl. China 3: 476. 2000; D. Z. Fu & G. Zhu in Fl. China 6: 296. 2001; W. T. Wang in High. Pl. China in Colour 3: 391. 2016; L. Xie in Med Fl. China 3: 266. 2016. ——*T. chelidonii* DC. var. *reniforme* (wall.) Hook. f. & Thoms. Fl. Ind. 13. 1855. Type: Nepal, Wallich 3716 (K, not seen).

T. neurocarpum Royle, Ill. Bot. Himal. 51. 1838.

多年生草本植物。茎高 0.8-1.5 m，有极短腺毛，上部分枝。中部茎生叶为 3 回羽状复叶，具短柄；叶片长约 20 cm；小叶草质，宽卵形，圆卵形或菱状宽倒卵形，1.2-2.5×1.5-3.2 cm，在基部圆形或浅心形，3 浅裂（裂片全缘或有少数浅圆齿），上面无毛，下面密被短腺毛，脉在下面稍隆起；小叶柄长 0.2-2 cm；叶柄长 2-6 cm，基部有狭鞘。聚伞圆锥花序长 20-30 cm，花序轴和花梗密被短腺毛（毛长约 0.1 mm）；苞片叶状；花梗长 0.8-2.5 cm。花：萼片 4，粉红色，卵形，9-13×5-10 m，在顶端微钝，无毛，每一萼片具 3 条基生脉（中脉 1 回二歧状分枝，2 侧生脉 3 回二歧状分枝）。雄蕊约 20，无毛；花丝狭条形，2.2-4×0.2-0.25 mm；花药黄色，条形，3-4×0.25-0.4 mm，顶端具长 0.3-0.4 mm 的短尖头。心皮 7-20；子房斜狭倒卵形，约 1×0.8 mm，被短腺毛，基部具长

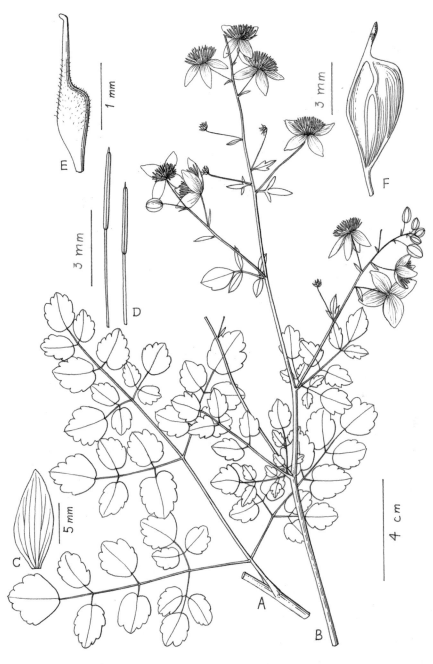

图版76. **美丽唐松草** A. 茎生叶，B. 聚伞圆锥花序，C. 萼片，D. 二雄蕊，E. 心皮（据青藏队75-1821），F. 瘦果（据青藏队74-2322）。

Plate 76. **Thalictrum reniforme** A. cauline leaf, B. thyrse, C. sepal, D. two stamens, E. carpel (from Qinghai-Xizang Exped. 75-1821), F. achene (from Qinghai-Xizang Exped. 74-2322).

0.3 mm 的柄；花柱钻形，长约 1.4 mm，腹面上部具柱头。瘦果扁平，斜倒卵形，约 6.5×3.5 mm，每侧具 3 条纵肋，基部具长 1.2 mm 的柄；宿存花柱长 1.5 mm；宿存柱头长 0.8 mm。7–10 月开花。

Perennial herbs. Stems 0.8–1.5 m tall, covered with very short glandular hairs, above branched. Middle cauline leaves 3-pinnate, shortly petiolate; blades ca. 20 cm long; leaflets herbaceous, broadly ovate, orbicular-ovate or broadly rhombic-obovate, 1.2–2.5×1.5–3.2 cm, at base rounded or subcordate, 3-lobed (lobes entire or sparsely crenate), adaxially glabrous, abaxially densely glandular-puberulous, nerves abaxially slightly prominent; petiolules 0.2–2 cm long; petioles 2–6 cm long, at base narrowly vaginate. Thyrses 20–30 cm long; rachis and pedicels densely glandular-puberulous(hairs ca. 0.1 mm long); bracts foliaceous; pedicels 0.8–2.5 cm long. Flower: Sepals 4, pink, ovate, 9–13×5–10 mm, at apex slightly obtuse, glabrous, each sepal with 3 basal nerves (midrib once dichotomizing, and 2 lateral nerves thrice dichotomizing). Stamens ca. 20, glabrous; filaments narrow-linear, 2.2–4×0.2–0.25 mm; anthers yellow, linear, 3–4×0.25–0.4 mm, apex with mucrones 0.3–0.4 mm long. Carpels 7–20; ovary obliquely narrow-obovate, ca. 1×0.5 mm, glandular-puberulous, base with a stipe 0.3 mm long; style subulate, ca. 1.4 mm long, adaxially above with a small stigma. Achene flattened, obliquely obovate, ca. 6.5×3.5 mm, on each side longitudinally 3-ribbed, base with a stipe 1.2 mm long; persistent style 1.5 mm long; persisitent stigma 0.8 mm long.

分布于中国西藏南部和东南部；以及尼泊尔、不丹、印度北部。生山地草坡、灌丛或冷杉林中，海拔 3100–3700 m。

根有清热解毒之效，用于炭疽病、肠炎、痢疾。全草用于虫病，痢疾等症。（《中国药用植物志》, 2016）

标本登录：西藏：吉隆，青藏队 75-6358，75-7052，李渤生等 13241；聂拉木，李渤生等 14259；樟木，陈家瑞 92268；卡马河下游，？451；亚东，傅国勋 1101，青藏队 74-2322；帕里至蛤坞，钟补求 5880；春丕，青藏队 74-2542；错那，青藏队 74-2830，青藏补点队 75-1901；米林，青藏队 74-1989。

77. 堇花唐松草（《中国植物志》）　图版 77

Thalictrum diffusiflorum Marq. & Shaw in J. Linn. Soc. Bot. 48: 153. 1929; W. T. Wang & S. H. Wang in Fl. Reip. Pop. Sin. 27: 573. pl. 145: 11. 1979; W. T. Wang in Fl Xizang. 2: 71, fig. 20: 1. 1985; D. Z. Fu & G. Zhu in Fl. Chin+a 6: 288. 2001; X. T. Ma et al. Wild Flow. Xizang

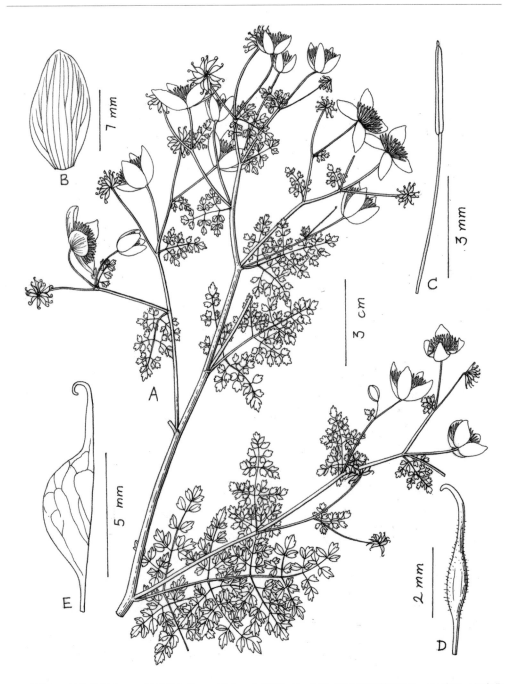

图版77. **堇花唐松草** A. 开花茎上部，B. 萼片，C. 雄蕊，D. 心皮（据倪志诚等2867），E. 瘦果（据青藏队74-3715）。

Plate 77. **Thalictrum diffusiflorum** A. upper part of flowering stem, B. sepal, C. stamen, D. carpel (from Z. C. Ni et al. 2867), E. achene (from Qinghai-Xizang Exped. 74-3715)

21, photo 52. 2016. Holotype: 西藏东南部，Tumbatse, 1924-07-07, F. Kingdon Ward 5899 (K, not seen).

多年生草本植物。茎高 60-100 cm，上部有短腺毛，自中部或上部分枝。叶为 3-5 回羽状复叶；叶片长 8-15 cm；小叶草质，圆菱形或宽卵形，4-12×3.5-10 mm，3-5 浅裂，上面无毛，下面疏被短腺毛，脉平或在下面稍隆起；小叶柄长 0.5-4 mm；叶柄长 1.5-4 cm，基部有狭鞘。聚伞圆锥花序顶生，长约 30 cm；花序轴和花梗疏被短腺毛（毛长约 0.05 mm）；苞片叶状；花梗长 1-3 cm。花：萼片 4-6，淡紫色，卵形或狭卵形，10-15×5-7 mm，在顶端钝，每一萼片具 7 条基生脉（中脉和 1 条侧生脉不分枝，3 条侧生脉 1 回二歧状分枝，2 条侧生脉 2 回二歧状分枝）。雄蕊 32-38，无毛；花丝丝形，长 3-4 mm；花药条形，2-2.8×0.4-0.5 mm，顶端具长 0.2-0.5 mm 的短尖头。心皮 7-10；子房狭椭圆形，约 1.6×0.5 mm，与花柱下部均被短腺毛（毛长 0.05 mm），基部具长 0.8 mm 的柄；花柱长 2 mm，直或上部稍弯曲；柱头不明显。瘦果扁平，半倒卵形，约 5×2.5 mm，在每侧有 3 条弯曲并网结的细纵肋，基部具长 1.2 mm 的柄；宿存花柱长 2.8-3 mm，顶端钩状弯曲。6-8 月开花。

Perennial herbs. Stems 60-100 cm tall, above covered with short glandular hairs, from the middle part or above branched. Leaves 3-5-pinnate; blades 8-15 cm long; leaflets herbaceous, orbicular-rhombic or broadly ovate, 4-12×3.5-10 mm, 3-5-1obed, adaxially glabrous, abaxially sparsely glandular-puberulous, nerves flat or abaxially slightly prominent; petiolules 0.5-4 mm long; petioles 1.5-4 cm long, at base narrowly vaginate, Thyrses terminal, ca. 30 cm long; rachis and pedicels sparsely glandular-puberulous(hairs ca. 0.05 mm long); bracts foliaceous; pedicels 1-3 cm long. Flower: Sepals 4-6, lilac. ovate or narrow-ovate, 10-15×5-7 mm, at apex obtuse, each sepal with 7 basal nerves(midrib and 1 lateral nerve simple, 3 lateral nerves once dichotomizing, and 2 laterl nerves twice dichotomizing). Stamens 32-38, glabrous; filaments filiform, 3-4 cm long; anthers linear. 2-.2.8×0.4-0.5 cm, apex with mucrones 0.2-0.5 mm long. Carpels 7-10; ovary narrow-elliptic, ca. 1.6×0.5 mm, with lower part of style glandular-puberulous(hairs 0.05 mm long), base with a stipe 0.8 mm long; style 2 mm long, straight or above slightly curved; stigma inconspicuous. Achene flattened, semi-obovate, ca. 5×2.5 mm, on each side with 3 curved and anastomosing thin ribs, base with a stipe 1.2 mm long; persistent style 2.8-3 mm long, apex hooked.

特产西藏东南部。生山地冷杉林中或沟边草丛，海拔 2900-4200 m。

标本登录：西藏：朗县，任再金 L194；工布江达，? 2121；米林，青藏队 74-3715，倪志诚等 2867，李渤生等 5766；林芝，西藏中草药队 3133，李渤生等 6221；鲁郎至色季拉山，张永田，郎楷永 997；波密，应俊生，洪德元 650，866，覃海宁等 332；札木，张永田，郎楷永 611；梭白拉，钟补求 6989。

78. 偏翅唐松草（《中国药用植物志》） 图版 78

Thalictrum delavayi Franch. in Bull. Soc. Bot. France 33: 367. 1886, nom. conserve.; et Pl Delav. 10.1889; Finet & Gagnep. in Bull. Soc. Bot. France 50: 623. 1903; Finet in J. de Bot. 21: 21. 1908; Hand.-Mazz. Symb. Sin. 7: 313. 1931; Boivin in J. Arn. Arb. 26: 113. 1945; Lauener & Green in Not. Bot. Gard. Edinburgh 23: 592. 1961; C. Pei & T. Y. Chou, Icon. Chin. Med. Pl. 7: fig. 333. 1964; Iconogr. Corm. Sin. 1: 680, fig. 1360. 1972; W. T. Wang in Fl. Xizang. 2: 70, fig. 20: 2-3. 1985; Y. K. Li in Fl. Guizhou. 3: 71, pl. 29: 5-11. 1990; W. T. Wang in Vasc. Pl. Hengduan Mount. 1: 502: 1993; in Acta Phytotax. Sin. 31(3): 213. 1993; in Fl. Yunnan. 11: 170, pl. 51: 1-6. 2000; et in High. Pl China 3: 475, fig. 758. 2000; D. Z. Fu & G. Zhu in Fl. China 6: 288. 2001; Y. M. Shui & W. H. Chen, Seed Pl. Honghe Reg. S.-E. Yunnan 77. 2003; F. W. Xing, Rare Pl. China 82, fig. 209. 2005; Z. Y. Li & M. Ogisu, Pl. Mount Emei 256. 2007; Y. Yang, X. M. & B. Liu, color Fl. Panzhihua cycas Nat. Reserve 48. 2011; W. T. Wang in High. Pl China in Colour 3: 390. 2016; L. Xie in Med. Pl. China 3: 264, with fig. & photo. 2016; X. T. Ma et al. Wild Flow. Xizang 26, photo 51. 2016. — *T. delavayi* var. *parviflorum* Franch. Pl. Delav. 11.1889. Type: 云南：大理，苍山，1882-07-04, Delavay Thalictrum 14 (holotype, P; isotype, PE).

T. thibeticum Franch. in Nouv. Arch. Mus. Hist. Nat. Paris, ser. 2,8: 186. 1885; et Pl. David. 2: 4. 1888; Lecoy. in Bull. Soc. Bot. Belg. 24: 153.1885; Finet & Gagnep. in Bull. Soc Bot. France 50: 614. 1903. Holotype：四川：宝兴，1869-08，A. David s. n. (holotype, P; photo, PE).

T. dipterocarpum Franch. in Bull. Soc Bot. France 33: 368. 1886; et Pl. Delav. 12. 1889; Finet & Gagnep. in l. c. 624. Syntypes: 云南：大理，Ta-pin-tze 之北，Pe-ngay-tze, 1882-08-22, Delavay Thalictrum 15 (P; photo, PE); Ta-pin-tze, Nien-kia-ke, 1885-09-06, Delavay 1852 (P; photo PE).

T. grandisepalum Lévl. in Bull. Acad. Int. Géogr. Bot 11: 297.1902; Boivin in J. Arn. Arb. 26: 113.1945.

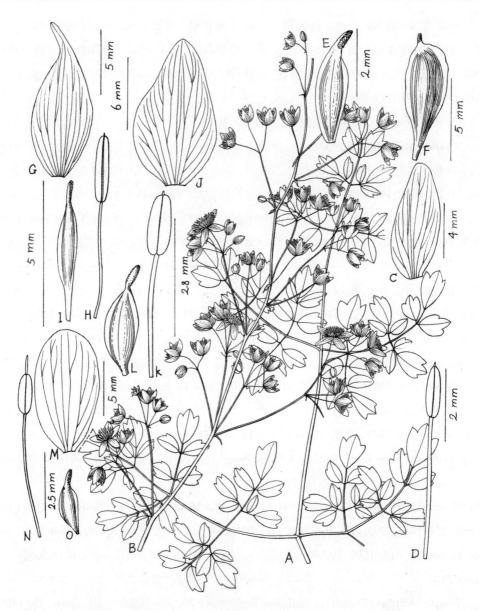

图版78. 偏翅唐松草 A–F 模式变种 A. 茎生叶，B. 聚伞圆锥花序，C. 萼片，D. 雄蕊，E. 心皮（据谢磊，毛建丰2），F. 瘦果（据刘慎谔21115）。 G–I **渐尖偏翅唐松草** G. 萼片，H. 雄蕊，I. 心皮（据王启无70560）。 J–L. **宽萼偏翅唐松草** J. 萼片，K. 雄蕊，L. 心皮（据植物研究所横断山队2027）。 M–O **短尖偏翅唐松草** M. 萼片，N. 雄蕊，O. 心皮（据税玉民等64082）。

Plate 78. **Thalictrum delavayi** A–F var. **delavayi** A. cauline leaf, B. thyrse, C. sepal, D. stamen, E. carpel (from L. Xie & J. F Mao 2), F. achene (from T. N. Liou 2115). G–I. var. **acuminatum** G. sepal, H. stamen, I. carpel (from C. W. Wang 70560). J–L var. **decorum** J.sepal, K. stamen, L. carpel (from Hengduan Maunt. Exped. Inst. Bot. 2027). M–O var. **mucronatum** M. sepal, N. stamen, O. carpel (from Y. M. Shui et al. 64082).

T. duclouxii Lévl. in Repert. Sp. Nov. Reg. Veg. 7: 98.1909. Syntypes：云南：昆明，黑龙潭，1904-07-23，ducloux 606 (not seen)；昆明，Tchong-Chan，1905-07-30，Ducloux 781 (not seen).

78a. var. **delavayi**

多年生草本植物，全部无毛。茎高 0.6-2 m，分枝。下部和中部茎生叶为 3-4 回羽状复叶；叶片长达 40 cm；小叶草质，圆卵形、倒卵形或椭圆形，0.5-3×0.3-2 (-2.5)cm，在基部圆形或楔形，3 浅裂或不分裂，边缘全缘或有 1-3 齿，脉平或下面稍隆起；小叶柄长 0.15-18 mm；叶柄长 1.4-8 cm，基部具鞘；托叶半圆形，边缘分裂或不分裂。聚伞圆锥花序长 15-40 cm，宽 10-20 cm，圆锥花序状，通常有多数花；苞片小，条形；花梗长 0.8-2.5 cm。花：萼片 4(-5)，紫色，卵形或狭卵形，5.5-9（-12）×2.2-4.5 (-5)cm，顶端微钝，每一萼片具 5 条基生脉（中脉和 1 条侧生脉 3 回二歧状分枝，1 条侧生脉 1 回二歧状分枝，其他 2 条侧生脉 2 回二歧状分枝）。雄蕊 24-28，长 5-7 mm；花丝丝形，长 2-5 mm；花药长圆形，1-1.5×0.5-0.6 mm，顶端具短尖头（短尖头长 0.2-0.4 mm）。心皮 9-22；子房椭圆形，约 2.2×1 mm，沿腹、背缝线有狭翅；花柱长约 1 mm，腹面密生柱头组织。瘦果扁，斜倒卵形，有时近镰刀形，5-8×2.5-3.2 mm，沿腹、背缝线有狭翅（翅宽 0.4-0.6 mm），在每侧有 3 条纵肋，基部具长 1-3 mm 的柄；宿存花柱长约 1 mm。6-9 月开花。

Perennial herbs, totally glabrous. Stems 0.6-2 m tall, branched. Lower and middle cauline leaves 3-4-pinnate; blades up to 40 cm long; leaflets herbaceous, orbicular-ovate, obovate or elliptic, 0.5-3×0.3-2(-2.5)cm, at base rounded or cuneate, 3-lobed or undivided, margin entire or 1-3-dentate, nerves flat or abaxially slightly prominent; petiolules 0.5-18 mm long; petioles 1.4-8 cm long, at base vaginate; stipules semi-orbicular, margin divided or undivided. Thyrses 15-40 cm long, 10-20 cm broad, paniculiform, usually many-flowered; bracts small, linear; pedicels 0.8-2.5 cm long. Flower: Sepals 4(-5), purple, ovate or narrow-ovate, 5.5-9(-12) ×2.2-4.5(-5)cm, at apex slightly obtuse, each sepal with 5 basal nerves(midrib and 1 lateral nerve thrice dichotomizing, 1 lateral nerve once dichotomizing, and the other 2 lateral nerves twice dichotomizing). Stamens 24-28, 5-7 mm long; filaments filiform, 2-5 mm long; anthers oblong, 1-1.5×0.5-0.6 mm, apex mucronate(mucrones 0.2-0.4 mm long). Carpels 9-22; ovary elliptic, ca. 2.2×1 mm, along ventral and dorsal sutures narrowly winged; style ca. 1 mm long, adaxially with dense stigmatic tissue. Achene compressed, obliquely obovate, sometimes subfalcate, 5-8×2.5-3.2 mm, along ventral and dorsal sutures narrowly winged (wings 0.4–0.6 mm broad). On each side longitudinally 3-ribbed, base with a stipe 1–3 mm long; persistent style ca. 1 mm long.

分布于西藏东南部、云南、四川西部和贵州西部。生山地林边、沟边、灌丛或疏林中，海拔 1050–3400 m。

根状茎和根有发汗、消瘀、解毒等效，用于胃肠热症、湿热痢疾、黄疸等症。(《中国药用植物志》，2016）

标本登录：西藏：米林，青藏队 74-4634，倪志诚等 3104；林芝，青藏补点队 75-1174；通麦，张永田，郎楷永 934；波密，具古乡，应俊生，洪德元 65-0358；察隅，倪志诚等 528；察瓦龙，王启无 66382；芒康，任再金 s.n.

云南：泸水，Gaoligong Shan Biodiversity Survey 25648；镇康，俞德浚 17132；大理，苍山，蔡希陶 53834，秦仁昌 22959，刘慎谔 21115，谢磊，毛建丰 2；漾濞，金效华，刘冰等 1252，1344；宾川，鸡足山，谢磊 24；洱源，云南药检所队 58-022；兰坪，蔡希陶 53720，56076；剑川，秦仁昌 23074；鹤庆，金沙江队 4746；宁蒗，姜恕 6066，韩裕丰等 81-1071；丽江，玉龙山，G. Forrest 5599，植物所横断山队 81-2477，罗毅波 47；丽江，石鼓，植物所横断山队 81-1861；中甸，俞德浚 12011，13623，中甸队 62-1510；维西，王启无 64277，68052，植物所横断山队 81-1768；德钦，王启无 64705，70072，俞德浚 9458，10029，青藏队 81-305；贡山，王启无 71793，俞德浚 19543，22460，22816，冯国楣 7315；景东，李鸣岗 1232；楚雄，李鸣岗 149；禄劝，毛品一 1460；昆明，刘慎谔 20817；嵩明，张英伯 214；东川，蓝顺彬 298；屏边，蔡希陶 62530。

四川：攀枝花，刘冰 366；德昌，朱水法 20088，四川植物队 12299；盐边，金效华 5，高信芬等 3819；盐源，青藏队 83-12137；雷波，俞德浚 3862；西昌，俞德浚 1234；越西，经济植物队 59-3558；洪溪，经济植物队 59-1234；冕宁，赵佐成 4197，4245，4268；甘洛，经济植物队 59-4163；木里，俞德浚 6593；稻城，四川植物队 2231；乡城，四川植物队 1286；巴塘，青藏队 81-3136；雅江，郎楷永，李良千，费勇 2986，D. E. Boufford, B. Bartholomew et al. 35957；泸定，关克俭，王文采等 1555，D. E. Boufford, et al. 27446；康定，王德全 4809，郎楷永，李良千，费勇 1068；峨眉山，杜大华 475，杨光辉 56285；天全，曲桂龄 9730；乾宁，四川植物队 5469；小金，张秀实 6456，郎楷永，李良千，费勇 1516；抚边，汪发缵 21340；金川，李馨 75588，77743，道孚，应俊生 9506；新龙，D. E. Boufford, et al. 34131；色达，D. E. Boufford, et al. 34320；埌塘，汤宗孝 1820；绰斯甲，第八森林大队 2354；马尔康，李馨 71866。

贵州：兴义，张志松，张永田 6397；威宁，蒋英 9179；纳雍，毕节队 662。

78b. 渐尖偏翅唐松草（《植物分类学报》，变种）

var. **acuminatum** Franch. Pl. Delav. 11.1889; W. T. Wang in Acta Phytotax. Sin. 31 (3): 214. 1993; et in Fl Yunnan. 11: 172. 2000; D. Z. Fu & G. Zhu in Fl. China 6: 288. 2001; L. Xie in Med. Fl. China 3: 265. 2016. Holotype：云南：Ta-pin-tze, Hia-ma-ti, alt. 1800 m, 1887-08-04, Delavay s. n. (P; photo. PE).

T. delavayi var. *acuminatum* Franch. f. *appendiculatum* W. T. Wang in Acta Phytotax. Sin. 31 (3): 214. 1993. Holotype：四川：德昌，alt. 1800 m, 1959-04-06，朱水法 20232（PE）.

与偏翅唐松草的区别：萼片顶端尾状渐尖。聚伞圆锥花序圆锥花序状。萼片长椭圆形或卵形，9.5-14×4-5 mm，顶端尾状渐尖，每一萼片具 5 条基生脉（中脉和 1 侧生脉 3 回二歧状分枝，1 侧生脉不分枝，其他 2 侧生脉均 1 回二歧状分枝）。花药长圆形，约 2×0.25 mm，顶端具长 0.1-0.4（-0.7）mm 的短尖头。

This variety differs from var. *delavayi* in its sepals with caudate-acuminate apexes. Thyrses paniculiform. Sepals long elliptic or ovate, 9.5-14×4-5 mm, at apex caudate-acuminate, each sepal with 5 basal nerves (midrib and 1 lateral nerve thrice dichotomizing. 1 lateral nerve simple, and another 2 lateral nerves all once dichotomizing). Anthers oblong, ca. 2×0.25 mm, apex mucronate [mucrones 0.1-0.4 (-0.7 mm) long].

分布于云南西北部和四川西南部。

标本登录：云南：丽江，王启无 70560，71301，71419；宁蒗，谢磊 52。

四川：木里，俞德浚 7589。

78c. 宽萼偏翅唐松草（《中国植物志》，变种）

var. **decorum** Franch. Pl. Delav. 11.1889; Hook. f. in Curtis's Bot. Mag. 116: pl. 7152. 1890; W. T. Wang & S. H. Wang in Fl. Reip. Pop. Sin. 27: 571. 1979; W. T. Wang in Acta Phytotax. Sin. 31(3):214. 1993; in Vasc. Pl. Hengduan Mount. 1: 503. 1993; in Fl. Yunnan. 11: 172, pl.51: 7-9. 2000; et in High. Pl. China 3: 475. 2000; D. Z. Fu & G. Zhu in Fl. China 6: 288. 2001; L. Xie in Med Fl. China 3: 265. 2016. Syntypes：云南：洱源，Yen-tze-hay, alt. 3000 m, 1887-07-08, Delavay 2377(P); Mo-so-yn 之北，Koutoui, alt. 3000 m, 1887-07-19, Delavay s. n. (P; photo, PE); Hee-chan-men, 1887-06-14, Delavay s. n. (P).

? *T. delavayi* auct. non Franch.: L. H. Bailey, Man. Cult. Pl. 391. 1949.

与偏翅唐松草的区别：萼片较大，宽卵形。聚伞圆锥花序圆锥花序状。萼片宽卵形，

7–11×4.8–7 mm，顶端稍钝，具 5 条基生脉（中脉和 1 条侧生脉 3 回二歧状分枝，其他 3 条侧生脉分别是 1, 2 和 5 回二歧状分枝）。花药宽长圆形，约 1.8×1 mm，顶端具长 0.1–0.2 mm 的短尖头。

This variety differs from var. *delavayi* in its broadly ovate sepals. Thyrses paniculiform. Sepals broad-ovate, 7–11×4.8–7 mm, at apex slightly obtuse, each sepal with 5 basal nerves (midrib and 1 lateral nerve thrice dichotomizing, and the another 3 lateral nerves 1, 2 or 5 times dichotomizing respectively). Anthers broad-oblong, ca. 1.8×1 mm, apex mucronate (mucrones 0.1–02 mm long).

分布于云南西北部和四川西南部。生山坡灌丛中，海拔 1600–3400 m。

本变种的花美丽，植株高达 2 m 或更高，在法国、瑞士、瑞典等国的植物园和私人花园栽培供观赏。

标本登录：云南：洱源，秦仁昌 23331；丽江，玉龙山，俞德浚 15479，秦仁昌 20739，30271，植物所横断山队 81-2027，谢磊 95；丽江，文笔山，中甸队 62-653；中甸，哈巴乡，中甸队 62-1510。

四川：稻城，四川植被队 2042。

78d. 角药偏翅唐松草（《中国植物志》，变种）

var. **mucronatum** (Finet & Gagnep.) W. T. Wang & S. H. Wang in Fl. Reip. Pop. Sin. 27: 571, Pl. 144: 8–10. 1979; W. T. Wang in Acta Phytotax. Sin. 31(3):215. 1993; et in Fl. Yunnan. 11: 172. 2000; Y. K. Li in Fl. Guizhou. 3: 71. 1990; D. Z. Fu & G. Zhu in Fl. China 6: 288. 2001; Y. M. Shui & W. H. Chen, Seed Pl. Honghe Reg. S. -E. Yunnan 78. 2003; L. Xie in Med. Fl. China 3: 266. 2016. — *T. dipterocarpum* Franch. var. *mucronatum* Finet & Gagnep. in J. de Bot. 21: 21. 1908. Syntypes: 四川：峨眉山 ,1904–06, E. H. Wilson 4701；同地，1903–07,E. H. Wilson 3083 (not seen). 云南：昆明，1904–08–10，Ducloux 2702; Pan-Kiao-ho, 1904–07, Ducloux 2875 (not seen). 贵州，无详细地点，1899–09–04，Beauvais 713 (not seen).

与偏翅唐松草的区别：聚伞圆锥花序较狭，似总状花序。萼片具 3 条基生脉。花药顶端的短尖头较长，长 0.9–1.8 mm。萼片椭圆形，宽椭圆形或卵形，6–9×3–7 mm，顶端微钝，具 3 条基生脉（中脉 2 回二歧状分枝，2 侧生脉 3 回二歧状分枝）。花药宽长圆形，约 1.8×1 mm。

This variety differs from var. *delavayi* in its narrow, raceme-like thyrses, in its sepal with 3 basal nerves, and its anthers at apex with larger mucrones 0.9–1.8 mm long. Sepals elliptic, broad-elliptic or ovate, 6–9×3–7 mm, at apex slightly obtuse, with 3 basal nerves (midrib twice dichotomizing, and 2 lateral nerves all thrice dichomizing). Anthers broad-oblong, ca. 1.8×1 mm.

分布于云南中部和东南部，以及贵州西部。生山谷灌丛、草坡或林边，海拔 1000-2200 m。

根有清热解毒之效，用于痢疾、肠类等症。(《中国药用植物志》，2016)

标本登录：云南：石林，税玉民等 64082、64326、64351、65009；罗平，张彩飞 1579；砚山，王启无 84052；文山，税玉民等 71125。

贵州：威宁，毕节队 35；盘县，钟补勤 1733、1738；安龙，蒋英 7357；Tating，蒋英 8832。

组 10. 二翅唐松草组

Sect. **Dipterium** Tamura in Acta Phytotax. Geobot. 43: 56. 1992; et in Hiepko, Nat. Pflanzenfam., Zwei. Aufl., 17a IV: 485. 1995. Type: *T. elegans* Wall. ex Royle.

小叶小。花组成分枝的聚伞圆锥花序；每一分枝似一总状花序，具 1-6 个无梗的，由 2-6 朵具短梗并密集花组成单歧聚伞花序。每朵花具 4 萼片；每一萼片具 3 条不分枝的基生脉。花丝丝形。心皮具长柄，具 1 条直花柱和 1 明显、无翅、狭条形的柱头。瘦果具长柄，扁平，近圆形，每侧具 1 条稍弯曲的纵脉，沿腹、背缝线各有 1 条狭翅。

Leaflets small. Flowers in branched thyrses; each branch raceme-like, with 1-6 sessil monochasia cosisting of 2-6 dense shortly pedicellate flowers. Sepals 4 per flower; each sepal with 3 simple basal nerves. Filaments filiform. Carpel long stipitate, with a straight style and a conspicuous not winged narrow-linear stigma. Achene long stipitate, flattened, suborbicular, on each side with a longitudinal slightly curved nerve, and along ventral and dorsal sutures narrowly winged.

1 种，分布于中国西南部，以及喜马拉雅山区。

79. **二翅唐松草** 小叶唐松草（《中国植物志》） 图版 79

Thalictrum elegans Wall. [Cat. n. 7428. 1832] ex. Royle, Ill. Bot. Himal. 51. 1838; Hook. f. & Thoms. Fl Ind. 13. 1855; et in Hook. f. Fl. Brit. Ind. 1: 10. 1872; Lecoy. in Bull. Soc. Bot. Belg. 24: 177. 1885; Frinet & Gagnep. in Bull. Soc. Bot. France 50: 623. 1903; W. T. Wang & S. H. Wang in Fl. Reip. Pop. Sin. 27: 566, fig. 24. 1979; W. T. Wang in Bull. Bot. Lab. E.-N. Forest. Inst. 8: 31. 1980; Grierson, Fl. Bhutan 1(2): 296. 1984; W. T. Wang in Fl.

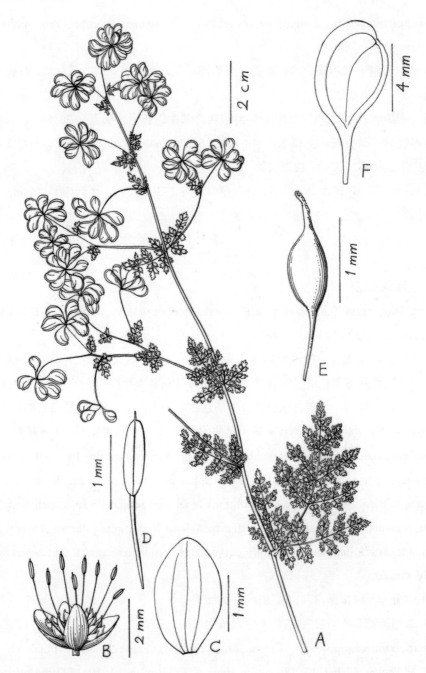

图版79. **二翅唐松草** A. 开花茎上部，B. 花，C. 萼片，D. 雄蕊，E. 心皮（据吴征镒等75-683），F. 瘦果（据青藏队10534）。

Plate 79. **Thalictrum elegans** A. upper part of flowering stem, B. flower, C. sepal, D. stamen. E. carpel (from C. Y. Wu et al. 75-683). F. achene (from Qinghai-Xizang Exped. 10534)

Xizang. 2: 69. 1985; in Vasc. Pl. Hengduan Mount. 1: 502. 1993; in Fl. Yunnan. 11: 167. 2000; et in High. Pl. China 3: 475, fig. 757. 2000; Rau in Sharma et al. Fl. India 1: 135. 1993; D. Z. Fu & G. Zhu in Fl. China 6: 289. 2001. Holotype: Himalaya: Choor Mountain, Wallich 7428 (not seen).

T. samariferum Boivin in J. Arn. Arb. 26: 114. 1945. Type: 云南：德钦，alt. 2700 m，1935-09，王启无 70156（holotype, GH; isotype, PE）. Paratypes：四川：稻城，贡嘎岭，1937-08-29，俞德浚 12993（GH，PE）. 西藏：察隅，察瓦龙，1935-8-9，王启无 64772，64885，65720，65913，66468（GH，PE）.

多年生草本植物，全部无毛。茎高 20-90 cm，细，不分枝或上部分枝。下部和中部茎生叶为 3-4 回羽状复叶，有短柄；叶片长 1.3-5 cm；小叶小，草质，近圆形，圆卵形、倒卵形或菱状倒卵形，1.5-4×1.5-4 mm，基部圆形，浅心形或楔形，3 浅裂至 3 全裂，有时不分裂，全缘，脉平或在下面稍隆起；小叶柄长 0-2 mm；叶柄长 5-15 mm；托叶狭，膜质。聚伞圆锥花序顶生，长约 15 cm；分枝约 10 条，长 1-8 cm，每一分枝似一总状花序，具 1-6 个无梗、由 2-6 朵具短梗并密集的花组成的单歧聚伞花序；苞片叶状；花梗长 0.5-1 mm，果期增长。花：萼片 4，宽椭圆形，2×1.2-1.5 mm，在顶端圆钝，每一萼片具 3 条不分枝的基生脉。雄蕊 9-11；花丝丝形，长 0.8-4 mm，顶端稍宽；花药长圆形，1×0.25-0.3 mm，顶端具长 0.2 mm 的短尖头。心皮 6-20；子房椭圆形，约 0.75×0.5 mm，基部具长 0.7 mm 的细柄；花柱长 0.5 mm；柱头狭条形，长 0.5 mm。瘦果扁平，斜宽倒卵形，4.5-7×2.6-4.2 mm，每侧各有 1 条稍弯曲的纵脉，基部具 2-4 mm 的柄，腹缝线翅宽 0.3 mm，背缝线翅宽 0.5 mm。6-7 月开花。

Perennial herbs, totally glabrous. Stems 20-90 cm tall, slender, simple or above branched. Lower and middle cauline leaves 3-4-pinnate, shortly petiolate; blades 1.3-5 cm long; leaflets small, herbaceous, suborbicular, orbicular-ovate, obovate or rhombic-obovate, 1.5-4×1.5-4 mm, at base rounded, subcordate or cuneate, 3-lobed to 3-sect, sometimes undivided, entire, nerves flat or abaxially slightly prominent; petiolules 0-2 mm long; petioles 5-15 mm long; stipules narrow, membranous. Thyrses teminal, ca. 15 cm long; branches ca. 10, each branch raceme-like, 1-8 cm long, with 1-6 sessile monochasia consisting of 2-6 dense shortly pedicellate flowers; bracts foliaceous; pedicels 0.5-1 mm long, after anthesis elongate. Flower: Sepals 4, broad-elliptic, 2×1.2-1.5 mm, at apex rounded-obtuse, each sepal with 3 simple basal nerves. Stamens 9-11; filaments filiform, 0.8-4 mm long, near apex slightly dilated;

anthers oblong, 1×0.25–0.3 mm, apex with mucrones 0.2 mm long. Carpels 6–20; ovary elliptic, ca. 0.75×0.5 mm, base with a slender stipe 0.7 mm long; style 0.5 mm long; stigma narrow-linear, 0.5 mm long. Achene flattened, obliquely broad-obovate, 4.5–7×2.6–4 mm, on each side with a slightly curved longitudinal nerve, at base with a stipe 2–4 mm long, and with wing along ventral suture 0.3 mm broad, and wing along dorsal suture 0.5 mm broad.

分布于中国西藏南部和东南部、云南西北部（中甸、德钦）和四川西南部（稻城）；以及不丹、尼泊尔、印度北部。生山地林边或草坡上，海拔 2700–4100 m。

标本登录：西藏：吉隆，吴征镒、陈书坤等 75–683；聂拉木，西藏中草药队 1633；帕里至蛤坞，钟补求 5879；察隅，青藏队 73–1081，81–10534。

云南：中甸，哈巴山，中甸队 62–2036；德钦，冯国楣 5995，谢磊，毛建丰 154。

四川：稻城，青藏队 81–4393，王利松，叶建飞 191。

组 11. 腺毛唐松草组

Sect. **Thalictrum.** DC. Syst. 1: 172. 1817; Tamura in Sci. Rep. Osaka. Univ. 17: 51. 1968; in Acta Phytotax. Geobot. 43: 56. 1992; et in Hiepko, Nat. Pflanzenfam., Zwei. Aufl., 17a IV: 483. 1995, p. p.; Emura in J. Fac. Sci. Univ. Tokyo, Sect. 3. Bot. 9: 107. 1972; W. T. Wang & S. H. Wang in Fl. Reip. Pop. Sin. 27: 558. 1979, p. p. Type: *T. foetidum* L.

Subsect. *Genuium* DC. Syst. 1: 172: 1817.

Subsect. *Filiformes* Lecoy. in Bull. Soc. Bot. Belg. 24: 112. 1885.

Subsect. *Flexuosa* Prantl in Bot. Jahrb. 9: 273. 1887.

Sect. *Homothalictrum* Boivin in Rhodora 46: 350. 1944. Type: *T. minus* L.

花小，组成聚伞圆锥花序。萼片 4(–5)，比雄蕊短，通常淡黄绿色或白绿色，具 3 条基生脉。花丝狭条形或丝形；花药狭长圆形或条形，顶端具短尖头。心皮无柄；柱头明显，无翅或具 2 侧生翅；花柱直。瘦果无柄，稍扁，通常沿腹、背缝线多少凸起，在每侧具 2–3 条明显纵肋，顶端具明显宿存柱头。

Flowers small, in thyrses. Sepals 4(–5), shorter than stamens, usually yellow-greenish or white-greenish, with 3 basal nerves. Filaments narrow-linear or filiform; anthers narrow-oblong or linear, at apex mucronate. Carpel sessile; stigma conspicuous, not winged or with 2 lateral wings; style straight. Achene sessile, slightly compressed, usually along ventral and dorsal sutures more or less convex, on each side with 2–3 cospicuous longitudinal ribs, at apex with a conspicuous persistent stigma.

本组在中国有 18 种，隶属 2 系，在全国多数省、区广布。

系 1. 川鄂唐松草系

Ser. **Osmundifolia**. W. T. Wang, ser. nov. Type: *T. osmundifolium* Finet & Gagnep.

Series nova haec est affinis Ser. *Thalictro*, a qua differt stigmatibus haud alatis. Filaments narrow-linear (*T. trichopus*, *T. honanense* and *T. osmundifolium*) or filiform. Style adaxially on apex or below apex with a not winged ellipsoidal, ovate or narrow-linear conspicuous stigma. Achenes not winged, rarely along dorsal suture narrowly winged (*T. finetii*).

本系与近缘的腺毛唐松草系的区别在于本系的心皮柱头无翅。花丝狭条形（毛发唐松草，河南唐松草，川鄂唐松草）或丝形。花柱在腹面顶端或顶端之下具 1 无翅、椭圆体形、卵形或狭条形的明显柱头。瘦果无翅，稀沿背缝线具狭翅（滇川唐松草）。

6 种，均为中国特有种，分布于中国西南部或中部。

80. 毛发唐松草（《中国药用植物志》，1964） 图版 80

Thalictrum tricopus Franch. in Bull. Soc. Bot. France 33: 370. 1886; et Pl. Delav. 11.1889; Finet & Gagnep. in Bull. Soc. Bot. France 50: 622. 1903; Finet in J. de Bot. 21: 20. 1908; Boivin in J. Arn. Arb. 26: 114. 1945; Laucner & Green in Not. Bot. Gard. Edinburgh 23(4):593. 1961; C. Pei & T. Y. Chou, Icon. Chin. Med. Pl. 7: fig. 360. 1964; Iconogr. Corm. Sin. 1: 681,fig. 1361. 1972; W. T. Wang & S. H. Wang in Fl. Reip. Pop. Sin. 27: 574, pl. 141: 6–10. 1979; W. T. Wang in Vasc. Pl. Hengduan Mount. 1: 503. 1993; et in Fl. Yunnan. 11: 172, pl. 50: 6–9. 2000; D. Z. Fu & G. Zhu in Fl. China 6: 299. 2001; W. T. Wang in High. Pl. China in colour 3: 392. 2016; L. Xie in Med. Fl. China 3: 267, with fig. & photo. 2016. Holotype: 云南：Tapin-tze 之北，Mao-kou-tchang, alt. 2000 m, 1885-07-04, Delavay 1854(P).

T. tenii Lévl. in Repert. Sp. Nov. Reg. Veg. 7: 98. 1909; et Cat. Pl. Yunnan 227. 1917. Holotype: 353 云南：Lou-Fou, prés Tong-Tchouan, 1907-07, Ten 584 (E, not seen).

多年生草本植物，全部无毛。茎高达 1.2 m，上部有长分枝。下部茎生叶为 3 回羽状复叶，长达 30 cm；叶片长约 20 cm；小叶草质，菱状卵形、卵形、椭圆形或倒卵形，0.4–2.1×0.3–1.6 cm，顶端微钝或圆形，基部圆形、宽楔形或浅心形，3 浅裂或不分裂，全缘、脉平、不明显；叶柄长达 9 cm，基部有鞘；托叶狭窄，全缘或分裂。聚伞圆锥花序具近平

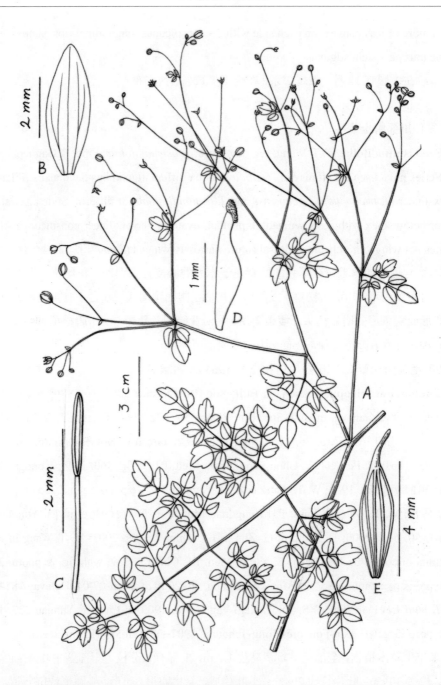

图版80. **毛发唐松草** A. 上部茎生叶和腋生聚伞圆锥花序，B. 萼片，C. 雄蕊，D. 心皮（据俞德浚16593），E. 瘦果（据周仕顺2123）。

Plate 80. **Thalictrum trichopus** A. upper cauline leaf and its axillary thyrse, B. sepal, C. stamen, D. carpel (from T. T. Yü 16593), E. achene (from S. S. Zhou 2123).

展细分枝；苞片叶状，或为三出复叶或单叶；花梗丝形，长 1.4–3.5 cm，果期长达 6 cm。

花：萼片 4，早落，白色，狭椭圆形或倒卵形，3–4×1.2–1.6 mm，顶端微尖或钝，每一萼片具 3 条基生脉（中脉不分枝，2 侧生脉 1 回二歧状分枝）。雄蕊约 17；花丝狭条形，2.6–2.8×0.15–0.2 mm；花药条形或狭长圆形，1.3–2×0.3 mm，顶端钝。心皮 2–5；子房斜狭倒卵形，约 1.5×0.5 mm；花柱长 0.5 mm；柱头生花柱腹面顶端，狭椭圆体形，长 0.5 mm。瘦果稍扁，狭椭圆体形，3–4×1.8 mm，每侧有 3 条纵肋；宿存花柱长约 1.4 mm。6–7 月开花。

Perennial herbs, totally glabrous. Stems up to 1.2 m tall, above long branched. Lower cauline leaves 3-pinnate, up to 30 cm long; blades ca. 20 cm long; leaflets herbaceous, rhombic-ovate, ovate, elliptic or obovate, 0.4–2.1×0.3–1.6 cm, at apex slightly obtuse or rounded, at base rounded, broad-cuneate or subcordate, 3-lobed or wndivided, entire, nerves flat and inconspicuous; petioles up to 9 cm long, at base vaginate; stipules narrow, entire or divided. Thyrses with nearly spreading slender branches; bracts foliaceous, ternate or simple. Flower: Sepals 4, caducous, white, narrow-elliptic or obovate, 3–4×1.2–1.6 mm, at apex slightly acute or obtuse, each sepal with 3 basal nerves (midrib simple, and 2 lateral nerves once dichotomizing). Stamens ca. 17; filaments narrow-linear, 2.6–2.8×0.15–0.2 mm; anthers linear or narrow-oblong, 1.3–2×0.3 mm, apex obtuse. Carpels 2–5; ovary obliquely narrow-obovate, ea. 1.5×0.5 mm; style 0.5 mm long; stigma borne on apex of adaxial surface of style, ellipsoidal, 0.5 mm long. Achene slightly compressed, narrow-ellipsoidal, 3–4×1.8 mm, on each side longitudinally 3-ribbed; persistent sytle ca. 1.4 mm long.

分布于云南西部和南部，四川南部。生山地灌丛草地，海拔 1400–2500 m。

根有清热解毒、消炎止泻之效，用于肠炎、痢疾、百日咳等疾病。(《中国药用植物志》，2016)

标本登录：云南：宾川，蔡希陶 52903；凤庆，俞德浚 16593；思茅，王洪 4809，周仕顺 2123；祥云，倪兆林 253。

四川：盐边，武素功 230；德昌，赵清盛等 5262。

81. 河南唐松草(《中国植物志》) 图版 81

Thalictrum honanense W. T. Wang & S.H.Wang in Fl Reip. Pop. Sin. 27: 576, 620, pl. 146: 3–7. 1979; B. Z. Ting et al. Fl Henan. 1: 468, fig.599. 1981; W. T. Wang in Acta Bot. Yunnan. 6(4): 378. 1984; et in High. Pl China 3: 476, fig. 761. 2000; D. Z. Fu & G. Zhu in Fl. China 6: 291. 2001; L. Xie in Med Fl. China 3: 369, with fig. & photo. 2016. Type: 河南：嵩

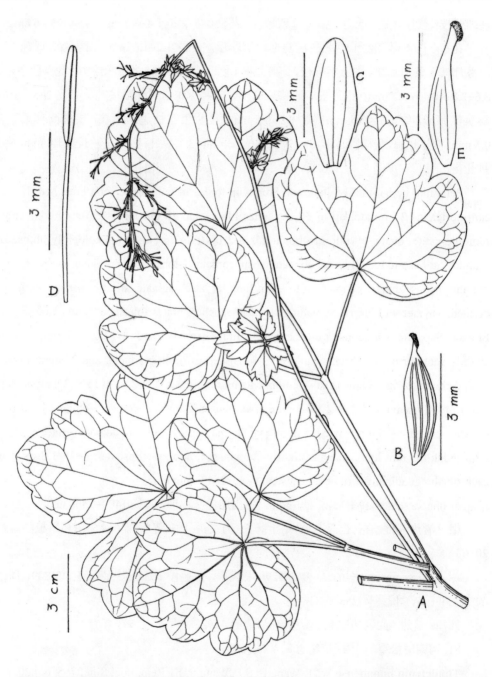

图版81. 河南唐松草　A. 结果茎上部，B. 瘦果（据河南林业厅队727），C. 萼片，D. 雄蕊，E. 心皮（据？940）。

Plate 81. **Thalictrum honanense**　A. upper part of fruiting stem, B. achene (from Exped. Bureau Forest. Henan Prov. 727), C. sepal, D. stamen, E. carpel (from ? 940).

县，西礓河，alt. 840 m，山谷林下，1956-09-12，河南森林调查队 940（holotype, PE）；栾川，老君山，alt. 1800 m，1959-08，王遂义 21012（paratype, PE）.

多年生草本，全部无毛。茎高 80-150 cm，上部分枝。中部茎生叶为 2-3 回三出复叶，有短柄；叶片长约 25 cm；小叶纸质，近圆形，宽卵形，心形或倒卵形，2.4-6.5×2.4-8.5 cm，顶端钝，基部圆形、心形或浅心形，3 浅裂，裂片有少数牙齿，脉隆起，脉网明显；小叶柄长 0.5-7 cm；叶柄长 0.9-4 cm。聚伞圆锥花序似总状花序，长约 30 cm，分枝稀疏，长 2-3 cm，有密集 5-8 花，有时基部分枝长 8 cm；基部苞片叶状，上部苞片狭卵形或三角形，长 0.2-1.6 cm；花梗长 4-6 mm。花：萼片 4，脱落，淡红色，狭椭圆形，约 4.6×2 mm，顶端钝，每一萼片具 3 条不分枝基生脉。雄蕊 30-35；花丝狭条形，3-5×0.2-0.25 mm；花药条形或狭长圆形，2-2.6×0.4-0.5 mm，顶端钝。心皮 3-9；子房狭卵球形，约 3×0.8 mm；花柱长 1 mm；柱头生花柱腹面顶端，正三角形，0.5×0.5 mm。瘦果狭椭圆体形，约 3×1 mm，在每侧有 3 条纵肋；宿存花柱长约 1 mm。8-9 月开花。

Perennial herbs, totally glabrous. Stems 80–150 cm tall, above branched. Middle cauline leaves 2–3, ternate, shortly petiolate; blades ca. 25 cm long; leaflets papery, suborbicular, broad-ovate, cordate or obovate, 2.4–6.5×2.4–8.5 cm, at apex obtuse, at base rounded, cordate or subcordate, 3-lobed, lobes with a few rotund teeth, nerves prominent and nervous nets conspicuous; petiolules 6.5–7 cm long; petioles 0.9–4 cm long. Thyrses raceme-like, ca. 30 cm long; branches sparse, 2–3 cm long, densely 5–8–flowered, sometimes basal branch up to 8 cm long; basal bract foliaceous, apper bracts simple, narrow-ovate or triangular, 0.2–1.6 cm long; pedicels 4–6 mm long. Flower: Sepals 4, deciduous, reddish, narrow-elliptic, ca. 4.6×2 mm, at apex obtuse, each sepal with 3 simple basal nerves. Stamens 30–35; filaments narrow-linear, 3–5×0.2–0.25 mm; anthers linear or narrow-oblong, 2–2.6×0.4–0.5 mm, apex obtuse. Carpels 3–9; ovary narrow-ovoid, ca. 3×0.8 mm; style 1 mm long; stigma borne on apex of adaxial surface of style, deltoid, 0.5×0.5 mm. Achene narrow-ellipsoidal, ca. 3×1 mm, on each side longitudinally 3-ribbed; persistent style ca. 1 mm long.

分布于河南和湖北。生山地灌丛或疏林中，海拔 840-1800 m。

根和全草有清热解毒之效，用于痈疖、肿毒等疾病。根含小檗碱（berberine）、药根碱（jatrorrhizine）等生物碱。（《中国药用植物志》，2016）

标本登录：河南：桐柏，伏牛山，河南林业厅队 727；登封，河南植物普查队 111，377。

湖北：枣阳，木德义 245（EDC）；宜城，苏明才 274（EDC）；钟祥，蒋传新 174；黄陂，王世木 3（EDC）。

82. 川鄂唐松草《中国植物志》》 图版 82

Thalictrum osmundifolium Finet & Gagnep. in Bull. Soc Bot. France 50: 615. pl. 19: A. 1903; S. H. Fu, Fl. Hupeh. 1: 359. 1976; W. T. Wang & S. H. Wang in Fl. Reip. Pop. Sin. 27: 578, pl. 146: 8–10. 1979; D. Z. Fu & G. Zhu in Fl. China 6: 294. 2001; Z. Y. Li & M. Ogisu, Pl. Mount Emei 255. 2001; Q. L. Gan, Fl Zhuxi Suppl. 246, fig. 295. 2011; L. Q. Li in Y. Jia et al. High. Pl. Daba Mount. 118. 2014. Syntypes：湖北西部，1901–07, E. H. Wilson 2393 (P, not seen). 重庆，城口，P. Farges s. n. (P, not seen).

多年生草本植物，只小叶柄有疏柔毛，其他部分无毛。茎高 80-100 cm，上部分枝。中部茎生叶为 3-4 回羽状复叶，有短柄；叶片长 15-26 cm；小叶草质，卵形、宽卵形、菱形或楔状倒卵形，0.8-3×0.5-1.7（-2.3）cm，顶端有短尖，基部浅心形或圆形，3 浅裂，裂片全缘或有少数牙齿，脉平或近平，脉网不明显；小叶柄长 0.5-12 mm；叶柄长 1.8-2.3 cm。聚伞圆锥花序长达 29 cm；苞片多叶状，顶端苞片小，长 2-3 mm；花梗细，长 0.8-2.5 cm。花：萼片 4（-5），脱落，椭圆状卵形，3-5×2-2.4 mm，顶端钝，每一萼片具 3 条不明显、不分枝的基生脉。雄蕊 30-35；花丝狭条形，长约 3.2 mm，上部宽 0.2 mm，下部稍变狭；花药长圆形，约 2.2×0.5 mm，顶端微尖。心皮 8-14；子房狭倒卵形，约 2.6×0.5 mm；花柱长 0.5 mm；柱头狭卵形，0.4×0.25 mm。瘦果斜椭圆形，约 3×1.5 mm，每侧有 2-3 条纵肋；宿存花柱长 1 mm。6 月开花。

Perennial herbs, only on petiolules sparsely puberulous, else-where glabrous. Stems 80–100 cm tall, above branched. Middle cauline leaves 3–4-pinnate, shortly petiolate; blades 15–26 cm long; leaflets herbaceous, ovate, broad-ovate, rhombic or cuneate-obovate, 0.8–3×0.5–1.7(–2.3) cm, at apex mucronate, at base subcordate or rounded, 3-lobed, lobes entire or few-dentate, nerves flat or nearly flat, nervous nets inconspicuous; petiolules 0.5–12 mm long; petioles 1.8–2.3 cm long. Thyrses up to 29 cm long; bracts mostly foliaceous, apical bracts small, 2–3 mm long; pedicels 0.8–2.5 cm long. Flower : Sepals 4, deciduous, elliptic-ovate, 3–5×2–2.4 mm, at apex obtuse, each sepal with 3 inconspicuous simple basal nerves. Stamens 30–35; filaments narrow-linear, ca. 3.2 mm long, above 0.2 mm broad, below slightly narrowed; anthers oblong, ca. 2.2×0.5 mm, apex slightly acute. Carpels 8–14; ovary narrow-obovate, ca. 2.6×0.5 mm; style 0.5 mm long; stigma narrow ovate, 0.4×0.25 mm. Achene

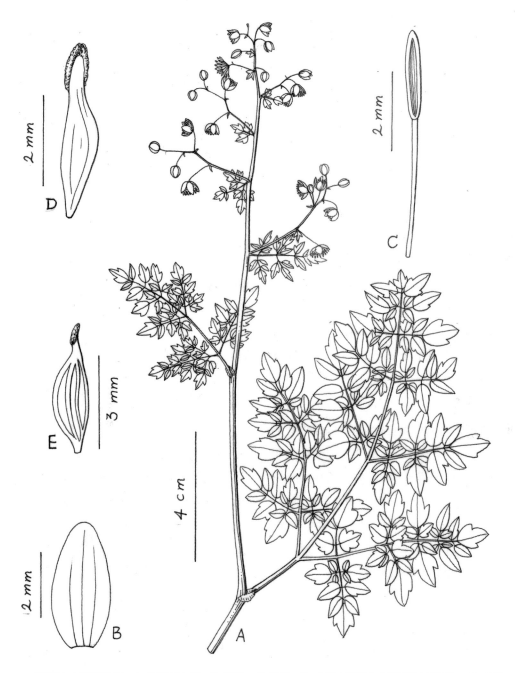

图版82. 川鄂唐松草 A. 开花茎上部，B. 萼片，C. 雄蕊，D. 心皮（据川大生物系队105625），E. 瘦果（据川大生物系队105720）。

Plate 82. **Thalictrum osmundifolium** A. upper part of flowering stem, B. sepal, C. stamen, D. carpel (from Exped. Depart. Biol. Univ. Sichuan 105625), E. achene (from Exped. Depart. Biol. Univ. Sichuan 105720).

obliquely elliptic, ca. 3×1.5 mm, on each side longitudinally 2-3-ribbed; persistent style 1 mm long.

分布于四川、重庆和湖北西部。生山地灌丛中或溪边，海拔 1400-2900 m。

标本登录：四川：小金，D. E. Boufford et al. 38260；峨眉山，贺贤育 6687，徐洪贵 4987；宝兴，宋滋圃 390442；平武，蒋兴麐 10601。

重庆：巫溪，杨光辉 58742；城口，戴天伦 100963，105720，105268，105625，巴山调查队 1950。

83. 亚东唐松草　　图版 83

Thalictrum yadongense W. T. Wang, sp. nov. Type：西藏（Xizang）；亚东县，上亚东乡（Yadong Xian, Shang-yadong Xiang），alt. 4017 m，山坡，花紫色（on slope, fls. purple），2013-07-06，PE- 西藏队（PE-Xizang Exped.）2589

Ob flores parvos in thyrsum dispositos, staminum filamenta regularia, et stigmata haud alata species nova haec affinis *T. osmundifolio* Finet & Gagnep. est, a quo filamentis filiformibus, antheris apice mucronatis, stigmatibus longioribus (1.6 mm longis) distinguitur. In *T. osmundifolio*, filamenta anguste linearia, antherae apice acutae, haud mucronatae, et stigmata breviora 0.4 mm longa sunt.

Perennial herbs, totally glabrous. Stems ca. 1 m tall, branched, near the middle ca. 8 mm in diam., smooth. Middle and upper cauline leaves shortly petiolate or sessile, 3-4-ternate-pinnate; blades 6-11×5-10 cm; leaflets thinly papery, broad-obovate, obovate or broad-elliptic, 3-10×2-11 mm, at base rounded or obtuse, 3-lobulate or undivided (lobes at apex rounded, entire or 1-denticulate), 3-5-nerved, nerves adaxially flat, abaxially prominent; petiolules 0-4 mm long; petioles short or wangting, narrowly vaginate. Terminal thyrses large, up to 40 cm long, axillary thyrses smaller, 15-20 cm long; bracts foliaceous, 0.2-3 cm long; pedicels filiform, 2-12 mm long. Flower: Sepals 4, purple, broad-elliptic, ca. 4×2.6 mm, apex rounded, each sepal with 3 simple basal nerves. Stamens 12-22; filaments filiform, ca. 8 mm long; anthers narrow-oblong, 2.5-3×0.5 mm, at apex with mucrones 0.4 mm long. Carpels ca. 6; ovary stipitate, ellipsoid, ca. 1.5×0.8 mm, base with a stipe 0.5 mm long; style 2 mm long; stigma narrow-linear, 1.6 mm long.

多年生草本植物，全部无毛。茎高约 1 m，分枝，近中部直径约 8 mm，光滑。中部和上部茎生叶具短柄或无柄，为 3-4 回三出羽状复叶；叶片 6-11×5-10 cm；小叶薄纸

图版83. 亚东唐松草 A. 茎生叶，B. 聚伞圆锥花序下部，C. 萼片，D. 雄蕊，E. 心皮（据isotype）。
Plate 83. **Thalictrum yadongense** A. cauline leaf, B. lower part of a thyrse, C. sepal. D. stamen, E. carpel (from isotype).

质，宽倒卵形、倒卵形或宽椭圆形，3-10×2-11 mm，基部圆形或钝，3 微裂或不分裂（裂片顶端圆形，全缘或具 1 小齿），具三出或五出脉，脉上面平，下面隆起；小叶柄长 0-4 mm；叶柄短或不存在，具狭鞘。顶生聚伞圆锥花序大，长达 40 cm，腋生聚伞圆锥花序较小，长 15-20 cm；苞片叶状，长 0.2-3 cm；花梗丝形，长 2-12 mm。花：萼片 4，紫色，宽椭圆形，约 4×2.6 mm，顶端圆形，每一萼片具 3 条不分枝的基生脉。雄蕊 12-22；花丝丝形，长约 8 mm；花药狭长圆形，2.5-3×0.5 mm，顶端具长 0.4 mm 的短尖头。心皮约 6；子房具柄，椭圆体形，约 1.5×0.5 mm，柄长 0.5 mm；花柱长 2 mm；柱头狭条形，长 1.6 mm。6-7 月开花。

特产西藏亚东县。生山坡上，海拔 4017 m。

本种与川鄂唐松草 *T. osmundifolium* Finet & Gagnep. 在亲缘关系上相近，区别特征为本种的花丝丝形，花药顶端具短尖头，以及柱头较长。

84. 陕西唐松草（《秦岭植物志》） 图版 84

Thalictrum shensiense W. T. Wang & S. H. Wang in Fl. Tsinling. 1 (2): 245, fig. 211. 1974; et in Fl. Reip. Pop. Sin. 27: 582, pl. 146: 11-12. 1979; D. Z. Fu & G. Zhu in Fl. China 6: 297. 2001. Holotype：陕西：旬阳，庙岭，1959-08-04，李培元 9608（WUK; photo, PE）.

多年生草本植物。茎高约 70 cm，中部以上被短柔毛，不分枝，约有 7 叶。中部茎生叶为 3-4 回三出复叶，长约 12 cm，具短柄；小叶纸质，楔状倒卵形、宽菱形或宽卵形，1.2-2.2×0.7-2 cm，顶端圆形或钝，有短尖头，基部圆形、楔形或近心形，3 浅裂（裂片全缘或 1-2 圆齿），上面变无毛，下面被短毛，并有明显的脉网；小叶柄长 1-6 mm，疏被短柔毛；叶柄长 1.5 cm。聚伞圆锥花序长约 13 cm，有密集的花；花梗长 2-4 mm，密被短柔毛（毛长 0.1-0.2 mm）。花：萼片 4，狭椭圆形，2.8-3×1-1.8 mm，顶端钝，外面疏被柔毛，每一萼片具 3 条不分枝的基生脉。雄蕊 5-9，无毛；花丝丝形，长 1-5 mm；花药条形或狭长圆形，2.4-2.6×0.4-0.5 mm，顶端具长 0.2 mm 的短尖头。心皮 2-3，无毛；子房近长圆形，长约 0.6 mm；花柱长 0.4 mm，腹面具不明显柱头。瘦果近长圆形，长约 1 mm，每侧有 2 条纵肋；宿存花柱长 0.4 mm。7-8 月开花。

Perennial herbs. Stem ca. 70 cm tall, above the middle part puberulous, simple, ca. 7-leaved. Middle cauline leaves 3-4-ternate, ca. 12 cm long, shortly petiolate; leaflets papery, cuneat-obovate, broad-rhombic or broad-ovate, 1.2-2.2×0.7-2 cm, at apex rounded or obtuse, mucronate, at base rounded, cuneate or subcordate, 3-lobed (lobes entire or 1-2-rotund-dentate), adaxially glabrescent, abaxially puberulous and with conspicuous nervous nets;

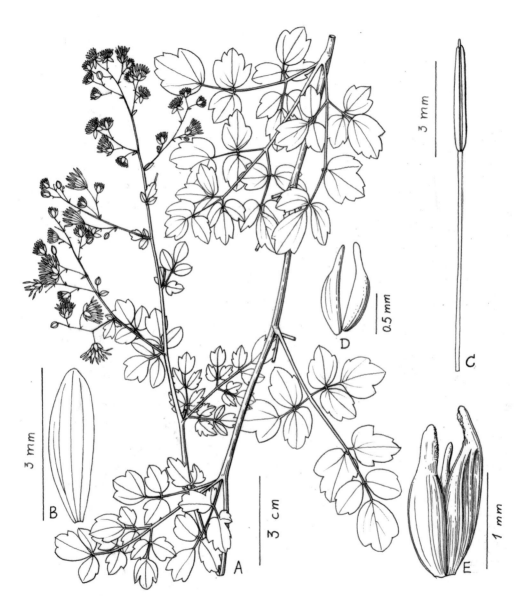

图版84. **陕西唐松草**　A. 开花茎上部，B. 萼片，C. 雄蕊，D. 雌蕊群，E. 二瘦果和一败育心皮（据 holotype）。

Plate 84. Thalictrum shensiense　A. upper part of flowering stem, B. sepal, C. stamen, D. gynoecium, E. two achenes and one abortive carpel (from holotype).

petiolules 1−6 mm long, sparsely puberulous; petioles 1.5 cm long. Thyrse ca. 13 cm long, with dense flowers; pedicels 2−4 mm long, densely puberulous (hairs 0.1−0.2 mm long). Flower: Sepals 4, narrow-elliptic, 2.8−3×1−1.8 mm, at apex obtuse, abaxially sparsely puberulous, each sepal with 3 simple basal nerves. Stamens 5−9, glabrous; filaments filiform, 1−5 mm long; anthers linear or narrow-oblong, 2.4−2.6×0.4−0.5 mm, apex with mucrones 0.2 mm long. Carpels 2−3, glabrous; ovary suboblong, ca. 0.6 mm long; style 0.4 mm long, adaxially with an inconspicuous stigma. Achene suboblong, ca. 1 mm long, on each side longitudinally 2-ribbed; persistent style 0.4 mm long.

特产陕西旬阳庙岭。生山地草坡。

85. 滇川唐松草（《药学学报》） 图版 85

Thalictrum finetii B. Boivin in J. Arn Arb. 26: 113, pl. 1: 1−3. 1945; Iconogr, Corm. Sin. 1: 682, fig. 1363. 1972; W. T. Wang & S. H. Wang in Fl Reip. Pop. Sin. 27: 578, pl. 539: 5−9. 1979; W. T. Wang in Fl. Xizang. 2: 72, fig. 21: 1−3. 1985; in Vasc. Pl. Hengduan Mount. 1: 503. 1993;in Fl. Yunnan. 11: 173. 2000; et in High. Pl. China 3: 477, fig. 762. 2000; D. Z. Fu & G. Zhu in Fl. China 6: 289. 2001; Z. Y. Li & M. Ogisu, Pl. Mount Emei 255. 2007; L. Xie in Med. Fl. China 3: 270, with fig. & photo. 2016. Holotype：四川：峨眉山，alt. 2200 m，1938−08−01，K. N. Yin 117 (GH, not seen). Paratypes：四川：峨眉山，alt. 3000 m,1931−07，汪发瓒 23458（GH. PE）；九龙, 1937−07−03, 俞德浚 6915（GH，PE）；Liang-he-kou, 1938−08，T. S. Wen 563 (GH, not seen). 云南：澜沧江－怒江分水岭，alt. 4000 m, 1938−08−11 俞德浚 22298（GH, PE）；镇康，1936−07−24，俞德浚 10930（GH, PE）；德钦，1935−07，王启无 64790（GH, PE）；知子罗，碧洛山，1934−08−15，蔡希陶 58014（GH, PE）；近维西，立地坪，1939−10−09，秦仁昌 22072（GH, not seen）；维西，叶枝，1935−07，王启无 68033（GH, PE）；中甸，哈巴山，1939−08−20，冯国楣 2100（GH, PE）；大理，1933−07−30，蔡希陶 53943（GH, PE）.

T. deciternatum B. Boivin in J. Arn. Arb. 26: 112, 1945, p. p. min. quoad specim. T. T. Yü 9891.

多年生草本植物，茎高 50-200 cm，变无毛，分枝。中部茎生叶为 3−4 回三出或羽状复叶，具短柄；叶片长约 18 cm；小叶草质，菱状倒卵形、宽卵形或近圆形，0.9−2×0.7−2 cm，顶端圆形，有短尖头，基部宽楔形、圆形或圆截形，3 浅裂，边缘有少数钝齿或全缘，上面无毛，下面脉上有短硬毛（毛长 0.1−0.3 mm）或变无毛，脉上面稍凹陷，下面稍

六、分类学处理　　　　　　　　　　　　　　　　307

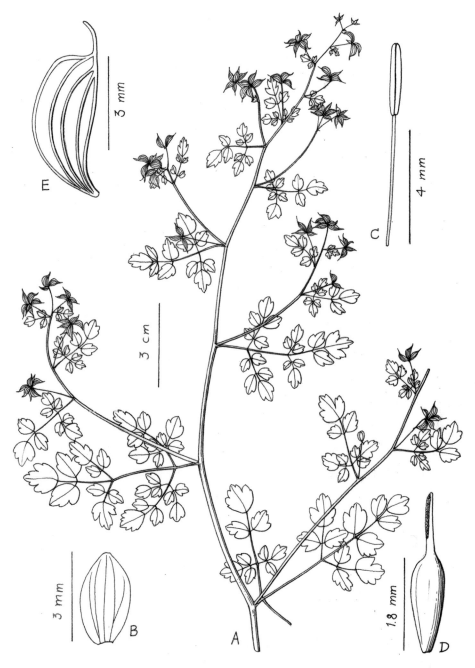

图版85. **滇川唐松草**　A. 开花茎上部，B. 萼片，C. 雄蕊，D. 心皮（据汪发缵23458），E. 瘦果（据 Gaoligone Shan Biodiv. Survey 32066）。

Plate 85. **Thalictrum finetii**　A. upper part of flowering stem, B. sepal, C. stamen, D. carpel (from F. T. Wang 23458), E. achene (from Gaoligone Shan Biodiv. Survey 32066)

隆起或平；小叶柄长 1–6 mm；叶柄长约 2 cm。聚伞圆锥花序长达 30 cm；苞片多为叶状，顶部苞片小，为三出复叶，小叶长 2–3 mm；花梗细，长 4–18 mm。花：萼片 4–5，白色或淡黄绿色，卵形或宽椭圆形，约 3×2 mm，顶端钝，每一萼片具 3 条基生脉（中脉不分枝，2 侧生脉均 1 回二歧状分枝）。雄蕊约 16；花丝丝形，长约 4.5 mm；花药狭长圆形，1.5–2.2×0.3–0.4 mm，顶端有长 0.1 mm 的短尖头。心皮 5–14；子房狭倒卵形，约 1.8×0.35 mm；花柱长 1 mm；柱头狭条形，长 0.8 mm。瘦果扁平，半圆形，3–4×1.2–3.2 mm，在每侧有 1–3 条稍弯曲的细纵肋，沿背缝线上部有狭翅；宿存花柱长 1.7–1.9 mm。7–8 月开花。

Perennial herbs. Stems 50–200 cm tall, glabrescent, branched. Middle cauline leaves 3–4-ternate or 3–4-pinnate, shortly petiolate; leaflets herbaceous, rhombic-obovate, broad-ovate or suborbicular, 0.9–2×0.7–2 cm, at apex rounded, mucronate, at base broad-cuneate, rounded or rounded-truncate, 3-lobed, at margin few-crenate or entire, adaxially glabrescent, abaxially on nerves hirtellous (hairs 0.1–0.3 mm long) or glabrescent, nerves adaxially impressed, abaxially slightly prominent or flat; petiolules 1–6 mm long; petioles ca. 2 cm long. Thyrses terminal, up to 30 cm long; bracts mostly foliaceous, apical bracts small, ternate, with leaflets 2–3 mm long; pedicels slender, 4–18 mm long. Flower: Sepals 4–5, white or yellowish-greenish, ovate or broad-elliptic, ca. 3×2 mm, at apex obtuse, each sepal with 3 basal nerves (midrib simple, and 2 lateral nerves once dichotomizing). Stamens ca. 16; filaments filiform, ca. 45 mm long; anthers narrow-linear, 1.5–2.2×0.3–0.4 mm, apex with mucrones 0.1 mm long. Carpels 5–14; ovary narrow-obovate, ca. 1.8×0.35 mm; style 1 mm long; stigma narrow-linear, 0.8 mm long. Achene flattened, semi-orbicular, 3–4×1.2–3.2 mm, on each side with 1–3 slightly curved thin longitudinal ribs, along above dorsal suture narrowly winged; persistent style 1.7–1.9 mm long.

分布于西藏东南部、云南西部和四川西部。生山地草坡、林边或林中，海拔 2200–4000 m。

根有清热解毒、泻火燥湿之效，用于湿热痢疾、黄胆等疾病。可代黄连、马尾连药用。（《中国药用植物志》，2016）

标本登录：西藏：米林，多雄拉，倪志诚等 3047；墨脱，青藏队 74-3840，生态室高原组 11144，程树志、李渤生 305；察隅，资源综考组 894。

云南：泸水，Gaoligong Shan Biodiv. Survey 25910；宾川，鸡足山，谢磊，毛建丰 27，28；丽江，植物所横断山队 81-2357；中甸，中甸队 62-914，62-1436，62-1719，谢磊，

毛建丰 75；维西，俞德浚 9015；福贡，Gaoligong Shan Biodiv. Survey 26741；贡山，俞德浚 22394，冯国楣 8348，Gaoligong Shan Biodiv. Survey 31363，32066，34518；德钦，俞德浚 9130，9372，9766，9891，10233。

四川：木里，青藏队 85-13075，85-14425；稻城，郎楷永，李良千，费勇 2589，2691；乡城，四川植物队 2853，青藏队 81-3824，D. E. Boufford, B. Bartholomew et al. 28467；九龙，D. E. Boufford et al. 33172，33446，42610；巴塘，郎楷永，李良千，费勇 2317，2353；新龙，姜恕 6434；雅江，郎楷永，李良千，费勇 2880；康定，关克俭，王文采等 317，郎楷永，李良千，费勇 686，920；金阳，杜大华 4527；冕宁，武素功 2287，赵佐成 4201；峨眉山，方文培 3005，汪发缵 23458. T. C. Lee 2881，熊济华，张秀实 31464，31814；泸定，J. I. Jeon 等 S11348；宝兴，杜大华 4390，宋滋圃 39028。

系 2. 腺毛唐松草系

Ser. **Thalictrum**, st. nov. —Sect. *Genuina* (DC.) Boivin in Rhodora 46: 360. 1944. —Subsect. *Thalictrum*. Tamura in Acta Phytotax. Geobot. 43: 56. 1992; et in Hiepko, Nat. Pflanzenfam., Zwei. Aufl., 17a Ⅳ : 484. 1995. Type: *T. foetidum* L.

花丝丝形。柱头完全与花柱贴生，狭长圆形或狭三角形，具 2 侧生狭翅，或宽三角形、正三角形或箭头状正三角形，具 2 侧生宽翅。瘦果无翅。

Filaments filiform. Stigma entirely adnate with style, narrow-oblong or narrow-triangular, with 2 narrow lateral wings, or broad-triangular, deltoid or sagittate-deltoid, with 2 broad lateral wings. Achenes not winged.

本系在中国有 12 种（4 种为特有种），广布中国多数省区，只在海南、广西、江西、浙江、福建、台湾等六省、区无分布。

86. 星毛唐松草（《中国植物志》） 图版 86

Thalictrum cirrhosum Lévl. in Repert. Sp. Nov. Reg. Veg. 7: 97. 1909; Boivin in J. Arn. Arb. 26: 112. 1945; Lauener & Green in Not. Bot Gard. Edinburgh 23(4): 591. 1961; W. T. Wang & S. H. Wang in Fl. Reip. Pop. Sin. 27: 579. 1979; W. T. Wang in Acta Bot. Yunnan. 6(4):379. 1984; et in Fl. Yunnan. 11: 174. 2000; D. Z. Fu & G. Zhu in FL. China 6: 287. 2001; L. Xie in Med . Fl. China 3: 272, with photo. 2016. Holotype：云南：昆明，Mont Tchong-chan, 1905-07-19，Ducloux 604 (E, not seen).

多年生草本植物。茎高 28-55 cm，上部和花序有极短的腺毛，分枝。中部茎生叶

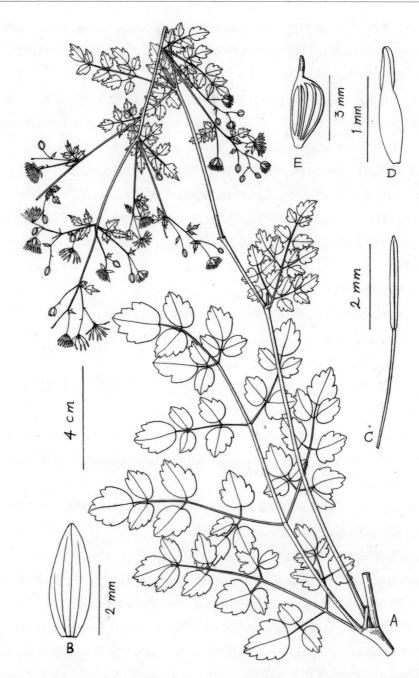

图版86. **星毛唐松草** A. 开花茎上部，B. 萼片，C. 雄蕊，D. 心皮（据刘慎谔16566），E. 瘦果（据税玉民等64620）。

Plate 86. **Thalictrum cirrhosum** A. upper part of flowering stem, B. sepal, C. stamen, D. carpel (from T. N. Liou 16566), E. achene (from Y. M. Shui et al. 64620).

为 3 回羽状复叶，具短柄；叶片长约 10 cm，具 4 对羽片；小叶纸质，圆菱形、宽菱形或菱状倒卵形，1–1.3×0.8-1.8 cm，基部圆截形或浅心形，3 浅裂（裂片全缘或有 1–2 圆齿），上面无毛，下面在脉网上密被灰白色短分枝毛，脉在上面凹陷，在下面隆起；小叶柄长 1–9 mm；叶柄长 0.5–1 cm。聚伞圆锥花序长约 15 cm；下部苞片叶状，上部苞片简单，狭披针形或条形，长 2–4 mm；花梗长 5–10 mm。花：萼片 4，早落，淡黄色，狭卵形，3–4×1.2–1.6 mm，背面脉上有短柔毛，每一萼片具 3 条不分枝的基生脉。雄蕊 10–17，无毛；花丝丝形，长约 3 mm；花药条形或狭长圆形，约 2.2×0.25 mm，顶端具长 0.1 mm 的短尖头。心皮 3–5，无毛；子房狭倒卵形，约 0.8×0.45 mm；花柱长约 0.7 mm；柱头狭三角形，约 0.7×0.3 mm，具 2 条膜质狭侧生翅。瘦果斜倒卵形，2.2–3×1–2 mm，每侧有 2–3 条纵肋；宿存花柱长约 0.7 mm。6 月开花。

Perennial herbs. Stems 28–55 cm tall, branched, above with thyrses with very short glandular hairs. Middle cauline leaves 3-pinnate, shortly petiolete; blades ca 10 cm long, with 4 pairs of pinnae; leaflets papery, orbicular-rhombic, broad-rhombic or rhombic-obovate, 1–1.3×0.8–1.8 cm. at base rounded-truncate or subcordate, 3-lobed (lobes entire or 1–2-rotund-dentate), adaxially glabrous, abaxially on nervous nets with dense grey branched hairs, nerves adaxially impressed, abaxially prominent; petiolules 1–9 mm long; petioles 0.5–1 cm long. Thyrses terminal, ca. 15 cm long; lower bracts foliaceous, upper bracts simple, narrow-lanceolate or linear, 2–4 mm long; pedicels 5–10 mm long. Flower : Sepals 4, caducous, yellowish, narrow-ovate, 3–4×1.2–1.6 mm, abaxially on nerves puberulous, each sepal with 3 simple basal nerves. Stamens 10–17, glabrous; filaments filiform, ca. 3 mm long; anthers linear or narrow-oblong. ca. 2.2×0.25 mm, apex with mucrones 0.1 mm long. Carpels 3–5, glabrous; ovary narrow-obovate, ca. 0.8×0.45 mm; style ca. 0.7 mm long; stigma narrow-triangular, ca. 0.7×0.3 mm, with 2 narrow membranous lateral wings. Achene obliquely obovate, 2.2–3×1–2 mm, on each side longitudinally 2–3-ribbed; persistent style ca. 0.7 mm long.

分布于云南西部至东北部、贵州西部和四川南部。生山地沟旁灌丛或山坡上，海拔 2200–3000 m。

根有清热燥湿、泻火解毒之效，用于腹泻、痢疾、呕吐等症。(《中国药用植物志》，2016)

标本登录：云南：大理，凤阳村，秦仁昌 22989；石林，税玉民等 64265，64620，

64811；昆明，刘慎谔 16376，16566，税玉民等 60208；嵩明，邱炳云 51243；东川，蓝顺彬 62。

贵州：威宁，符开国 81178。

四川：普格，1979-09，普格队 s. n. (IMC)。

87. 札达唐松草（《广西植物》） 图版 87

Thalictrum zhadaense W. T. Wang in Guihaia 37 (6): 684, fig. 6. 2017. Holotype：西藏：札达，马阳村，alt. 3500 m, 1976-03-20，青藏队 12911（PE）.

多年生草本，全部无毛。茎高约 90 cm，分枝。上部茎生叶为 3 回三出羽状复叶，具短柄；叶片三角形，约 20×12 cm；小叶纸质，宽倒卵形、倒卵形、宽椭圆形或近方形，1-1.8×0.7-1.5 cm，顶端急尖，基部圆形或截形，3 浅裂或 3 中裂（裂片顶端急尖，具 1-2 尖齿或全缘），具 3-5 脉，叶脉上面平，下面稍隆起；小叶柄丝形，长 0.2-2.2 cm；叶柄长约 1.5 cm，基部具狭鞘。聚伞圆锥花序顶生，长约 50 cm；分枝约 6 条，长 7-15 cm，疏具 3-8 花；苞片叶状；花梗丝形，长 0.4-1.5 cm。花：萼片 4，狭卵形，5-5.6×2.2-2.8 mm，顶端渐狭，每一萼片具 3 条淡紫色基生脉（中脉不分枝，2 侧生脉 1 回二歧状分枝）。雄蕊 14-20；花丝丝形，长 2-2.5 mm；花药狭长圆形，1.8-2.8×0.4-0.6 mm，顶端具长 0.1-0.2 mm 的短尖头。心皮 6-7；子房斜椭圆形，约 1×0.55 mm，在每侧有 3 条不明显纵肋；花柱长 0.7 mm；柱头狭披针形，约 0.6×0.3 mm，具 2 条侧生狭翅。3 月开花。

Perennial herbs, totally glabrous. Stem ca. 90 cm tall, branched. Upper cauline leaves 3-ternate-pinnate, shortly petiolate; blades triangular, ca. 20×12 cm; leaflets papery, broad-obovate, obovate, broad-elliptic or subquadrate, 1-1.8×0.7-1.5 cm, at apex acute, at base rounded or truncate, 3-lobed or 3-fid (lobes at apex acute, with 1-2 acute teeth or entire), 3-5-nerved, nerves adaxially flat, abaxially slightly prominent; petiolules filiform, 0.2-2.2 cm long; petioles ca. 1.5 cm long, at base narrow-vaginate. Thyrse terminal, ca. 50 cm long; branches ca. 6, 7-15 cm long, sparsely 3-8-flowered; bracts foliaceous; pedicels filiform, 0.4-1.5 cm long. Flower : Sepals 4, narrow-ovate, 5-5.6×2.2-2.8 mm, at apex attenuate, with 3 purplish basal nerves (midrib simple, and 2 lateral nerves once dichotomizing). Stamens 14-20; filaments filiform, 2-2.5 mm long; anthers narrow-oblong, 1.8-2.8×0.4-0.6 mm apex with mucrones 0.1-0.2 mm long. Carpels 6-7; ovary obliquely elliptic, ca. 1×0.55 mm, on each side with 3 inconspicuous longitudinal ribs; style 0.7 mm long; stigma narrow-lanceolate, ca. 0.6×0.3 mm, with 2 narrow lateral wings.

特产西藏西南部札达。生山地河滩上，海拔 3500 m。

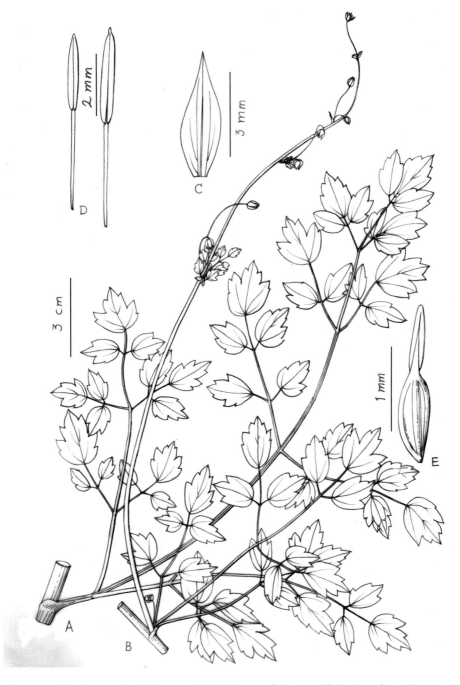

图版87. 札达唐松草　A. 下部茎生叶，B. 开花茎的中部，C. 萼片，D. 二雄蕊，E. 心皮。（据holotype）
Plate 87. **Thalictrum zhadaense**　A. lower cauline leaf, B. middle part of flowering stem, C. sepal, D. two stamens, E. carpel. (from holotype)

88. 藏南唐松草　图版 88

Thalictrum austrotibeticum J. Y. Li, L. Xie & L. Q. Li in Phytotaxa 207 (3): 281, figs. 1–2. 2015. Type：西藏：聂拉木，樟木，707 大桥附近，alt. 3205 m，河边灌草丛，2013-07-28，PE 西藏队 4387（holotype and isotypes, PE）；聂拉木，立新村，alt. 2308 m，2013-07-29，PE 西藏队 4289（paratype, PE）；吉隆，扎村，alt. 2800–3970 m，2013-07-28，PE 西藏队 4130 (Paratype, PE)；亚东，县城西北部，2013-07-07，PE 西藏队 2848（paratype, PE）。

多年生草本植物。茎高 1.2–2 m，无毛，不分枝或分枝。茎生叶为 3 回三出羽状复叶，长 40–60 cm；小叶蓝绿色，纸质，圆卵形、卵形、倒卵形或宽倒卵形，2–5.5(–7)×2.2–5.7(–7.5) cm，顶端圆形或钝，具小短尖头，基部心形、圆形或近截形，边缘有粗齿，3 浅裂，上面近无毛，下面在隆起的脉网上被短粗毛（毛长 0.15–0.3(–0.4) mm），具 3–5 条脉，脉上面平，下面隆起；小叶柄长 1.6–3.5 cm；叶柄长 2–7.5 cm，基部具鞘。聚伞圆锥花序顶生，长约 20 cm，3 回分枝，具多数花；花序轴、分枝和花梗被短柔毛（毛长 0.5–1 mm）；下部苞片叶状，长 2–2.5 cm，上部苞片简单，狭条形或钻形，长 4–8 mm；花梗长 0.4–1.8 cm。花：萼片 4，白色或粉红，椭圆状卵形，4×2.4 mm，无毛，顶端圆形，每一萼片具 3 条不分枝的基生脉。雄蕊 18–24；花丝紫色，丝形，长 5.5–7 mm，无毛；花药淡黄色，狭长圆形，2.5–3.2×0.4–0.5 mm，密被小毛（毛长 0.05 mm），顶端具长 0.2–0.4 mm 的短尖头。心皮 7–14；子房斜倒卵形，约 1.4×0.5 mm，密被短柔毛；花柱长 2 mm，无毛；柱头狭三角形，约 1.5×0.4 mm，具 2 条侧生狭翅。瘦果扁，斜倒卵形，长约 4.6 mm，基部具长 0.5 mm 的柄；宿存花柱长约 1.6 mm。7–8 月开花。

Perennial herbs. Stems 1.2–2 m, glabrous, simple or branched. Cauline leaves 3-ternate-pinnate, 40–60 cm long; leaflets blue-green, papery, orbicular-ovate, ovate, obovate or broad-ovate, 2–5.5(–7)×2.2–5.7(–7.5) cm, at apex rounded or obtuse, minutely mucronate, at base cordate, rounded, or subtruncate, at margin coarsely dentate, 3-lobed, adaxially subglabrous, abaxially on prominent nervous nets covered with short thick hairs 0.15–0.3(–0.4)mm long, 3–5-nerved, nerves adaxially flat, abaxially prominent; petiolules 1.6–3.5 cm long; petioles 2–7.5 cm long, at base vaginate. Thyrses terminal, ca. 20 cm long, thrice branched, many-flowered; rachis, branches and pedicels puberulous (hairs 0.5–1 mm long); lower bracts foliaceous, 2–2.5 cm long, upper bracts simple, narrow-linear or subulate, 4–8 mm long; pedicels 0.4–1.8 cm long. Flower : Sepals 4, white or pink, elliptic-ovate, 4×0.4 mm, glabrous, at apex rounded, each sepal with 3 simple basal nerves. Stamens 18–24; filaments

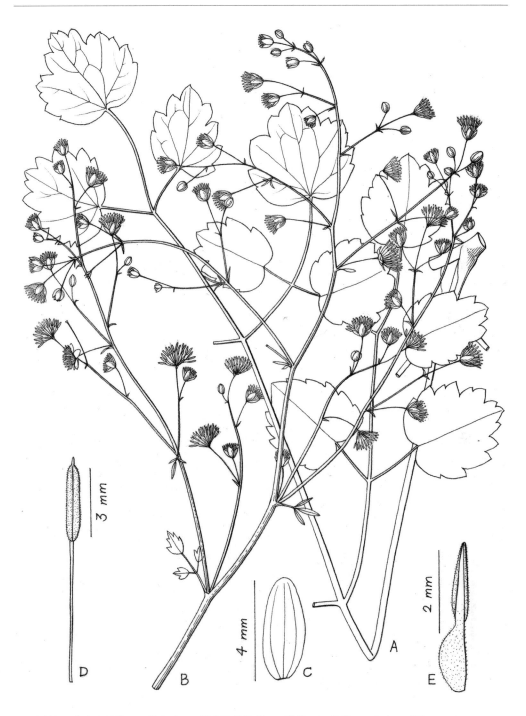

图版88. **藏南唐松草** A. 茎生叶，B. 聚伞圆锥花序，C. 萼片，D. 雄蕊，E. 心皮。（据isotype）
Plate 88. **Thalictrum austrotibeticum** A. cauline leaf, B. thyrse, C. sepal, D. stamen, E. carpel. (from isotype)

purple, filiform, 5.5–7 mm long, glabrous; anthers yellowish, narrow-oblong, 2.5–3.2×0.4–0.5 mm, densely with little hairs 0.05 mm long, apex with mucrones 0.2–0.4 mm long. Carpels 7–14; ovary obliquely obovate, ca. 1.4×0.5 mm, densely minutely puberulous; style 2 mm long, glabrous; stigma narrow-triangular, ca. 1.5×0.4 mm, with 2 narrow lateral wings. Achene compressed, obliquely obovate, ca. 4.6 mm long, base with a stipe 0.5 mm long; persistent style ca. 1.6 mm long.

特产西藏南部。生山地林边或草坡上，海拔 2300–3900 m。

89. **多叶唐松草**（《中国药用植物志》，1964） 图版 89

Thalictrum foliolosum DC. Syst. 1: 175. 1818; Prodr. 1: 12. 1824; D. Don, Prodr. Fl Nepal. 192. 1825; Wall. Cat. n. 3711. 1832; Royle, Ill. Bot. Himal. 51. 1858; Hook. f. & Thoms. Fl. Ind. 16. 1855; et in Hook. f. Fl. Brit. Ind. 1: 14. 1872; Finet & Gagnep. in Bull. Soc. Bot. France 50: 619. 1903; C. Pei & T. Y. Chou, Icon. Chin. Med. Pl. 7: fig. 331. 1964; Iconogr. Corm. Sin. 1: 681, fig. 1362. 1972; W. T. Wang & S. H. Wang in Fl. Reip. Pop. Sin. 27: 575, pl.140: 4–8. 1979; Grierson, Fl. Bhutan 1(2): 298. 1984; W. T. Wang in Fl. Xizang. 2: 72. 1985; in Vasc. pl. Hengduan Mount. 1: 503. 1993; in Fl Yunnan. 11: 172. 2000; et in High. Pl. China 3: 476, fig. 759, 2000; Rau in Sharma et al. Fl. India 1: 135. 1993; D. Z. Fu & G. Zhu in Fl. China 6: 290. 2001; Y. M. Shui & W. H. Chen, Seed Pl. Honghe Reg. S–E. Yunnan 78. 2003; W. T. Wang in High. Pl. China in Colour 3: 393. 2016; L. Xie in Med. Fl. China 3: 368, with fig. 2016. Holotype : Nepal. near Suembo, F. Buchanan s. n. (not seen).

T. smithii B. Boivin in J. Arn. Arb. 26: 114. 1945, p.p. quoad paratypos H. T. Tsai 54653 et 58383.

多年生草本植物，全部无毛。茎高 90–200 cm，多分枝。中部茎生叶为 3 回三出羽状复叶；叶片长达 36 cm；小叶草质，菱状椭圆形、卵形或宽卵形，1–2.5×0.5–1.5 cm，顶端圆形或钝，基部浅心形或圆形，3 浅裂，边缘有少数圆齿，叶脉平或下面稍隆起，脉网稍明显；小叶柄长 1–16 mm；叶柄长 1.5–5 cm， 基部有狭鞘。聚伞圆锥花序顶生，长 12–28 cm，有多数花；基部苞片叶状，上部苞片小，为三出复叶或单叶；花梗长 1.5–6 cm。花：萼片 4，早落，淡黄绿色，近长圆形，3–.4.5×1.2–1.5 mm，顶端圆钝，每一萼片具 3 条不分枝的基生脉。雄蕊 16–27；花丝长 2.2–2.5 mm，上部狭条形，宽 0.2 mm，下部变为丝形；花药条形或狭长圆形，2.5–3×0.2–0.3 mm，顶端有长 0.2 mm 的短尖头。心皮 3–7；子房半倒卵形，约 1.8×0.5 mm；花柱长 1.2 mm；柱头条形，约 1×0.25 mm，

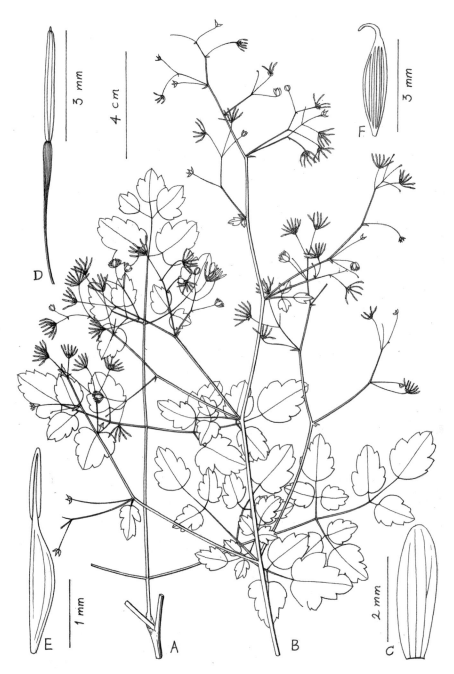

图版89. **多叶唐松草** A. 上部茎生叶，B. 聚伞圆锥花序，C. 萼片，D. 雄蕊，E. 心皮（据俞德浚23039），F. 瘦果（据王洪6505）。

Plate 89. **Thalictrum foliolosum** A. upper cauline leaf. B. thyrse, C. sepal, D. stamen, E. carpel (from T. T. Yü 23039), F. achene (from H. Wang 6505).

有 2 条狭侧生翅。瘦果纺锤形，3–3.5×1–1.2 mm，每侧有 3 条纵肋；宿存花柱长 0.8–1.2 mm。8–9 月开花。

Perennial herbs, totally glabrous. Stems 0.9–2 m tall, much branched. Middle cauline leaves 3-ternate-pinnate; blades up to 36 cm long; leaflets herbaceous, rhombic-orbicular, ovate or broad-ovate, 1–2.5×0.5–1.5 cm, at apex rounded or obtuse, at base subcordate or rounded, 3–lobed, at margin few-rotund-dentate, nerves flat, or abaxially slightly prominent, nervous nets slightly conspicuous; petiolules 1–16 mm long; petioles 1.5–5 cm long, at base narrow-vaginate. Thyrses terminal, 12–28 cm long, many-flowered; basal bracts foliaceous, upper bracts small, ternate or simple; pedicels 1.5–6 cm long. Flower : Sepals 4, caducous, yellowish-green, suboblong, 3–4.5×1.2–1.5 mm, at apex rounded-obtuse, each sepal with 3 simple basal nerves. Stamens 16–27; filaments 2.2–2.5 mm long, above linear, 0.2 mm broad, below narrowed to filiform; anthers linear or narrow-oblong, 2.5–3×0.2–0.3 mm, apex with mucrones 0.2 mm long. Carpels 3–7; ovary semi-obovate, ca. 1.8×0.5 mm; style 1.2 mm long; stigma linear, ca. 1×0.25 mm, with 2 narrow lateral wings. Achene fusiform, 3–3.5×1–1.2 mm, on each side longitudinally 3–ribbed; persistent style 0.8–1.2 mm long.

分布于中国西藏南部和东南部、云南、四川南部；以及不丹、尼泊尔、印度北部。生山地草坡上或林中；海拔在西藏南部为 1700–3500 m，自西藏东南部至云南大部和四川南部为 1300–2400 m，在云南南部为 580–950 m。

根状茎和根有清热解毒、杀菌止痢之效，用于湿热泻痢、肠炎、黄疸等疾病。根含芬氏唐松草定碱（thalifendine）、小檗碱（berberine）等生物碱。(《中国药用植物志》，2016)

标本登录：西藏：聂拉木，青藏队 75–7580，75–6002；墨脱，生态室高原组 11144，程树志，李渤生 1462。

云南：贡山，俞德浚 23039，青藏队 8313，Gaoligong Shan Biodiv. Survey 23191 (GH,PE)，15549，33226，34329（GH）；福贡，alt. 1500 m，山坡上，花淡黄色，1933-09-27，蔡希陶 54683；福贡，知子罗，alt. 3200 m，林中，花黄色，1934-09-07，蔡希陶 58383（以上二号标本均为 *T. smithii* B. Boivin 的 paratypes）；漾濞，刘慎谔 22495，22843；腾冲，Gaoligong Shan Biodiv. Survey 18088（GH,PE），29568；永平，宗庭疾 160；景东，哀牢山，杨国平 203；澜沧，钱义咏 3016；普洱，王洪 9967；思茅，昆明植物所队 6008；勐腊，勐仑，周仕顺 638；易门，六街队 254；蒙自，王启无 81504，83264，税玉民等 40400；屏边，王启无 82131。

四川：会理，俞德浚 1415。

90. 高原唐松草（《药学学报》） 图版 90

Thalictrum cultratum Wall. [Cat. n.3715. 1828.] Pl. As. Rar. 2: 26. 1831; Hook. f. & Thoms. in Hook. f. Fl. Brit. Ind. 1: 11. 1872; Lecoy. in Bull. Soc. Bot. Belg. 24: 178. 1885; Franch. Pl. Delav. 13. 1889; Finet & Gagnep. in Bull. Soc. Bot. France 50: 617. 1903; Finet in J. de Bot. 21: 20. 1908; Hand.-Mazz. Symb. Sin. 7: 311. 1931; Fl. Tsinling. 1(2): 245, fig. 210. 1974; W. T. Wang & S. H. Wang in Fl. Reip. Pop. Sin. 27: 579, pl.145: 13–15. 1979; W. T. Wang in Bull. Bot. Lab. N.-E Forest. Inst. 8: 34. 1980; et in Iconogr. Corm. Sin. Suppl. 1: 929, fig. 8617. 1982; Grierson, Fl. Bhutan 1(2):297. 1984; W. T. Wang in Fl. Xizang. 2: 72. 1895; in Vasc. Pl. Hengduan Mount. 1: 503. 1993; in Fl. Yunnan. 11: 173. 2000; et in High Pl. China 3: 477, fig. 763. 2000; Rau in Sharma et al. Fl. India 1: 134. 1993; D. Z. Fu & G. Zhu in Fl. China 6: 288. 2001; W. T. Wang in High. Pl. China in Colour 3: 393. 2016; L. Xie in Med. Fl. China 3: 271, with fig. & photo. 2016; X. T. Ma et al. Wild Flow, Xizang 26, photo 50. 2016. Syntype: Nepal, Wallich 3715A (K, not seen).

T. deciternatum B. Boivin in J. Arn. Arb. 26: 112, pl. 1: 4–7. 1945, p. p. min. excl. specim. T. T. Yu 9891. Holotype：云南：丽江, alt. 3200 m, 1914-06-16, C. Schneider 1805（GH, not seen）. Paratypes: 丽江, 1937-08-27, 俞德浚 15480（GH, PE）；丽江, 1922-05-24, J. F. Rock 3801（GH, not seen）；丽江, 1938-07, 王启无 70955（GH, PE）；维西，叶枝, 王启无 68341（GH, PE）; Chi-ling Shan, Cheng-king, 1939-08-25, H. Wang 41681（GH, not seen）.

T. yui B. Boivin in l. c. 115. Type: 四川：木里, Kulu, alt. 3300 m, 1937-09-14, 俞德浚 14273（holotype. GH; isotype, PE）.

多年生草本植物，全部无毛或茎上部、小叶背面和子房被短毛。茎高50–120 cm，分枝。中部茎生叶为3–4回羽状复叶，具短柄；叶片长9–20 cm，具4–6对一回羽片；小叶薄革质，菱状倒卵形、宽菱形或近圆形，5–10(–14)×3–10(–14)mm，顶端急尖，基部钝、圆形或浅心形，3浅裂，裂片全缘或1–2小齿，脉上面凹陷，下面隆起；小叶柄长0.5–6 mm；叶柄长1–4 cm。聚伞圆锥花序长10–24 cm；下部苞片叶状，上部苞片小；花梗长4–14 mm。花：萼片4，脱落，绿白色，狭椭圆形或狭卵形，3–3.7×1.2–2 mm，顶端圆钝，每一萼片具3条不分枝的基生脉。雄蕊12–15；花丝丝形，长(1.3–)3–5 mm；花药狭长圆形或条形，2.2–3×0.4–0.5 mm，顶端具长0.05–0.15 mm的短尖头。心皮4–9；子房狭倒卵形，1.2–2×0.6–1 mm，常有短毛，基部常有短柄；花柱长约1 mm；柱头狭三角形，长约0.8 mm，

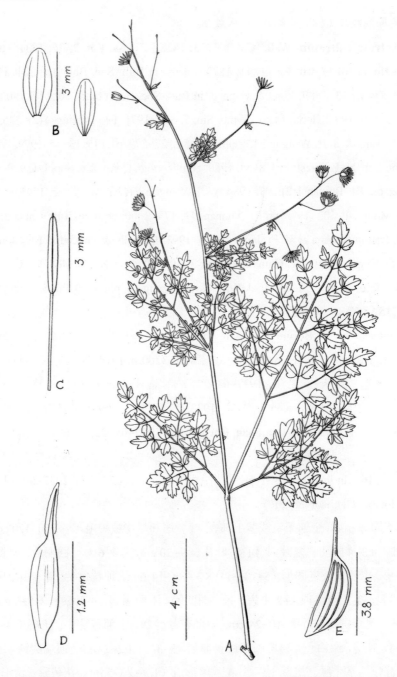

图版90. 高原唐松草　A. 开花茎上部，B. 二萼片，C. 雄蕊，D. 心皮（据傅国勋1178），E. 瘦果（据D. E. Boufford et al. 32390）。

Plate 90. **Thalictrum cultratum**　A. upper part of flowering stem, B. two sepals, C. stamen, D. carpel (from G. X. Fu 1178), E. achene (from D. E. Boufford et al. 32390).

有 2 条侧生狭翅。瘦果扁，斜倒卵形或半圆形，2.8-3.5×1.5-1.7 mm，每侧有 3 条纵肋，无柄或有极短柄；宿存花柱长约 1 mm。6-7 月开花。

Perennial herbs, totally glabrous, or stems above, leaflets abaxially and ovary puberulous. Stems 50-120 cm tall, branched. Middle cauline leaves 3-4-pinnate, shortly petiolate; blades 9-20 cm long, with 4-6 pairs of primary pinnae; leaflets subcoriceous, rhombic-obovate, broad-rhombic or suborbicular, 5-10(-14)×3-10(-14)mm, at apex acute, at base obtuse, rounded or subcordate, 3-lobed, lobes entire or 1-2-denticulate, nerves adaxially impressed, abaxially prominent; petiolules 0.5-6 mm long; petioles 1-4 cm long. Thyrses 10-24 cm long; lower bracts foliaceous, upper bracts small; pedicels 4-14 mm long. Flower: Sepal 4, deciduous, greenish-white, narrow-elliptic or narrow-ovate, 3-3.7×1.2-2 mm, at apex rounded-obtuse, each sepal with 3 simple basal nerves. Stamens 12-15; filaments filiform, (1.3-)3-5 mm long; anthers narrow-oblong or linear, 2.2-3×0.4-0.5 mm, apex with mucrones 0.05-0.15 mm long. Carpels 4-9; ovary narrow-obovate, 1.2-2×0.6-1 mm, often with very short hairs, base offen shortly stipitate; style ca. 1 mm long; stigma narrow-triangular, ca. 0.8 mm long, with 2 narrow lateral wings. Achene compressed, obliquely obovate or semi-orbicular, 2.8-3.5×1.5-1.7 mm, on each side longitudinally 3-ribbed, sessile or base with a very short stipe; persistent style ca. 1 mm long.

分布于中国西藏东南部、云南西北部和中部、四川西部、青海南部、甘肃南部；以及不丹、尼泊尔、印度北部。生山地草坡、灌丛、沟边草地或林中，海拔 1700-4300 m。

根状茎和根有清热燥湿、杀菌止痢等效，用于湿热泻痢、传染性肝炎、黄疸等疾病。根含巴马汀（palmatine）、药根碱等生物碱。(《中国药用植物志》，2016)

标本登录：西藏：错那，任再金 L361；隆子，傅国勋 1178；墨竹工卡，FLPH Tibet Exped. 12-2209；左贡，青藏队 76-12146；贡觉，杨竞生 91-481，91-484；江达，D. E. Boufford et al. 31233，33809，33842；昌都，D. E. Boufford et al. 32390，B. Bartholomew et al. 41351；类乌齐，D. E. Boufford et al. 32225。

云南：昆明，太华山，钟观光 2063；大理，苍山，金效华，刘冰等 1512；丽江，G. Forrest 5654，5939，姜恕 6316，陈家瑞 92463，杨亲二 9604；中甸，中甸队 62-806，青藏队 81-1325，杨亲二，孔宏智 98-257；维西，王启无 64698，68341。

四川：越西，经济植物队 59-3543；乡城，D. E. Boufford, B. Bartholomew et al. 28384；康定，D. E. Boufford et al. 27614；炉霍，赵清盛等 111254，D. E. Boufford et al. 27784；色达，

D. E. Boufford et al. 34638。

青海：囊谦，藏药队 1176。

甘肃：舟曲，紫盖山，王作宾 14207。

91. **腺毛唐松草**(《中国植物志》) 图版 91

Thalictrum foetidum L. Sp. Pl. 545. 1753; DC. Prodr. 1: 13. 1824; Ledeb. Fl. Alt. 2: 349. 1830; et Fl. Ross. 1: 7. 1842; Lecoy. in Bull. Soc Bot. Belg. 24: 181. 1885; Finet & Gagnep. in Bull. Soc. Bot. France 50: 618. 1903; Finet in J. de Bot. 21: 20. 1908; Kom. & K.-Alisova, Key Pl. Far East USSR 1: 554. 1931; Nevski in Fl. URSS 7: 520, t. 35: 2. 1937; Iconogr. Corm. Sin. 1: 683, fig. 1365. 1972; Fl. Tsinling. 1(2): 246. 1974; S. H. Li in Fl. Pl. Herb. Chinae Bor.-Or. 3: 211, pl.93. 1975; W. T. Wang & S. H. Wang in Fl. Reip. Pop. Sin. 27: 580, pl. 147: 4–5. 1979; B. Z. Ding et al. Fl. Henan. 1: 474. 1981; S. Y. He, Fl. Beijing, rev. ed., 1: 239, fig. 298. 1984; Grierson, Fl. Bhutan 1(2):297. 1984; W. T. Wang in Fl. Xizang. 2: 74, fig. 21: 10–11. 1985; J. W. Wang in Fl. Hebei 1: 444, fig. 442. 1986; D. Z. Ma, Fl. Ningxia, 1: 175, fig. 153. 1986; Y. Z. Zhao in Fl. Intramongol. 2: 457, pl. 184: 3–4. 1990; W. T. Wang in Vasc. Pl. Hengduan Mount. 1: 504. 1993; J. G. Liu in Fl. Xinjiang. 2(1):271. 1994; S. H. Li in Clav. Pl. Chinae Bor.-Or., ed. 2,207, pl. 97: 4. 1995; L. H. Zhou in Fl. Qinghai. 1: 333. 1997; W. T. Wang in Fl. Yunnan. 11: 174. 2000; et in High. Pl. China 3: 478, fig. 764. 2000; Rau in Sharma et al. Fl. India 1: 136. 1993; D. Z. Fu & G. Zhu in Fl. China 6: 290. 2001; W. T. Wang in High. Pl. China in Colour 3: 394. 2016; L. Xie in Med. Pl. China 3: 273, with fig. & photo. 2016. — *T. minus* L. var. *foetidum* (L) Hook. f. & Thoms. in Hook. f. Fl. Brit. Ind. 1: 14. 1872. Described from S. Europe.

91a. var. foetidum

多年生草本植物。茎高 15-100 cm，无毛或幼时被短柔毛，变无毛，上部分枝或不分枝。中部茎生叶为 3 回羽状复叶，具短柄；叶片长 5.5-12 cm；小叶草质，菱状卵形或宽倒卵形，4-15×3-15 mm，顶端急尖或钝，基部宽楔形或圆形，有时浅心形，3 浅裂，（裂片全缘或有 1-2 小齿），下面在脉网上被短柔毛和短腺毛，有时无毛，脉上面稍凹陷，下面稍隆起；小叶柄长 1-15 mm；叶柄长 1.8-6 cm，基部有鞘；托叶膜质，褐色。聚伞圆锥花序顶生，长 10-22 cm，有少数或多数花；下部苞片叶状，上部苞片小，为三出复叶或简单，狭卵形；花梗细，长 5-12 mm，常被短柔毛和极短腺毛。花：萼片 4-5，早落，淡黄绿色，长圆形或卵形，约 4×1.6 mm，顶端圆形，外面常被疏柔，每一萼片具 3 条基生脉（中

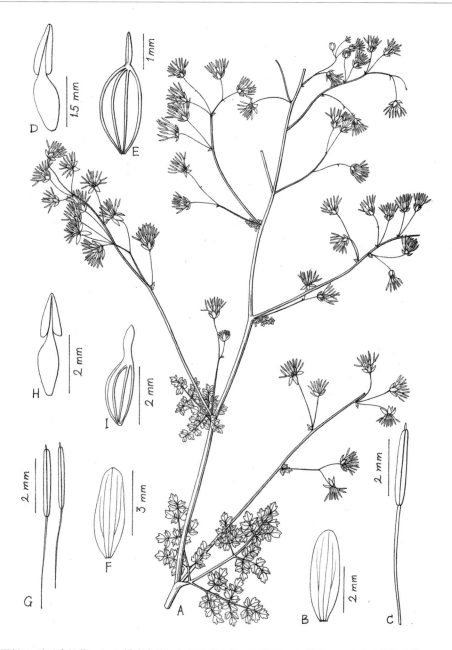

图版91. 腺毛唐松草　A–E. 模式变种　A. 开花茎上部，B. 萼片，C. 雄蕊，D. 心皮（据张建武H93113）E. 瘦果（据关克俭3362）。　F–I. 扁果唐松草（变种）　F. 萼片，G. 二雄蕊，H. 心皮（据刘继孟2651），I. 瘦果（据刘继孟2726）。

Plate 91. **Thalictrum foetidem**　A–E. var. **foetidum**　A. upper part of flowering stem, B. sepal, C. stamen, D. carpel (from J. W. Zhang H93113), E. achene (from K. C. Kuan 3362).　F–I. var. **glabrescens** F. sepal, G. two stamens, H. carpel (from J. M. Liou 2651), I. achene (from J. M. Liou 2726).

脉不分枝，2 侧生脉 1 回二歧状分枝）。雄蕊 15-20，无毛；花丝丝形，上部稍宽呈狭条形，长 4-4.5 mm；花药狭长圆形，2.5-3.5×0.35-0.5 mm，顶端具长 0.2-0.3 mm 的小尖头。心皮 4-12；子房椭圆形，约 1.5×0.8 mm，常有疏柔毛；花柱长约 1.5 mm；柱头狭三角形，长约 1.6 mm，具 2 条侧生狭翅，瘦果扁平，椭圆形或半倒卵形，长 2.2-5 mm，常有短柔毛，每侧有 2-3 条纵肋；宿存花柱长约 1 mm。5-7 月开花。

Perennial herbs. Stems 15-100 cm tall, glabrous or when young puberulus, glabreseent, above branched or simple. Middle cauline leaves 3-pinnate, shortly petiolate; blades 5.5-12 cm long; leaflets herbaceous, rhombic-ovate or broadly ovate, 4-15×3-15 mm, at apex acute or obtuse, at base broad-cuneate or rounded, sometimes subcordate, 3-lobed (lobes entire or 1-2-denticulate), abaxially on nervous nets puberulous and glandular-puberulous, sometimes glabrous, nerves adaxially slightly impressed, abaxially slightly prominent; petiolules 1-15 mm long; petioles 1.8-6 cm long, at base vaginate; stipules membranous, brown. Thyrses terminal, 10-22 cm long, with few or many flowers; lower bracts foliaceous, upper bracts small, ternate or simple; pedicels slender, 5-12 mm long, often puberulous and glandular-puberulous. Flower : Sepals 4-5, caducous, yellowish-greenish, oblong or ovate, ca. 4×1.6 mm, at apex rounded, abaxially often sparsely, puberulous, each sepal with 3 basal nerves (midrib simple, and 2 lateral nerves once dichotomizing). Stamens 15-20; filaments filiform, above broader and narrow-linear, 4-4.5 mm long; anthers narrow-oblong, 2.5-3.5×0.35-0.5 mm, apex with mucrones 0.2-0.3 mm long. Carpels 4-12; ovary elliptic, ca. 1.5×0.8 mm, often with sparse hairs; style ca. 1.5 mm long; stigma narrow-triangular, ca. 1.6 mm long, with 2 narrow lateral wings. Achene flattened, elliptic or semi-obovate, 2.2-5 mm long, often with short hairs, on each side longitudinally 2-3-ribbed; persistent style ca. 1 mm long.

分布于中国西藏、云南西北部、四川西部、青海、新疆、甘肃、陕西、河南西北部、山西、河北、内蒙古；以及蒙古、亚洲西部和欧洲。生山地草坡或高山多石砾处；海拔高度：在西藏 3500-4500 m，在云南西北、四川西部和青海 2200-3500 m，在新疆 1800-2700 m，在甘肃至内蒙古南部 900-1800 m，在内蒙古北部 350-950 m。

根状茎和根有清热燥湿、杀菌止痢之效，用于湿热痢疾、病毒性肝炎、麻疹等疾病。根含香唐松草碱（thalfoetidine）、氧化小檗碱（oxyberberine）等生物碱。(《中国药用植物志》，2016）

标本登记：西藏：札达，青藏队 76-8047；普兰，西北高原生物所队 74-4108；阿里、嘎尔，青藏队 76-8329；吉隆，青藏队 75-7181；聂拉木，张永田，郎楷永 3859，青藏队 75-5825；浪卡子，唐宇丹，林秦文 2010-2；南木林，西藏中草药队 995；拉萨，张永田，郎楷永 1773，2593；当雄，资源综考会队 47；林周，西藏中草药队 1952；墨竹工卡，青藏队 74-2045；错那，吴征镒等 75-1109；米林，倪志诚等 2924；加查，青藏补点队 75-1434；林芝，西藏中草药队 3158；察隅，青藏队 73-246；左贡，青藏队 76-12064，D. E. Boufford, B. Bartholomew et al. 40701，40708，40748；易贡，钟补求 6689；邦达，谢磊 BD08；芒康，西藏队 76-11991. 76-12033；比如，陶德定 11301；江达，D. E. Boufford, B. Bartholomew et al. 41975；贡觉，青藏队 76-12599，D. E. Boufford, B. Bartholomew et al. 41615。云南：中甸，杨竞生 88-108；德钦，杨亲二，袁琼 2535；

四川：得荣，青藏队 81-3171，81-3287；乡城，四川植被队 3530，D. E. Boufford. B. Bartholomew et al. 28710，D. E. Boufford et al. 30687；稻城，四川植被队 2064，郎楷永，李良千，费勇 2535；九龙，应俊生 3875；巴塘，四川植物队 4342；德格，贾慎修 183；甘孜，钟补求 5101，姜恕 9664；道孚，四川植物队 5591；雅江，D. E. Boufford, B. Bartholomew et al. 35865，35949；康定，胡文光，何铸 11282，13122；小金，郎楷永、李良千、费勇 1507，1512；理县，资源综考队 1920；刷经寺，张泽云，周洪富 22583。

重庆：石柱，谭士贤 1364。

青海：囊谦，杨永昌 1185，何廷农等 2526；玛沁，何廷农等 635；贵德，刘尚武 3117；共和，郭本兆 11912；门源，刘继孟 6821；乌兰，郭本兆 11490；海北，郭本兆 12272；德令哈，甘青队 59-883，59-3186，郭本兆 11652；铁卜加，? 077；柴达木，甘青队 59-774。

新疆：塔什库尔干，青藏队 87-0536；东帕米尔，Wündicsh 293，皮山，李渤生等 11680；和静，李安仁，朱家柟 6285，6429，王庆瑞 3955，西北植物所队 2346；和硕，李安仁，朱家柟 6968，7697；特克斯，关克俭 3362；乌鲁木齐，秦仁昌 177，周太炎等 65-1206；温泉，关克俭 4661；托里，关克俭 2759，张建武 H9313；青河，秦仁昌 1452；哈密，王成智 87118。

甘肃：卓尼，西北大学队 8；平凉，崆峒山，王作宾 13326；莲花山，张学刚 2226；永登，张学刚 6364；山丹，? 810；肃南，河西队 132，539；嘉峪关，青甘队 60-3036。

陕西：佛坪，傅坤俊 5040；太白山，Zhu, Chen, Xu, Wang 1352；华阴，王作宾 19668。

河南：林州，经济植物队 59-4145。

山西：沁源，刘继孟 1998；保德，黄河二队 2878。

河北：内丘，刘瑛 13051，刘心源 653；阜平，刘继孟 3222；易县，狼牙山，河北农大队 3453；张家口，黄秀兰等 4937；张北，黄秀兰等 8547。

北京：门头沟，中草药队 162。

内蒙古：阿拉善右旗，陈又生等 14-0966；贺兰山，于兆英 2389；近磴口，大青山，王作宾 2403；巴林右旗，孙英宝等 2-024；乌兰浩特，孙英宝等 09070304。

91b. 扁果唐松草（《中国植物志》，变种）

var. **glabrescens** Takeda in J. Bot. 48: 266. 1910; Tamura in Acta Phytotax. Geobot. 15: 86. 1953; W. T. Wang & S. H. Wang in Fl. Reip. Pop. Sin. 27: 582. 1979; X. Y. Yu et al. in Fl. Shanxi. 1: 604. 1992; D. Z. Fu & G. Zhu in Fl. China 6: 290. 2001; Y. Kadota in Iwatsuki et al. Fl. Japan 2a:338. 2006. —*T. foetidum* L. ssp. *glabrescens* (Takeda) T. Shmizu in Acta Phytotax. Geobot. 17: 149. 1958. Holotype：日本：Yezo: prov. Shiribeshi, Jozankei et al. s. n. (not seen).

与腺毛唐松草的区别在于本变种的植株全部无毛，柱头稍宽，呈三角形。茎高 23-42 cm。小叶草质，4-16×3-15 mm。

This variety differs from var. *foetidum* in its totally glabrous plants and its slightly broader triangular stigma. Stems 23-42 cm tall. Leaflets herbaceous, 4-16×3-15 mm.

分布于中国山西南部、河北西部；以及日本。生山地草坡，海拔 1700-2200 m。

标本登录：山西：沁源，刘继孟 1485。

河北：涞源，刘继孟 2651, 2726；小五台山，李建藩 10793，王启无 61521。

北京：门头沟区，西灵山，杨朝广 1589。

92. 亚欧唐松草（《中国植物志》） 图版 92

Thalictrum minus L. Sp. Pl. 546. 1753; DC. Syst. 1: 178. 1818; et Prodr. 1: 13. 1824; Lecoy. in Bull. Soc. Bot. Belg. 24: 199,293. 1885; Finet & Gagnep. in Bull. Soc. Bot. France 50: 620. 1903; Nevski in Fl. URSS 7: 524. 1937; Hand. -Mazz. Symb. Sin. 7: 313. 1931; et in Acta Hort. Gotob. 13: 172. 1939; W. T. Wang & S. H. Wang in Fl. Reip. Pop. Sin. 27: 583. 1979; W. T. Wang in Fl. Xizang. 2: 74. 1985; et in Vasc. Pl. Hengduan Mount. 1: 504. 1993; Y. Z. Zhao in Fl. Intramongol., ed. 2, 2: 460, pl. 186: 1-2. 1990; J. G. Liu in Fl. Xinjiang. 2(1): 272. 1994; S. H. Li in Clav. Pl. Chinae Bor. -Or., ed. 2: 207. 1995; L. H. Zhou in Fl. Qinghai. 1: 334. 1997; D. Z. Fu & G. Zhu in Fl. China 6: 293. 2001; Y. Z. Zhao, Vasc. Pl. Inn. Mongol. 161. 2012; L.Q. Li in High. Pl. Daba Mount. 118. 2014; L.Q. Li et al. Pl. Baishuijiang st. Nat. Reserve Gansu Prov. 71. 2014; H. Y. Feng & J. B. Pan, Field Guid, Wild Pl. China: Qilian Mount. 364. 2016; L. Xie in Med. Fl. China 3: 276. 2016. Described from N. Europe.

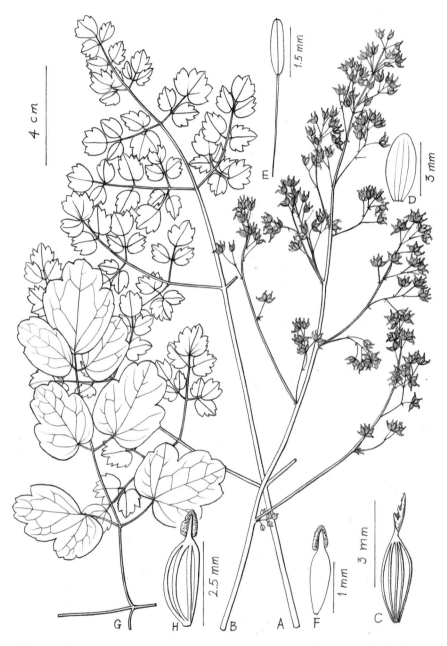

图版92. 亚欧唐松草 A–F. **模式变种** A. 茎生叶，B. 聚伞圆锥花序，C. 瘦果（据周太炎等651138），D. 萼片，E. 雄蕊，F. 心皮（据关克俭956）。 G–H. **圆齿亚欧唐松草** G. 茎生叶的一个羽片，H. 瘦果（据 holotype）。

Plate 91. **Thalictrum minus** A–F. var. **minus** A. cauline leaf, B. thyrse, C. achene (from T. Y. Zhou et al. 651138), D. sepal, E. stamen, F. carpel (from K. C. Kuan 956). G–H. var. **rotundidens** G. a pinna of cauline leaf, H. achene (from holotype).

92a. var. **minus**

多年生草本植物，全部无毛。茎高 35–70 cm，通常上部分枝。茎生叶为 4 回三出羽状复叶；叶片长达 20 cm；小叶纸质或薄革质，楔状倒卵形、宽倒卵形、近圆形或菱形，0.7–1.5×0.4–1.5 cm，基部楔形或圆形，3 浅裂或不分裂，边缘有少数小齿，脉多平，只中脉下面稍隆起，脉网不明显；小叶柄长 1–6 mm；叶柄长达 4 cm，基部有狭鞘。聚伞圆锥花序长 15–20 cm；花梗长 3–8 mm。花：萼片 4，脱落，淡黄绿色，狭椭圆形，3.4–4×1–1.2 mm，顶端近圆形，每一萼片具 3 条不分枝的基生脉。雄蕊 15–20；花丝丝形，长约 3.2 mm；花药长圆形，1.5–2×0.5 mm，顶端有长 0.1 mm 的短尖头。心皮 3–7；子房狭倒卵形，长约 1.2 mm；花柱长约 0.4 mm；柱头正三角形，约 0.5×0.6 mm，具 2 条侧生宽翅。瘦果倒卵形或狭椭圆体形，3–3.5×1.4 mm，每侧有 3 条纵肋；宿存柱头长约 1 mm。6–7 月开花。

Perennial herbs, totally glabrous. Stems 35–70 cm tall, usually above branched. Cauline leaves 4-ternate-pinnate; blades up to 20 cm long; leaflets papery or subcoriaceous, cuneate-obovate, broad-ovate, suborbicular or rhombic, 0.7–1.5×0.4–1.5 cm, at base cuneate or rounded, 3-lobed or undivided, at margin few-denticulate, nerves mostly flat, only midrib abaxially slightly prominent, nervous nets inconspicuous; petiolules 1–6 mm long; petioles up to 4 cm long. Thyrses 15–20 cm long; pedicels 3–8 mm long. Flower: Sepals 4, deciduous, yellowish-greenish, elliptic, 3.4–4×1–1.2 mm, at apex subrounded, each sepal with 3 simple basal nerves. Stamens 15–20; filaments filiform, ca. 3.2 mm long; anthers oblong, 1.5–2×0.5 mm, apex with mucrones 0.1 mm long. Carpels 3–7; ovary narrow-obovate, ca. 1.2 mm long; style ca. 0.4 mm long; stigma deltoid, ca. 0.5×0.6 mm, with 2 broad lateral wings. Achene obovate or narrow-ellipsoid, 3–3.5×1.4 mm, on each side longitudinally 3-ribbed; persistent stigma ca. 1 mm long.

分布于中国西藏西南部、四川西部、青海、新疆、甘肃、山西、河北、内蒙古、吉林、黑龙江；以及亚洲北部、西部和欧洲。生于山地草坡、田边、灌丛或林中，海拔 1400–2700(–3500) m。

根有清热凉血、理气消肿之效，用于痢疾等症。根含亚欧唐松草林碱（thalmineline）等生物碱。(《中国药用植物志》, 2016)

标本登录：西藏：札达，青藏队 76-8080。

四川：康定，胡文光、何铸 10342，关克俭、王文采等 67；道孚，伍煜庭 111660；炉霍，应俊生 9367；甘孜，钟补求 5130；小金，张秀实，任有铣 5878；金川，李馨 75108，78009；壤塘，四川植物队 9432；马尔康，四川植物队 9549；黑水，李馨 73578，73686；茂县，张超 07-1134；松潘，傅坤俊 1998；南坪，李金喜 2714，2743；若尔盖，四川植物队 10342。

青海：玛沁，何廷农 305；同德，吴玉虎 6380，何廷农 98；贵德，何廷农 979；民和，何廷农 532；乐都，吴玉虎 4267；互助，刘继孟 5877；大通，刘继孟 6310；祁连，青甘队 60-2768。

新疆：阿克苏，新疆队 78-689；伊犁，关克俭 3098；温泉，陈家瑞 86337；托里，张建武 93298；新源，周太炎 650464；阿勒泰，秦仁昌 2409；乌鲁木齐，李安仁，朱家柟 6138；石河子，关克俭 560；沙湾，关克俭 956，周太炎 651138；焉耆，李安仁，朱家柟 6210。

甘肃：岷县。王作宾 4680；卓尼，J. F. Rock 12567.16589；临潭，夏伟瑛 8644；夏河，王作宾 7047；兰州，崔友文 10301；永登，张学刚 6269；张掖，钟补求 9109。

山西：沁源，刘继孟 1998；宁武，管岑山，？304。

河北：小五台山，段俊喜 633。

内蒙古：兴和，赵一之 77-108；翁牛特旗，王薇 3328。

吉林：通榆，叶居新 179；安图，白河队 1051。

92b. 东亚唐松草（《北京植物志》） 图版 92a

var. **hypoleucum** (Sieb. & Zucc.) Miq. in Ann. Mus. Bot. Lugd. –Bat. 3:3. 1867; Tamura in Acta Phytotax. Geobot. 15: 87. 1953; W. T. Wang & S. H. Wang in Fl. Reip. Pop. Sin. 27: 584. pl. 149: 5–9. 1979; S. Y. He, Fl. Beijing, rev. ed., 1: 241, fig. 402. 1984; Tamura & Shimizu in Satake et al. Wild Flow. Pl. Japan 2: 84, pl. 83: 3–4. 1984; J. W. Wang in Fl. Hebei 1: 447, fig. 446. 1986; X. W. Wang in Fl. Anhui 2: 325. 1987; Y. K. Li in Fl. Guizhou. 3: 71, pl. 30: 1–5. 1990; Y. Z. Zhao in Fl. Intramongol., ed. 2, 2: 462, pl. 186: 3–5. 1990; X. Y. Yu et al. In Fl. Shanxi. 1: 604, fig. 383. 1992; W. T. Wang in Vasc. Pl. Hengduan Mount. 1: 504. 1993; et in Keys Vasc. Pl. Wuling Mount. 165. 1995; S. H. Li in Clav. Pl. Chinae Bor. –Or., ed. 2: 207, pl. 97: 2. 1995; L. H. Zhou in Fl. Qinghai. 1: 656, fig. 2–512. 1997; Y. J. Zheng in Fl. Shandong. 2: 31, fig. 25. 1997; K. M. Liu in Fl. Hunan 2: 656, fig. 2–512. 2000; W. T. Wang in High. Pl. China 3: 478, fig. 765. 2000; D. Z. Fu & G. Zhu in Fl. China 6: 293. 2001; R. J. Wang in Fl. Guangdong 5: 20. 2003; Y. Kadota in Iwatsuki et al. Fl. Japan 2a: 336. 2006;

图版92a. A-F. **东亚唐松草**　A. 开花茎顶部, B. 萼片, C. 二雄蕊, D. 心皮（据李建藩10720), E. 瘦果（据PE考察队2828)。 F-J. **长梗亚欧唐松草**　F. 聚伞圆锥花序, G. 萼片, H. 雄蕊, I. 心皮（据刘慎谔3039), J. 瘦果（据陈家瑞86145)。

Plate 92a. A-E. **Thalictrum minus** var. **hypoleucum**　A. apical part of flowering stem, B. sepal, C. two stamens, D. carpel (from J. F. Li 10720), E. achene (from PE Exped. 2828), F-J. **T. minus** var. **stipellatum** F. thyrse, G. sepal, H. stamen, I. carpel (from T. N. Liou 3039), J. achene (from C. J. Chen 86145).

C. X. Yang et al. Keys Vasc. Pl. Chongqing 216. 2009; Bing Liu, Field Guid. Wild Pl. China: Shandong Prov. 316. 2009; Q. L. Gan, Fl. Zhuxi Suppl. 246, fig. 296. 2011; Y. Z. Zhao, Vasc. Pl. Inn. Mongol. 161. 2012; M. B. Deng & K. Ye in Fl. Jiangsu 2: 83, fig. 2−133. 2013; L. Q. Li in High. Pl. Daba Mount. 118. 2014; L. Q. Li et al. Pl. Baishuijiang St. Nat. Reserve Gansu Prov. 71. 2014; F. H. Li et al. Pl. Yanqing 143. 2015; L. Xie in Med. Fl. China 3: 279, with photo. 2016. —*T. hypoleucum* Sieb & Zucc. In Abh. Akad. Muench. 4: 178. 1845. Described from Japan.

T. thunbergii DC. Syst. 1: 183. 1818; et Prodr. 1: 15. 1824; Nakai in Bot. Mag. Tokyo 42: 2. 1928; Fl. Beijing 1: 295, fig. 242. 1962; Iconogr. Corm. Sin. 1: 684, fig. 1367. 1972; Fl. Tsinling. 1(2):246. 1974; S. H. Li in Fl. Pl. Herb. Chinae Bor. –Or. 3: 215. Pl. 95: 57. 1975; S. H. Fu, Fl. Hupeh. 1: 359. 1976; B. Z. Ding et al. Fl. Henan. 1: 473, fig. 607. 1981; D. Z. Ma, Fl. Ningxia. 1: 176. 1986. Holotype：日本，C. P. Thunberg s. n. (UPS, not seen).

T. minus L. var. *elatum* Lecoy. in Bull. Soc. Bot. Belg. 24: 202. 1885; Finet & Gagnep. in Bull. Soc. Bot. France 50: 620. 1903; Boivin in J. Arn. Arb. 26: 117. 1945; Lauener & Green in Not. Bot. Gard. Edinburgh 23(4):593. 1961.

T. amplissimum Lévl. & Vant. In Bull. Acad Geogr. Bot. 11: 51. 1902. —*T. minus* L. var. *amplissimun* (Lévl. & Vant.) Lévl. Fl. Koug−Tcheu 339. 1915. Holotype：贵州：Mont du Collège, 1897−07−29, Bodinier 1733 (E, not seen).

T. chinense Freyn in Oesterr. Bot. Zeitschr. 51: 377. 1901; Kom. & K.-Alisova., Key Pl. Far East. Reg. USSR 1: 563, t.170. 1931.

T. purdomii J. J. Clarke in Kew Bull. 1913: 39. 1913,p.p.

T. thunbergii DC. var. *majus* (Miq.)Nakai in Bot. Mag. Tokyo 42: 4. 1928; Kitagawa, Lineam. Fl. Mansh. 228. 1939.

T. flavum auct. non L.:Thunb, Fl. Jap. 241. 1784, fide DC.

T. minus auct. non L.: Forbes & Hemsl. in J. Linn. Soc. Bot. 23: 8. 1886.

T. acteaefolium auct. non Sieb. & Zucc.: Chun in Sunyatsenia 1: 232. 1934.

与亚欧唐松草的区别在于本变种的茎高达 150 cm，小叶较大，长和宽均为 1.5–4（–5）cm，下面具白粉，并具隆起的叶脉和明显的脉网，以及较长的狭长圆形或条形花药。花梗长 3–10 mm。花药 3.2–3.6×0.4–0.5 mm。

This variety differs from var. *minus* in its stems up to 150 cm tall, larger leaflets which are 1.5–4(–5)cm long and broad, and abaxially glaucous and with prominent nerves and conspiouous nervous nets, and longer narrow-oblong or linear anthers. Pedicels 3–10 mm long. Anthers 3.2–3.6×0.4–0.5 mm.

分布于中国四川、重庆、贵州、广东北部、湖南、湖北、江苏北部、安徽、河南、陕西、甘肃、青海东部、宁夏、山西、山东、河北、内蒙古、辽宁、吉林、黑龙江；以及朝鲜、日本。生山地或丘陵林边或山谷沟边；海拔 260-2000 m。

根有清热解毒之效，用于牙痛、急性皮炎、湿疹等症。根含秋唐松草替定碱（thalmelatidine）等生物碱。(《中国药用植物志》，2016)

标本登录：四川：汉源，王作宾 8833；宝兴，关克俭，王文采 2881；理县，姜恕 1594，李馨 46544；茂县，方文培 1511；马尔康，谢磊 20010900；黑水，姜恕 1521；刷经寺，赵清盛 1453；松潘，傅坤俊 1999；九寨沟，李沛琼 60；万源，李培元 4516。

重庆：南川，金佛山，刘正宇 4523；武隆，刘正宇 181575；酉阳，谭士贤 186；彭水，曹亚玲，溥发鼎 136；丰都，曹亚玲，溥发鼎 657；石柱，曹亚玲，溥发鼎 785；忠县，三峡队 2189；开县，戴天伦 101746；奉节，周洪富 110545，张泽荣 25767；巫山，杨光辉 59951；巫溪，曲桂龄 1747；城口，戴天伦 101746，102167，107145。

贵州：兴义，张志松，张永田 7205；兴仁，张志松，张永田 7577；普安，安顺队 1662；纳雍，毕节队 662；赫章，禹平华 1342；仁怀，王光莲 2180；习水，安明态 1310；遵义，A. N. Steward 等 254；息烽，阮红 4-0050；修文，安明态 350；贵阳，曹子余 301；德江，赵佐成 88-2295；沿河，黔北队 2413；石阡，武陵山队 2383；万山，武陵山队 2149。

广东：乳源，高锡朋 52984。

湖南：洪江，李泽棠 3014；新晃，武陵山队 851；芷江，武陵山队 1943，2402；保靖，刘林翰 9933；永顺，湖南队 53-528；慈利，席光银 196。

湖北：鹤峰，李洪钧 6150；恩始，戴伦膺 611；建始，周鹤昌 1493；长阳，王作宾 11407；秭归，王作宾 11732；巴东，赵常明 Ex2773；兴山，陈之端 961310；神农架，鄂植物科考队 25270；房县，刘继孟 8941，9333；武当山，刘克荣 138；宜昌，赵常明 Ex0006。

江苏：连云港，云台山，刘昉勋 10823。

安徽：贵池，经济植物队 59-7217；繁昌，刘晓龙 583。

河南：内乡，刘继孟 5384；西峡，关克俭 1183；栾川，马成功 364；卢氏，刘继孟 4957，5020；桐柏，河南队 1695；修武，云台山，云台山队 1171，1302；嵩县，河南队 35002，关克俭 2259；嵩山，郝景盛 3649；登封，河南队 50010；禹州，河南农学院队

55144。

陕西：留坝，汪劲武 318；勉县，傅坤俊 5607；洋县，傅坤俊 5188；宁陕，张珍万 1780；高县，王作宾 16032；岚皋，巴山队 1561；眉县，王振华，钟补求 73；太白山，刘慎谔，钟补求 258，王作宾 1573；周至，秦岭队 555；户县，郭本兆 748；太华山，郝景盛 3830；终南山，白银元 1117；宜君，尚崇礼 194；延安，黄河队 990；志丹，崔友文 10858。

甘肃：文县，杨金祥 3381；康县，张志英 16416；舟曲，何业祺 698；临漳，廉永善 96834，临夏，孙学刚 20010187；榆中，黄河队 5952；连城，何业祺 5034；天祝，何业祺 4682；张掖，？209；天水，刘继孟 10326；华亭，仲世奇 114；平凉，黄河队-甘一队 2019；合水，黄河队 537。

青海：西宁，郑斯绪 350。

宁夏：银川，贺兰山，何业祺 7339；固原，黄河队-甘一队 2285；海原，黄河队 5284。

山西：伏牛山，关克俭，陈艺林 179；运城，黄河队 571；夏县，刘天慰 600；垣曲，包士英 620；晋城，包士英 1601；芮城，包士英 843；陵川，刘继孟 7629，7853；平顺，刘继孟 8022；霍县，山西队 709；阳城，孔冬梅 K0383；沁县，关克俭，陈艺林 930；介休，黄土高原队 2800；汾阳，王作宾 2775；中阳，中阳甲队 344；离山，黄河队 2310；孝义，王作宾 3385；平定，刘继孟 3940；五台，关克俭，陈艺林 2407；兴县，黄河二队 2488；芦芽山，H. Smith 8118；五寨，黄河二队 2319；宁武，山西队 258；太谷，姚长梓 114；隰县，王作宾 2859。

山东：曲阜，刘继孟 1689；泰山，郝景盛 1706，周太炎 7166；蒙阴，周太炎 6325；崂山，山东大学队 195；牟平，刘继孟 1225；烟台，刘继孟 1687。

河北：磁县，关克俭 6032；武安，关克俭 5848；正定，5 分队 5429；平山，？620；赞皇，刘鑫源 825；邢台，周汉藩 43501；阜平，阜平队 85；涞源，刘继孟 2750，2932；定兴，刘继孟 3895；兴隆，刘慎谔等 4713；涿鹿，杨朝广 1158；小五台山，王作宾 575，黄秀兰等 5018；围场，傅国勋 69；山海关，易新德 12；青龙，王忠涛 140314；杨家坪，王启无 61946；张家口，黄秀兰等 4771；张北，黄秀兰等 3844。

北京：动物园，王作宾 292；戒台寺，周汉藩 41424；西山，刘慎谔 1427；房山，上方山，夏纬瑛 3298，刘继孟 927；百花山，夏纬瑛 2253，王启无 60218；雾灵山，王文采，张敬 2695；昌平，夏口，张敬 2024；延庆，松山，傅德志，向秋云 83169；妙峰山，陈介 3812。

内蒙古：贺兰山，陈又生等 141117；土默特右旗，九峰山，陈又生等 141346；呼和浩特，夏纬瑛 2644，2925，2992；化德，崔友文 865；洛古河，朱有昌等 437；额尔古纳，王战等 1965；满洲里，王战等 1115。

辽宁：旅顺，老铁山队 621；丹东，王薇等 1018；鞍山，傅沛云 2544；铁岭，孔宪武 689；西丰，李文俊 s. n.；开原，傅国勋 15。

吉林：安图，王薇等 2590；汪清，王崇书 1383；蛟河，孔宪武 2005。

黑龙江：宁安，俞德浚 75283；依兰，张玉良 1867；海伦，东北队 344；伊春，刘慎谔等 7645；带岭，刘慎谔等 7205。

92c. 圆齿唐松草（变种）

var. **rotundidens** W. T. Wang, var. nov. Holotype：河南省 (Henan Province)：修武县，云台山，东岭（Xiuwu xian, Yuntai Shan, Dongling）。alt. 850–950 m，路边林下（under forest by road），2009-08-23，云台山采集队（Yuntai Shan Exped.）1301 (PE).

A var. *minore* differt foliolis majoribus usque ad 4 cm longis et 4.4 cm latis margine grosse rotundo-dentatis. Upper cauline leaf 3-pinnate; blade 25×22 cm; leaflets papery rotund-ovate, broadly ovate or subquadrate, 2.4–4×2–4.4 cm, at base subcordate, rarely rounded, 3-lobed, lobes at apex rounded, at margin coarsely and rotundly 1–2-dentate or entire. Fruiting thyrse ca. 30×22 cm; fruiting pedicels 3–10 mm long.

与亚欧唐松草的区别在于本变种的小叶较大，长达 4 cm，宽达 4.4 cm，边缘有粗圆齿。上部茎生叶为 3 回羽状复叶；叶片 25×22 cm；小叶纸质，圆卵形、宽卵形或近方形，2.4–4×2–4.4 cm，在基部近心形，稀圆形，3 浅裂，裂片顶端圆形，边缘有 1–2 粗圆齿或全缘。果期聚伞圆锥花序约 30×22 cm；果期花梗长 3–10 mm。

特产河南修武云台山。生低山路边林下，海拔 850–950 m。

92d. 长梗亚欧唐松草（《中国植物志》，变种）

var. **stipellatum** (C.A.Mey.) Tamura in Acta Phytotax. Geobot. 15: 87. 1953; W. T. Wang & S. H. Wang in Fl Reip. Pop. Sin. 27: 583. 1979; Y. Z. Zhao in Fl. Intramongol., ed. 2, 2: 462, pl. 186: 6–7. 1990; J. G. Liu in Fl. Xinjiang. 2(1): 272. 1994; S. H. Li in Clav. Pl. Chinae Bor. –Or., ed. 2: 207. 1995; Y. Kadota in Iwatsuki et al. Fl. Japan 2a: 339. 2006。— *T. kemense* Fries var. *stipellatum* C. A. Mey. ex Maxim. Prim. Fl. Amur. 16. 1859. Syntypes: Am untern Amur, 1854–1855, L. v. Schrenck s. n. (not seen).

T. minus L. var. *kemense* (Fries) Trel. in Proc. Bost. Soc. Nat. Hist. 23: 300. 1888; D. Z. Fu & G. Zhu in Fl. China 6: 294. 2001; Y. Z. Zhao, Vasc. Pl. Inn. Mongol. 161. 2012。— *T.*

kemense Fries, Fl. Halland. 95. 1817; Hara in J. Fac. Sci. Univ. Tokyo. Sect. 3, 4: 56. 1952.

与亚欧唐松草的区别在于本种的花梗长 15-30 mm，花药呈条形，顶端具长 0.4-0.5 mm 的较长短尖头，以及狭三角形的柱头。茎高 40-60 cm。小叶 0.8-2×0..6-2 cm；脉平或近平，不明显。

This variety differs from var. *minus* in its pedicels 15-30 mm long, linear anthers at apex with mucrones 0.4-0.5 mm long, and narrow-triangular stigma. Stems 40-60 cm tall. Leaflets 0.8-2×0.6-2 cm; nerves flat or nearly flat, inconspicuous.

分布于中国新疆，以及日本、亚洲北部和欧洲北部，生山坡草地，海拔 1100-2100 m。

标本登录：新疆：乌鲁木齐，刘慎谔 3039；博乐，王忠涛 399；察布查尔，王忠涛 129。

93. **细唐松草**（《北京植物志》） 图版 93

Thalictrum tenue Franch. in Nouv. Arch. Mus. Hist. Nat. Paris, ser. 2, 5: 168. 1883; et Pl. David. 1: 16, pl. 7. 1884; Lecoy. in Bull. Soc. Bot. Belg. 24: 175. 1885; Forbes & Hemsl. in J. Linn. Soc. Bot. 23: 9. 1886; Finet & Gagnep. in Ball. Soc. Bot. France 50: 617. 1903; Fl. Beijing. 1: 294. 1962; Iconogr. Corm. Sin. 1: 682, fig. 1364. 1972; W. T. Wang & S. H. Wang in Fl. Reip. Pop. Sin. 27: 575, pl. 140: 9-13. 1979; S. Y. He, Fl. Beijing, rev. ed., 1: 239, fig. 299. 1984; J. W. Wang in Fl. Hebei. 1: 446, fig. 444. 1986; D. Z. Ma, Fl. Ningxia 1: 175. 1986; Y. Z. Zhao in Fl. Intramongol., ed. 2, 2: 455, pl. 184: 5-6. 1990; X. Y. Yu et al. In Fl. Shanxi. 1: 600, fig. 381. 1992; L.H. Zhou in Fl.Qing hai. 1: 333. 1997, Y. Z. Zhao, Vasc. Pl. Inn. Mongol. 160. 2012; W. T. Wang in High. Pl. China 8: 476, fig. 760. 2000; D. Z. Fu & G. Zhu in Fl. China 6: 299. 2001; W. T. Wang in High. Pl. China in Colour 3: 392. 2016; Holotype：北京，San-yu, 1863-08, A. David 2343 (P, not seen).

多年生草本植物，全部无毛。茎高 27-70 cm，分枝。茎生叶为 3-4 回羽状复叶，具短柄；叶片长 5-17 cm；小叶草质，卵形、椭圆形或圆卵形，4-6×2-4 mm，顶端钝或圆形，基部圆形；不分裂，稀三浅裂，全缘，脉不明显；小叶柄长 0.5-6 mm；叶柄长 2-4 cm，基部有短鞘。聚伞圆锥花序长约 30 cm，具纤细分枝；苞片叶状；花梗纤细，长 0.7-3 cm。花：萼片 4，脱落，淡黄绿色，长圆状卵形或椭圆形，3-3.6×1-1.5 mm，顶端圆形，每一萼片具 3 条基生脉（中脉 1 回二歧状分枝，1 侧生脉不分枝，另 1 侧生脉 2 回二歧状分枝）。雄蕊约 10；花丝丝形，长 1-3 mm；花药条形或狭长圆形，2-2.8×0.25-0.4 mm，顶端具长 0.2-0.25 mm 的短尖头。心皮 3-6；子房宽倒卵形，约 1×0.9 mm；花柱长约 0.6 mm；柱头宽三角形，约 0.6×0.4 mm，有 2 条侧生膜质翅。瘦果扁，倒卵形，3.5-5×2-3 mm，

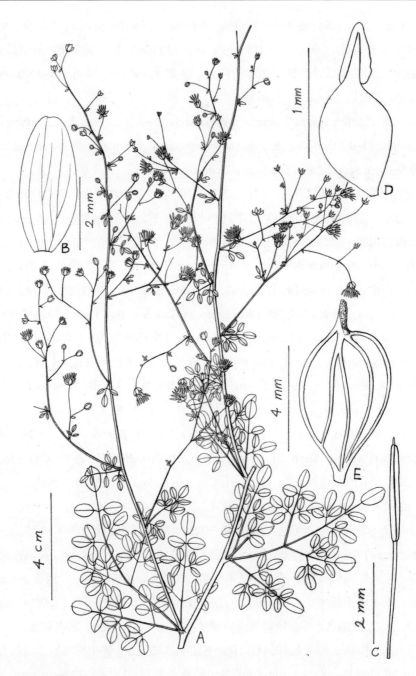

图版93. **细唐松草** A. 开花茎一部分，B. 萼片，C. 雄蕊，D. 心皮（据金德福355），E. 瘦果（据黄秀兰等3585）。

Plate 93. **Thalictrum tenue** A. part of flowering stem, B. sepal, C. stamen, D. carpel (from D. F. Jin 355), E. achene (from X. L. Huang et al. 3585).

每侧有 3 条纵肋，基部具极短柄；宿存柱头长约 0.7 mm。6 月开花。

Perennial herbs, totally glabrous. Stems 27–70 cm tall, branched. Cauline leaves 3–4-pinnate, shortly petiolate; blades 5–17 cm long; leaflets herbaceous, ovate, elliptic or orbicular-ovate, 4–6×2–4 mm, at apex obtuse or rounded, at base rounded, undivided, rarely 3–lobed, entire, nerves inconspicuous; petiolules 0.5–6 mm long; petioles 2–4 cm long, at base shortly vaginate. Thyrses ca. 30 cm long, with slender branches; bracts foliaceous; pedicels slender, 0.7–3 cm long. Flower: Sepals 4, deciduous, yellowish-greenish, oblong-ovate or elliptic, 3–3.6×1–1.5 mm, at apex rounded, each sepal with 3 basal nerves (midrib once dichotomizing. 1 lateral nerve simple and another lateral nerve twice dichotomizing). Stamens ca. 10; filaments filiform, 1–3 mm long; anthers linear or narrow-oblong, 2–2.8×0.25–0.4 mm, apex with mucrones 0.2–0.25 mm long. Carpels 3–6; ovary broad-obovate, ca. 1×0.9 mm; style ca. 0.6 mm long; stigma. broad-triangular, ca. 0.6×0.4 mm, with 2 membranous lateral wings. Achene compressed, obovate, 3.5–5×2–3 mm, on each side longitudinally 3–ribbed, base with a very short stipe; persistent stigma ca. 0.7 mm long.

分布于青海东部、四川东北部、甘肃南部、宁夏、陕西北部、山西、河北、内蒙古西南部。生低山或丘陵的干燥山坡或田边，海拔 400–1900 m。

标本登录：四川：昭化，郝景盛 306。

甘肃：迭部，杨亲二，孔宏智 98-92；贺岗山，黄河队 56-8840。

宁夏：银川，贺兰山，何业祺 7234，7309。

陕西：黄龙（石堡），傅坤俊 3199；延川，傅坤俊 7818；绥德，傅坤俊 6483，黄河队 826；黑郎山，黄河队 6952。

山西：永济，孔冬梅 307；五台，刘继孟 3688。

河北：涉县，关克俭 5547，邯郸队 72-138；小五台山，刘瑛 10965；张家口，黄秀兰等 3585。

北京：房山，上方山，刘继孟 903；房山，谢磊 2005002；门头沟，军响，门头沟队 71-322；百花山，金德福 355；西山，刘慎谔 7996；昌平，高崖，昌平队 72-208，72-499。

94. 紫堇叶唐松草（《中国植物志》）　图版 94

Thalictrum isopyroides C. A. Mey. in Ledeb. Fl. Alt. 2: 346. 1830; Lecoy. in Bull. Soc. Bot. Belg. 24: 196. 1885; Finet & Gagnep. in Bull. Soc. Bot. France 50: 619. 1903; Nevski in Fl. URSS 7: 523, t. 5: 6. 1937; W. T. Wang & S. H. Wang in Fl. Reip. Pop. Sin. 27: 582,

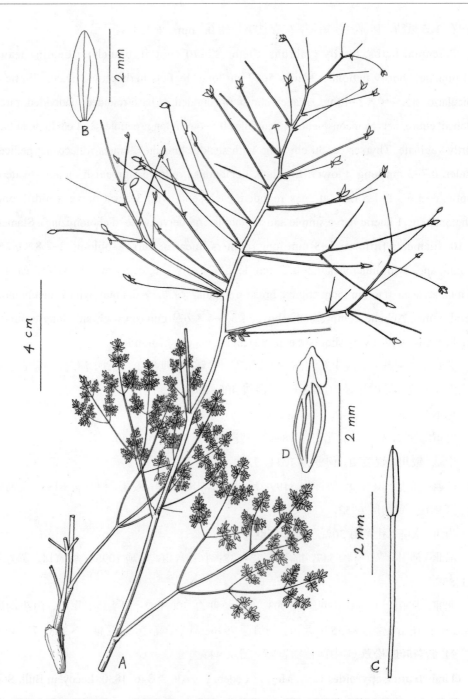

图版94. 紫堇叶唐松草 A.结果植物，B.萼片，C.雄蕊，D.瘦果（据周太炎等650285）。
Plate 94. **Thalictrum isopyroides** A. fruiting plant, B. sepal, C. stamen, D. achene (from T. Y. Zhou et al. 650285).

pl. 146: 1-2. 1979; J. G. Liu in Fl. Xinjiang. 2(1): 272, pl. 73. 1994; D. Z. Fu & G. Zhu in Fl. China 6: 292. 2001; L. Xie in Med. Fl. China 3: 275. with fig. 2016. Described from Mt. Lirmuchaika near Buchtarminsk in the Altay Range.

多年生草本植物，全部无毛。茎高 20-40 cm。基生叶约 3，与下部茎生叶为 4 回三出羽状复叶，有短柄；叶片长 4.5-8 cm；小叶厚，稍肉质，顶生小叶有短柄（小叶柄长约 2 mm），宽菱形，长和宽均约 4 mm，3 浅裂或 3 深裂（裂片狭椭圆形或倒披针形），全缘，脉不明显，侧生小叶较小，无柄，不等 2 深裂或不分裂而呈椭圆形；叶柄长 1-1.8 cm，基部有狭鞘。聚伞圆锥花序顶生，长 12-20 cm，有稀疏的花；苞片狭卵形，长 1-5 mm；花梗长 1-3 cm。花：萼片 4，脱落，淡绿色，狭卵形或卵形，2-2.4×1.2 mm，顶端钝，每一萼片具 3 条不分枝的基生脉。雄蕊 5-8；花丝丝形，长 2.5-3.4 mm；花药狭长圆形，约 1.6×0.4 mm，顶端有长 0.1 mm 的短尖头。心皮 3-5；子房椭圆形、柱头正三角形，约 0.6×0.6 mm，具 2 侧生膜质宽翅。瘦果椭圆形，2-4×0.8-1 mm，每侧有 2-3 条纵肋；宿存柱头长约 0.7 mm。6 月开花。

Perennial herbs, totally glabrous. Stems 20–40 cm. Basal leaves ca. 3, 4-ternate-pinnate, shortly petiolate; blades 4.5–8 cm long; leaflets thick, slightly carnose, terminal leaflets shortly petiolulate (petiolules ca. 2 mm long), broad-rhombic, ca. 4×4 mm, 3-lobed or 3-parted (lobes narrow-elliptic or oblanceolate), entire, nerves inconspicuous, lateral leaflets sessile, smaller, unequally 2-parted or undivided and elliptic in shape; petioles 1–1.8 cm long, at base narrowly vaginate. Thyrses terminal. 12–20 cm long, sparsely flowered; bracts narrow-ovate, 1–5 mm long; pedicels 1–3 cm long. Flower: Sepals 4 deciduous, greenish, narrow-ovate or ovate, 2–2.4×1.2 mm, at apex obtuse, each sepal with 3 simple basal nerves. Stamens 5–8; filaments filiform, 2.5–3.4 mm long; anthers narrow-oblong, ca. 1.6×0.4 mm, apex with mucrones 0.1 mm long. Carpels 3–5; ovary elliptic; stigma deltoid, ca. 0.6×0.6 mm, with 2 membranous lateral wings. Achene elliptic., 2–4×0.8–1 mm, on each side longitudinally 2–3-ribbed; persistent stigma ca. 0.7 mm long.

分布于中国新疆西部；以及亚洲西部。生于多石砾山坡，海拔 1200 m。

根和根状茎有清热燥湿、杀菌、止痢之效。（《中国药用植物志》，2016）

标本登录：新疆：和靖，李安仁，朱家柟 6285；巩留，周太炎等 65-285。

95. 箭头唐松草（《中国植物志》） 图版 95

Thalictrum simplex L. Mant. 1: 78. 1767; DC. Syst. 1: 183. 1818; et Prodr. 1: 15. 1824;

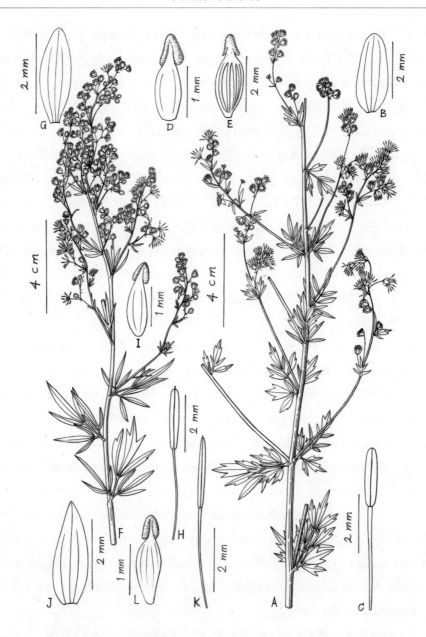

图版95. A–E. 箭头唐松草　A. 开花茎上部，B. 萼片，C. 雄蕊，D. 心皮（据周太炎等650836）。E. 瘦果（据林有润74759）。F–I. 短梗箭头唐松草　F. 开花茎上部，G. 萼片，H. 雄蕊，I. 心皮（据傅德志，向秋云3056）。J–L. 腺毛唐松草　J. 萼片，K. 雄蕊，L. 心皮（据holotype）。

Plate 95. **Thalictrum simplex**　A–E. var. **simplex**　A. upper part of flowering stem, B. sepal, C. stamn, D. carpel (from T. Y. Zhou et al. 650836), E. achene (from Y. R. Lin 74759).　F–I. var. **brevipes**　F. upper part of flowerig stem, G. sepal, H. stamen, I. carpel (from D. Z. Fu & Q. Y. Xiang 3056)　J–L. var. **glandulosum**　J. sepal, K. stamen, L. carpel (from holotype).

Ledeb. Fl. Alt. 2: 353. 1830; Lecoy. in Bull. Soc. Bot. Belg. 24: 204. 1885; Nevski in Fl. URSS 7: 526. 1937; S. H. Li in Fl. Pl. Herb. Chinae Bor.-Or. 3: 215, pl. 95: 1-4. 1975, p. p.; W. T. Wang & S. H. Wang in Fl. Reip. Pop. Sin. 27: 584. 1979; Y. Z. Zhao in Fl. Intramongol., ed. 2, 2: 458, pl. 185: 1. 1990; J. G. Liu in Fl. Xinjiang. 2(1): 274. 1994; S. H. Li in Clav. Pl. Chinae Bor.-Or., ed. 2: 207. 1995; L. H. Zhou in Fl. Qinghai. 1: 335. 1997; D. Z. Fu & G. Zhu in Fl. China 6: 298. 2001; Y. Z. Zhao, Vasc. Pl. Inn. Mongol. 160. 2012; L. Xie in Med. Fl. China 3: 280. 2016. Described from Europe.

95a. var. **simplex**

多年生草本植物，全部无毛。茎高54–100 cm，不分枝或下部分枝。茎生叶向上近直展，为2回羽状复叶；下部茎生叶的叶片长达20 cm；小叶纸质，圆菱形、菱状宽卵形或倒卵形，2–4×1.4–4 cm，基部圆形，3浅裂，裂片顶端钝或圆形，有圆齿，脉在下面隆起，脉网明显；上部茎生叶较小，小叶倒卵形或楔状倒卵形，基部圆形、钝或楔形，裂片顶端急尖；小叶柄长0–2 mm；下部茎生叶有长4–7 cm的柄，上部叶无柄。聚伞圆锥花序长9–30 cm，分枝与轴成45°斜上方开展；花梗长达7 mm。花：萼片4，早落，长卵形或狭椭圆形，2.2–2.6×1 mm，顶端钝，每一萼片具3条不分枝的基生脉。雄蕊9–15；花丝丝形，长约4 mm；花药长圆形，2×0.4–0.5 mm，顶端有长0.05 mm的小短尖头。心皮3–12；子房宽长圆形，约1×0.6 mm；花柱长约0.8 mm；柱头正三角形，约0.8×0.65 mm，具2膜质侧生翅。瘦果椭圆体形，约2×1 mm，在每侧均有3条纵肋；宿存柱头长约0.7 mm。7月开花。

Perennial herbs, totally glabrous. Stems 54–100 cm tall, simple or below branched. Cauline leaves upwards spreading, 2-pinnate; blades of lower cauline leaves up to 20 cm long; leaflets papery, orbicular-rhombic, broadly rhombic-ovate or obovate, 2–4×1.4–4 cm, at base rounded, 3-lobed, lobes at apex obtuse or rounded, at margin with a few rotund teeth, nerves abaxially prominent and nervous nets conspicuous; upper cauline leaves smaller, their leaflets obovate or cuneate-obovate, at base rounded, obtuse or cuneate, and with acute lobes; petiolules 0–2 mm long; petioles of lower cauline leaves 4–7 cm long and upper cauline leaves sessile. Thyrses 9–30 cm long, with branches upwards obliquely spreading; pedicels up to 7 mm long. Flower: Sepals 4, caducous, long ovate or narrow-elliptic, 2.2–2.6×1 mm, at apex obtuse, each sepal with 3 simple basal nerves. Stamens 9–15; filaments filiform, ca. 4 mm long; anthers oblong, 2×0.4–0.5, apex with minute mucrones ca. 0.05 mm long. Carpels 3–12; ovary broad-oblong, ca. 1×0.6 mm; style ca. 0.8 mm long; stigma deltoid, ca. 0.8×0.65 mm, with 2 membranous lateral wings. Achene ellipsoid, ca. 2×1 mm, on each side longitudinally 3-ribbed; persistent stigma ca. 0.7 mm long.

分布于青海东部、甘肃、宁夏、山西、内蒙古、新疆；以及亚洲北部和西部、欧洲。生山地草坡或沟边，海拔 1400–2600 m。

根和根状茎有清热燥湿、杀菌止痢之效，用于痢疾、肠炎、传染性肝炎等症。根含亚欧唐松草碱（thalicmine）等生物碱。(《中国药用植物志》，2016)

标本登录：青海：西宁，钟补求 8187；互助，杨竞生 97-879。

甘肃：平凉，陈学林 94-118。

山西：五台，唐进 1090。

内蒙古：锡林郭勒盟，嘎顺马拉山，综考会队 34；牙克石，傅德志 86-0293。

新疆：塔城，李安仁，朱家柟 10873；托里，张建武 93141；霍城，西北植物所队 3200；绥定，王庆瑞 3244；巩留，王庆瑞 2705；昭苏，林有润 74759；和静，王庆瑞 4121；沙湾，关克俭 1300；石河子，？205；玛纳斯，关克俭 1115；乌鲁木齐，八一农学院队 1-298；天山，关克俭 1413；富蕴，秦仁昌 2213；阿勒泰，秦仁昌 2630，朱格麟 7041；清河，秦仁昌 1240。

95b. 锐裂箭头唐松草（《中国植物志》，变种）

var. **affine** (Ledeb.) Regel in Bull. Soc. Nat. Mosc. 34: 44. 1861; W. T. Wang & S. H. Wang in Fl. Reip. Pop. Sin. 27: 585. 1979; Y. Z. Zhao in Fl. Intramongol., ed. 2, 2: 460, pl. 185: 5. 1990; S. H. Li in Clav. Pl. Chinae Bor.-Or., ed. 2: 207. 1995; D. Z. Fu & G. Zhu in Fl. China 6: 298. 2001; Y. Z. Zhao, Vasc. Pl. Inn. Mongol. 161. 2012; L. Xie in Med. Fl. China 3: 282. 2016. —— T. *affine* Ledeb. Fl Ross. 1: 10. 1842. Described from S. Siberia and Davuria.

T. *angustifolium* auct. non L.: C. Pei & T. Y. Chou, Icon. Chin. Med. Pl. 7: fig. 332. 1964.

与箭头唐松草的区别在于本变种的小叶楔形，具狭楔形的基部和狭三角形、顶端锐尖的裂片。花梗长 4–7 mm。

This variety differs from var. *simplex* in its cuneate leaflets with narrow-cuneate bases and narrow-triangular, pungent lobes. pedicels 4–7 mm long.

分布于山西北部、河北北部、内蒙古、吉林、黑龙江；以及俄罗斯西伯利亚地区。生山坡草地或湿草地，海拔 400-1470 m。

根有清湿热、解毒之效，用于黄疸、痢疾、腹泻等症。(《中国药用植物志》，2016)

标本登录：山西：宁武，山西队 53-305。

河北：张家口，沽源牧场，胡叔良 101。

内蒙古：乌兰坝林场，傅德志等 399。

黑龙江：孙吴县，J. Sato7731；克山，东北队 308；伊春，刘慎谔等 7713；洛古河自然林，朱有昌等 436。

吉林：通榆，东北师大队 260；甘河，东北师大队 615；安图，刘慎谔等 3717。

95c. 短梗箭头唐松草（《中国植物志》，变种）

var. **brevipes** Hara in J. Fac. Sci. Univ. Tokyo, Sect.3, 6(2): 56. 1952; Iconogr. Corm. Sin. 1: 683, fig. 1366. 1972; Fl. Tsinling. 1(2):247. 1974; S. H. Fu, Fl. Hupeh. 1: 358, fig. 502. 1976; W. T. Wang & S. H. Wang in Fl. Reip. Pop. Sin. 27: 585, pl. 147: 1–3. 1979; B. Z. Ding et al. Fl. Henan. 1: 474. 1981; S. Y. He, Fl. Beijing, rev. ed., 1: 240, fig. 301. 1984; Tamura & Shimizu in Satake et al. Wild Flow. Japan 2: 83, pl. 83: 1. 1984; J. W. Wang in Fl. Hebei 1: 446, fig. 445. 1986; D. Z. Ma, Fl. Ningxia. 1: 176, fig. 155. 1986; Y. Z. Zhao in Fl. Intramongol., ed. 2, 2: 460. 1990; X. Y. Yu et al. In Fl. Shanxi. 1: 607, fig. 384. 1992; S. H. Li in Clav. Pl. Chinae Bor. –Or., ed. 2: 207, pl. 97: 3. 1995; L. H. Zhou in Fl. Qinghai. 1: 335. 1997; Y. J. Zheng in Fl. Shandong. 2: 34, fig. 26.1997; W. T. Wang in High. Pl. China 3: 479, fig. 766: 1–3. 2000; D. Z. Fu & G. Zhu in Fl China 6: 298. 2001; Y. Kadota in Iwatsuki et al. Fl. Japan 2a:338. 2006; Y. Z. Zhao, Vasc. Pl. Inn. Mongol. 161. 2012; M. B. Deng & K.Ye in Fl. Jiangsu 2: 84, fig. 2-134. 2013; L. Xie in Med. Fl. China 3: 282. 2016. Holotype：日本：Kyushu. Prov. Bungo: mt. Tsurumi, 1926-11-03, S. Saito 3 (T1. not seen).

T. simplex auct. non L.: Forbes & Hemsl. in J. Linn. Soc. Bot. 23: 9. 1885; S. H. Li in Fl. Pl. Herb. Chinae Bor.-Or. 3: 215. 1975. p. p.

与箭头唐松草的区别在于本变种的小叶楔形，具狭三角形锐尖裂片，花梗较短，长 1–4(–5) mm，以及狭长圆形花药（长 2 mm，宽 0.3 mm）。

This variety differs from var. *simplex* in its cuneate leaflets with narrow-triangular, pungent lobes, shorter pedicels 1–4(–5) mm long, and narrow-oblong anthers 2 mm long and 0.3 mm broad.

分布于中国四川、重庆、湖北、江苏南部、安徽东南部、河南、陕西、甘肃中南部、青海东部、宁夏、山西、山东、河北、内蒙古、辽宁、吉林、黑龙江；以及朝鲜、日本。生平原或低山湿草地，沟边或山坡；海拔在本种分布区大部地区自海边达 1400 m，在甘肃和青海东部达 2300–2600 m。

根有清湿热、解毒之效，用于黄疸、痢疾、腹泻等症。(《中国药用植物志》，2016)

标本登录：四川：成都，王作宾 8261；彭山，经济植物队 59-5275。

重庆：江津，经济植物队 59-673；合川，黄志明 690；南川，刘正宇 14250；丰都，三峡植被组 139。

湖北：巴东，郝景盛 11；宜昌，钟观光 3502，郑重 115。

江苏：南京，灵谷寺，左景烈 1863。

安徽：繁昌，刘晓龙 583。

陕西：凤县，傅竞秋 2909；太白，杨金祥 1160；周至，白银元 1303；榆林，夏纬瑛 3658，傅坤俊 7152。

甘肃：连城，何业祺 4927，5030；永登，何业祺 4301，孙学刚 6335，6456。

青海：互助，刘继孟 5884；贵德，1957-06，采集人不明。

山西：五寨，夏纬瑛 1668；五台，关克俭，陈艺林 1429；应县，段长虹 678。

河北：涞源，河北农大队 1775；徐水，河北农大队 4173，4175；乐亭，侯学煜 10256；迁西，迁西队 284；围场，黄士坎 612。

北京：清华大学，? 572；琉璃河，房山队 659；昌平，昌平队 256；雾灵山，刘瑛 11724。

天津：茶淀，刘慎谔 s. n.

内蒙古：锡林浩特，陈又生 140680；昭盟克旗，张玉良 328；东乌珠穆新旗，李世英 284；阿巴嘎旗，李世英 328；多伦，崔友文 739；额尔古纳旗，王战 1240。辽宁：旅顺，孙英宝等 404；沈阳，北陵，孔宪武 897；东安，王薇 1294。

吉林：和龙，延边二组 773；汪清，傅沛云 889；安图，延边一组 664；额穆，孔宪武 2361；镇赉，白城组 126；通榆，叶居新 260。

黑龙江：镜泊湖，孔宪武 2076；阿城，东北队 415；安达，张玉良 726；依兰，张永良 1993。

95d. 腺毛箭头唐松草(《中国植物志》，变种)

Var. glandulosum W. T. Wang in Fl Reip. Pop. Sin-27: 585, 620. 1979; D. Z. Fu & G. Zhu in Fl. China 6: 298. 2001. Type：黑龙江，肇东，宋站试验林场，1959，郭树昌 74 (holotype, NEFI)，83 (paratype, NEFI)。

与箭头唐松草的区别在于本变种的叶和花序均有极短的腺毛，以及长 1-3 mm 的较短花梗。

This variety differs from var. *simplex* in its leaves and thyrses covered with very short glandular hairs and its shoter pedicels 1–3 mm long.

分布于甘肃东南部、北京、黑龙江南部，生于山地草坡。

标本登录：甘肃：平凉，崆峒山至蒿店途中，仲世奇，杨平礼 162。

北京：延庆，松山林场，傅德志，向秋云 83-056。

96. 黄唐松草（《中国植物志》）　图版 96

Thalictrum flavum L. Sp. Pl. 546. 1753; DC. Prodr. 1: 14. 1824; Ledeb. Fl. Alt. 2: 355. 1830; Lecoy. in Bull. Soc. Bot. Belg. 24: 208. 1885; Finet & Gagnep. in Bull. Soc. Bot. France 50: 622. 1903; Kom. & K.-Alisova. Key Pl. Far East. Reg. USSR 1: 563. 1931; Nevski in Fl. URSS 7: 527. 1937; W. T. Wang & S. H. Wang in Fl. Reip. Pop. Sin. 27: 386. 1979; J. G. Liu in Fl. Xinjiang. 2(1): 274. 1994; D. Z. Fu & G. Zhu in Fl. China 6: 289. 2001. Described from N. Europe.

多年生草本，全部无毛。茎高约 1.5 m，分枝。叶为 3 回羽状复叶；中部茎生叶长约 30 cm，有柄，顶生小叶楔状倒卵形或狭倒卵形，4–7×2.5–5.5 cm，基部圆形或钝，上部有 3 粗齿或 3 浅裂，侧生小叶稍斜，狭卵形或狭椭圆形，边缘有 1–2 齿或全缘；上部茎生叶长 9–15 cm，小叶楔形或楔状倒披针形，长达 4 cm，宽达 1.8 cm，基部楔形或钝，上部有 3 个三角形尖齿或小裂片；小叶柄长 1–1.9 cm，侧生小叶通常无柄。聚伞圆锥花序长约 25 cm，有多数密集的花；苞片狭条形或钻形，长约 2.5 mm；花梗长 5–10 mm。花：萼片 4，早落，卵形或卵状长圆形，3–4×1.2–1.5 mm，顶端急尖，每一萼片具 3 条不分枝的基生脉，或中脉和 1 侧生脉 1 回二歧状分枝。雄蕊 11–18；花丝丝形，长 3–6 mm；花药狭长圆形，1.5–2.2×0.4–0.8 mm，顶端钝或具长 0.05 mm 的小尖头。心皮 9–10；子房椭圆形，1×0.5 mm；花柱长 0.8 mm；柱头正三角形，约 0.8×0.6 mm，顶端微凹，具 2 宽侧生翅。瘦果椭圆形，约 2.4×1 mm，每侧有 3 条纵肋；宿存柱头长约 0.8 mm。7 月开花。

Perennial herbs, totally glabrous. Stems ca. 1.5 m tall, branched. Leaves 3-pinnate, middle cauline leaves ca. 30 cm long, petiolate, their terminal leaflets cuneate-obovate or narrow-obovate, 4–7×2.5–5.5 cm, at base rounded or obtuse. above coarsely 3-dentate or 3-lobed, lateral leaflets slightly oblique, narrow-ovate or narrow-elliptic, margin 1–2-dentate or entire; upper cauline leaves 9–15 cm long, their leaflets cuneate or cuneate-oblanceolate, up to 4 cm long and 1.8 cm broad, at base cuneate or obtuse, above with 3 triangular pungent teeth or 3 lobes, petiolules 1–1.9 cm long, lateral leaflets usually sessile. Thyrses ca. 25 cm long,

图版96. 黄唐松草 A. 茎生叶，B. 聚伞圆锥花序，C. 二萼片，D. 雄蕊，E. 心皮（据安争夕等722416），F. 瘦果（据P. C. Hoch & C. J. Chen 86145）。

Plate 96. **Thalictrum flavum** A. cauline leaf, B. thyrse, C. two sepals, D. stamen, E. carpel (from Z. X. An et al. 722416), F. achene (from P. C. Hoch & C. J. Chen 86145).

densely many-flowered; bracts narrow-linear or subulate, ca. 2.5 mm long; pedicels 5–10 mm long. Flower: Sepals 4, caducous, ovate or ovate-oblong, 3–4×1.2–1.5 mm, at apex acute. each sepal with 3 basal nerves (3 nerves all simple, or midrib and 1 lateral nerve once dichotomizing). Stamens 11–18; filaments filiform, 3–6 mm long; anthers narrow-oblong, 1.5–2.2×0.4–0.8 mm, apex obtuse or with minute mucrones 0.05 mm long. Carpels 9–10; ovary elliptic, 1×0.5 mm; style 0.8 mm long; stigma deltoid, ca. 0.8×0.6 mm, at apex emarginate, with 2 broad lateral wings. Achene elliptic, ca. 2.4×1 mm, on each side longitudinally 3-ribbed; persistent stigma ca. 0.8 mm long.

分布于中国新疆阿尔泰山区；以及亚洲北部和西部及欧洲。生于山地溪边草地，海拔 500–1750 m。

根和根状茎有清热燥湿，杀菌止痢之效，用于湿热泻痢、黄疸、目赤肿痛等病。根含前绿心樟碱（preocoteine）、亚美尼亚罂粟碱（armepovine）等生物碱。(《中国药用植物志》，2016）

标本登录：新疆：阿尔泰山，安争夕，杨昌友，李志仁 72-2416；阿勒泰，P. C. Hoch & C. J. Chen 86-145。

97. 展枝唐松草（《东北植物检索表》） 图版 97

Thalictrum squarrosum Steph. ex Willd. Sp. Pl. 2: 1279. 1799; DC. Prodr. 1: 13. 1824; Lecoy. in Bull. Soc. Bot. Belg. 24: 198. 1885; Forbes & Hemsl. in J. Linn. Soc. Bot. 23: 9. 1886; Kom. in Acta Hort. Petrop. 22: 309. 1903; Finet & Gagnep. in Bull. Soc. Bot. France 50: 619. 1903; Kom. & K.-Alisova, Key Pl. Far East Reg. USSR 1: 563. 1931; Nevski in Fl. URSS 7: 523. 1937; Kitag. Lineam. Fl. Mansh. 228. 1939; Fl. Beijing 1: 294. 1962; Iconogr. Corm. Sin. 1: 684, fig. 1368. 1972; S. H. Li in Fl. Pl. Herb. Chinae Bor. –or. 3: 211, Pl. 94. 1975; W. T. Wang & S.H. Wang in Fl. Reip. Pop. Sin. 27: 386, pl. 148. 1979; S. Y. He, Fl. Beijing, rev. ed., 1: 240, fig. 300. 1984; J. W. Wang in Fl. Hebei 1: 445, fig. 443. 1986; D. Z. Ma, Fl. Ningxia. 1: 175. 1986; Y. Z. Zhao in Fl. Intramongol., ed. 2, 2: 457, pl. 184: 1–2. 1990; X. Y. Yu et al. in Fl. Shanxi. 1: 607, fig. 385. 1992; S. H. Li in Clav. Pl. Chinae Bor. –Or., ed. 2: 207, pl. 97: 1. 1995; L. H. Zhou in Fl. Qinghai. 1: 335. 1997; W. T. Wang in High. Pl. China 3: 480, fig. 767. 2000; D. Z. Fu & G. Zhu in Fl. China 6: 299. 2001; Y. Zhou, Pl. Resour. Changbai Mount. China 140, pl. 79: 2, 3, 6. 2010; Q. Lu et al. Desert Pl. China 163. 2012; M. B. Deng & K. Ye in Fl. Jiangsu 2: 84, fig. 2-136. 2013; F. H. Li et al. Pl. Yanqing

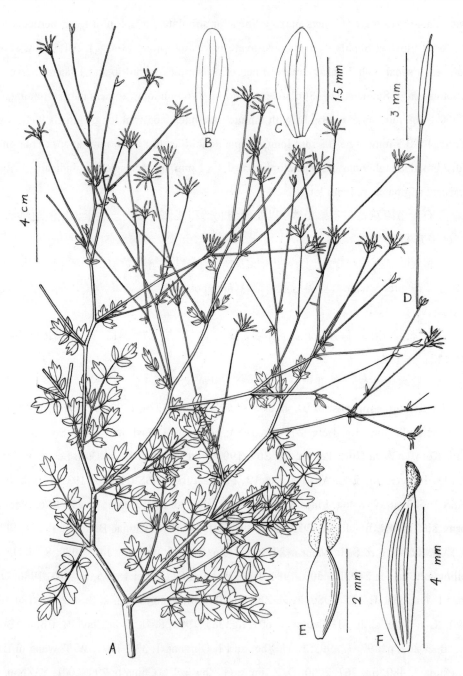

图版97. 展枝唐松草　A. 开花茎顶部，B、C. 二萼片，D. 雄蕊，E. 心皮（据傅坤俊7271），F. 瘦果（据夏纬瑛3869）。

Plate 97. **Thalictrum squarrosum**　A. apical part of flowering stem, B、C. two sepals, D. stamen, E. carpel (from K. Z. Fu 7271), F. achene (from W. Y. Xia 3869).

144. 2015; W. T. Wang in High. Pl. China in Colour 3: 394. 2016; L. Xie in Med. Fl. China 3: 283, with fig. 2016. Described from Siberia.

T. trigynum Fisch. ex DC. Prodr. 1: 14. 1824; Ledeb. Fl. Ross. 1: 11. 1842; Franch. Pl. David. 1: 16. 1884. Described from Dahuria.

T. purdomii J. J. Clarke in Kew Bull. 1913: 39. 1913, p. p.

多年生草本植物，全部无毛。茎高 60-100 cm，通常自中部二歧状分枝。下部和中部茎生叶为 2-3 回羽状复叶，有短柄；叶片长 8-18 cm；小叶纸质，楔状倒卵形，宽倒卵形，长圆形或圆卵形，0.8-2（-3.5）×0.6-1.5（-2.6）cm，顶端急尖，基部楔形或圆形，通常 3 浅裂（裂片全缘或有 2-3 小齿），下面有白粉，脉上面稍凹陷，下面平或稍隆起，脉网稍明显；小叶柄长 0.5-14 mm；叶柄长 1-4 cm。聚伞圆锥花序伞房状，宽约 30 cm，近二歧状分枝；花梗细，长 1.5-3 cm，果期稍增长。花：萼片 4，脱落，淡黄绿色，长椭圆形或近长圆形，3-3.8×1-1.6 mm，顶端钝，每一萼片具 3 条基生脉（3 条脉均不分枝，或中脉和 1 侧生脉 1 回二歧状分枝）。雄蕊 5-14；花丝丝形，长 2.5-7 mm；花药狭长圆形，2.2-3×0.3-0.4 mm，顶端具长 0.3-0.5 mm 的短尖头。心皮 1-3（-5）；子房狭倒卵形，约 2×0.8 mm；花柱长 1 mm；柱头近圆形，约 1.2×1 mm，具 2 宽侧生翅。瘦果纺锤形，约 4.2×1.6 mm，每侧具 3 条纵肋；宿存柱头长约 1.5 mm。7-8 月开花。

Perennial herbs, totally glabrous. Stems 60–100 cm tall, from the middle upwards densely dichotomously branched. Lower and middle cauline leaves 2–3-pinnate, shortly petiolate; blades 8–18 cm long; leaflets papery, cuneate-obovate, broad-obovate, oblong or orbicular-ovate, 0.8–2(–3.5)×0.6–1.5(–2.6)cm, at apex acute, at base cuneate or rounded. usually 3-lobed(lobes entire or 2–3-denticulate), abaxially glaucous, nerves adaxially slightly impressed, abaxially flat or slightly prominent, nervous nets slightly conspicuous; petiolules 0.5–14 mm long; petioles 1–4 cm long. Thyrses corymbiform, ca. 30 cm broad, densely dichotomously branched; pedicels slender, 1.5–3 cm long, slightly elongating in fruit. Flower: Sepals 4, deciduous, yellowish-greenish, long elliptic or suboblong, 3–3.8×1–1.6 mm, at apex obtuse, each sepal with 3 basal nerves (3 nerves all simpe, or 1 lateral nerve simple and midrib and the other lateral nerve once dichotomizing). Stamens 5–14; filaments filiform, 2.5–7 mm long; anthers narrow-oblong, 2.2–3×0.3–0.4 mm, apex with mucrones 0.3–0.5 mm long. Carpels 1–3(–5); ovary narrow-obovate, ca. 2×0.8 mm; style 1 mm long; stigma suborbicular, ca. 1.2×1 mm, with 2 broad lateral wings. Achene fusiform, ca. 4.2×1.6 mm, on each side longitadinally 3-ribbed; persistent stigma ca. 1.5 mm long.

分布于中国青海东部、甘肃东南部、宁夏、陕西北部、江苏东北部、山西、河北、内蒙古、辽宁、吉林、黑龙江；以及蒙古、俄罗斯西伯利亚和远东地区。生平原草地、田边或山地草坡，海拔200-1900 m。

全草有清热解毒、健胃等效，用于头痛、头晕、胃病吐酸等症。(《中国药用植物志》, 2016)

标本登录：甘肃：会宁，黄河队56-5733。

陕西：延安，夏纬瑛3561；吴起，黄河队56-8243；子长，傅坤俊7615；靖边，黄河队56-8010；米脂，黄河队56-6566；横山，傅坤俊6814，7167，黄河队56-7446；榆林，夏纬瑛3720，崔友文10611，黄河队56-7141。

山西：隰县，黄河队55-3010；交城，H. Smith7560；太原，Y. Yabe s.n.；兴县，黄河二队55-2178；岢岚，黄河队55-2966；山阴，樊润威296；五寨，夏纬瑛1665；保德，黄河队55-2873；河曲，黄河队55-3162；五台，刘继孟3698，关克俭、陈艺林2313；Kushehshan，王作宾3710。

河北：阜平，刘继孟3286；小五台山，李建藩10625，杨承元6103，黄秀兰等2437；张家口，崔友文426，黄秀兰等4821；张北，黄秀兰等3498，6054；康保，黄秀兰等8537；围场，承德队71-362，傅国勋32。

北京：天坛，S. M. Liang 8；昌平，昌平队72-304；百花山，刘长江195；东灵山，王启无60309。

天津：永清，中草药队77-21。

内蒙古：贺兰山，内大生物系队251；鄂尔多斯，白银元57，夏纬瑛3868；乌审旗，傅坤俊7271；包头，王作宾2391；丰镇，崔友文1040；多伦，农林处队171；正蓝旗，陈又生14-0549；赤峰，刘慎谔等5126；克什克腾旗，杨雅玲382，孙英宝等2-104；西联旗，李世英299；索伦，中德队8186；伊克昭盟，郎学忠280；海拉尔，刘慎谔8002，陈又生等14-1994；满洲里，中德队8337。

吉林：通榆，白城组182；四平，J. Sato 3115。

黑龙江：哈尔滨，王薇等416；汤原，刘慎谔等1329。

组 12. 石砾唐松草组

Sect. **Schlagintweitella** (Ulbr.) W. T. Wang & S. H. Wang in Fl. Reip. Pop. Sin. 27: 588, 620. 1979. —Gen. *Schlagintweitella* Ulbr. in Notizbl. Bot. Gart. Berl.10: 877. 1929. Type: *T. squamiferum* Lecoy.

花沿茎或总状花序轴向顶发育。萼片4，小，淡黄绿色，早落，每一萼片具3条不分枝或二歧状分枝的基生脉。花丝丝形；花药狭长圆形或条形，稀长圆形，顶端具短尖头。花柱完全与柱头贴生；柱头宽三角形或正三角形，具二侧生宽膜质翅。瘦果沿腹、背二缝线凸起，每侧具3条纵肋。

Flowers along stem or raceme rachis acropetally developed. Sepals 4, small, yellowisgreenish, caducous; each sepal with 3 simple or dichotomizing basal nerves. Filaments filiform; anthers narrow-oblong or linear, rarely oblong, apex muconate. Style entirely adnate with stigma; stigma broad-triangular or deltoid, with 2 broad membranous lateral wings. Achene convex along both ventral and dorsal sutures, on each side longitudinally 3-ribbed.

2种，隶属2系。

系 1. 石砾唐松草系

Ser. **Squamifera** W. T. Wang & S. H. Wang in Fl. Reip. Pop. Sin. 27: 588, 620. 1979. — Subsect. *Squamifera* (W. T. Wang & S. H. Wang) Tamura in Acta Phytotax. Geobot. 43: 56. 1992; et in Hiepko, Nat. Pflanzenfam., Zwei. Aufl., 17a Ⅳ : 484. 1995. Type: *T. squamiferum* Lecoy.

基生叶不存在。花单生于茎生叶腋部，沿茎向顶发育。瘦果无柄。

Basal leaves wanting. Flowers singly borne in axils of cauline leaves, and along stem acropetally developed. Achenes sessile.

1种。

98. 石砾唐松草（《药学学报》） 图版 98

Thalictrum squamiferum Lecoy. in Bull. Soc. Bot. Belg. 16: 327. 1880 et 24: 184, t. 4: 8. 1885; W. T. Wang & S. H. Wang in Fl. Reip. Pop. Sin 27: 588, pl. 147: 6-9. 1979; W. T. Wang in Bull. Bot. Lab. N.-E. Forest. Inst. 8: 36. 1980; et in Iconogr. Corm. Sin. Suppl. 1: 429, fig. 8618. 1982; Grierson, Fl. Bhutan 1(2):297. 1984; W. T. Wang in Fl. Xizang. 2: 74, fig. 21: 7-9. 1985; in Vasc. Pl. Hengduan Mount. 1: 504. 1993; in Fl. Yunnan. 11: 174. 2000; et in High.

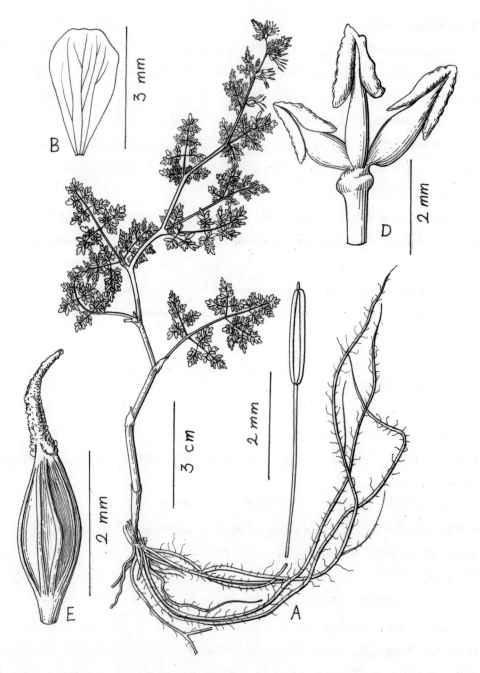

图版98. 石砾唐松草 A. 开花植株，B. 萼片，C. 雄蕊，D. 雌蕊群（据俞德浚9170），E. 瘦果（据Harry Smith 11954）。

Plate 98. **Thalictrum squamiferum** A. flowering plant, B. sepal, C. stamen, D. gynoecium (from T. T. Yü 9170), E. achene (from Harry Smith 11954).

Pl. China 3: 480, fig. 166: 4-8. 2000; Rau in Sharma et al. Fl. India 1: 143. 1993; D. Z. Fu & G. Zhu in Fl. China 6: 299. 2001; W. T. Wang in High. Pl. China in Colour 3: 395. 2016; L. Xie in Med Fl. China 3: 286, with fig. & photo. 2016; X. T. Ma et al. Wild Flow. Xizang 29, photo 56. 2016. Described from the Kyungar Pass, Tibet.

T. cultratum Wall. ssp. *tsangense* Bruhl in Ann. Bot. Gard. Culc. 5(2): 72, t. 102: 3, 6, 7. 11, 15, 18. 1896. Syntypes collected by King's collector from Tibet and Sikkim.

T. glareosum Hand.-Mazz. in Anseig. Akad. Wiss. Wien Math. Nat. Kl. 62: 218. 1925; et Symb. Sin. 7: 312. 1931. ——*Schlagintweitella glareosa* (Hand.-Mazz.) Ulbr. in Notizbl. Bot. Gart. Berl. 12: 355. 1935. Holotype: 云南: Gebirge Waha bei Yungning（永宁）, 1915-07-20, Handel-Mazzetti 7114（not seen; photo, PE）. Paratype: 云南: 德钦，白马山，1917-07, G. Forrest 14454（not seen）.

Schlagintweitella fumarioides Ulbr. in Notizbl. Bot. Gart. Berl. 10: 878, fig. 15. 1929. Syntypes: 西藏: Prov. Gnari Khorsum, south of the Satlej, 1855-07-16, Schlagintwett 7024 (not seen). 云南: 德钦之西的 Moting（山名）, alt. 14000 ft. 1923-07, J. F. Rock 10323（not seen）.

多年生小草本植物，全部无毛，有时具少数短腺毛。根状茎短。茎渐升或直立，长6-20 cm，下部常生石砾中，在节处有鳞片，分枝。中部茎生叶为3-4回羽状复叶，长3-9 cm，有短柄，上部茎生叶较小；叶片长2-4.5 cm；小叶多无柄，密集，近革质，顶生小叶卵形，三角状宽卵形或心形，1-2×0.6 mm，侧生小叶较小，卵形、椭圆形或狭卵形，边缘全缘，干时反卷，脉不明显；叶柄长3-15 mm，基部有狭鞘。花单生叶腋；花梗长1.2-6.5（-20）mm；萼片4，早落，淡黄绿色，常带紫色，倒卵形或椭圆状卵形，约3×2 mm，顶端圆形，每一萼片具3条基生脉（中脉3回二歧状分枝，2侧生脉1回二歧状分枝）。雄蕊7-20枚；花丝丝形，长4.2-5 mm；花药狭长圆形，1.8-2.2×0.3 mm，顶端具长0.2 mm 的短尖头。心皮2-6；子房椭圆形，约1.4×0.8 mm，花柱长1 mm；柱头宽箭头状三角形，1×0.8-1 mm，具2宽侧生翅。瘦果稍扁，椭圆形，2-3×1 mm，每侧有3条纵肋；宿存柱头长约1.4 mm。7月开花。

Small perenniall herbs, totally glabrous, sometimes with sparse short glandular hairs. Rhizome short. Stems ascendent or erect, 6-20 cm long, below often among gravels and on nodes with small scales, above this scaly portion branched. Middle cauline leaves 3-4-pinnate, 3-9 cm long, shortly petiolate, upper cauline leaves smaller; blades 2-4.5 cm long; leaflets mostly sessile, dense, subcoriaceous, terminal leaflets ovate, triangular-ovate or cordate, 1−2×0.6 mm, lateral leaflets smaller, ovate, elliptic or narrow-ovate, margin entire, revolute when drying, nerves inconspicuous; petioles 3−15 mm long, at base narrowly vaginate. Flowers singly borne in leaf axils; pedicels 1.2−6.5(−20)mm; sepals 4, caducous, yellowish-greenish, often purple-tinged, obovate or elliptic-ovate, ca. 3×2 mm, at apex rounded, each sepal with 3 basal nerves (midrib thrice dichotomizing, and 2 lateral herves once dichotomizing). Stamens 7−20; filaments filiform, 4.2−5 mm long; anthers narrow-oblong, 1.8−2.2×0.3 mm, apex with mucrones 0.2 mm long, Carpels 2−6; ovary elliptic, ca. 1.4×0.8 mm; style 1 mm long; stigma broadly sagittate-triangular, 1×0.8−1 mm, with 2 broad lateral wings. Achene slightly compressed, elliptic, 2−3×1 mm, on each side longitudinally 3-ribbed; persistent stigma ca. 1.4 mm long.

分布于中国西藏东南部至西南部、云南西北部、四川西部、青海南部；以及不丹、尼泊尔、印度北部。生山地多石砾山坡、流石滩、河岸石砾沙地或林边，海拔 3600−5000 m。

全草有清热解毒、健胃燥湿之效，用于感冒发热、咳嗽、目赤肿痛等症。(《中国药用植物志》, 2016)

标本登录：西藏：察隅，倪志诚等 1164；昌都，D. E. Boufford, B. Bartholomev et al. 41414；类乌齐，D. E. Boufford, B. Bartholomew 32003, 32271；索县，王金亭 3258；博古湖至马拉山，吴征镒等 75-471；江孜，青藏队 75-0388；仲巴，青藏队 75-6522；南木林，青藏队 75-7494；定日，郑度 29，中日队 1790；绒布寺至珠峰，张永田，郎楷永 3657；珠峰，登山队 59-601；江孜，青藏队 75-0388；吉隆，西藏中草药队 667；帕里，青藏队 74-2679；札达，青藏队 76-7812。

云南：丽江，G. Forrest 12771；中甸，俞德浚 12102，青藏队 81-1095, 81-1506，植物所横断山队 2890, 3292，D. E. Boufford et al. 42202；德钦，俞德浚 9170，青藏队 81-3102，横断山植被组 4604，谢磊 126。

四川：木里，俞德浚 6463；乡城，D. E. Boufford et al. 30728；稻城，青藏队 81-4394；九龙，D. E. Boufford et. al. 42583, 42620；康定，刘振书 1520，关克俭，王文采 1148；雅江，D. E. Boufford, B. Bartholomew et al. 36004；道孚，泰宁，H. Smith 11954；甘孜，D. E. Boufford et al. 34209。

青海：囊谦，杨永昌 1123。

系 2. 高山唐松草系

Ser. **Alpina** (Tamura) W. T. Wang & S. H. Wang in Fl. Reip. Pop. Sin. 27: 589. 1979. — Subsect. *Alpina* Tamura in Sci. Rep. Osaka Univ. 17: 31. 1968; et in Hiepko, Nat. Pflanzenfam., Zwei. Aufl. 17a IV : 484. 1995. Type: *T. alpinum* L.

叶均基生。花在花葶上组成顶生总状花序，沿花序轴向顶发育。瘦果无柄或具柄。

Leaves all basal. Flowers in terminal raceme on scape, along raceme rachis acropetally developed. Achenes sessile or stipitate.

1 种。

99. 高山唐松草（《中国高等植物图鉴》） 图版 99

Thalictrum alpinum L. Sp. Pl. 545. 1753; DC. Syst. 1: 175. 1818; et Prodr 1: 12. 1824; Lecoy. in Bull. Soc. Bot. Belg. 24: 193. 1885; Finet et Gagnep. in Bull Soc. Bot. France 50: 616. 1903, p.p.; Nevski in Fl. URSS 7: 519, fig. 35: 1. 1937; Iconogr. Corm. Sin. 1: 685, fig. 1369. 1972; W. T. Wang & S. H. Wang in Fl. Reip. Pop. Sin. 27: 589, pl. 119 et 120: 28. 1979; Grierson Fl. Bhutan 1(2):295. 1984; W. T. Wang in Fl. Xizang. 2: 75. 1985; Y. Z. Zhao in Fl. Intramongol., ed. 2, 2: 460, pl. 181: 1–3. 1990; J. G. Liu in Fl. Xinjiang. 2(1): 270. 1974; L. H. Zhou in Fl. Qinghai. 1: 336. 1997; W. T. Wang in High. Pl. China 3: 41, fig. 768. 2000; D. Z. Fu & G. Zhu in Fl. China 6: 285, 2001; Y. Z. Zhao, Vasc. Pl. Inn. Mongol. 158. 2012; L. Xie in Med. Fl. China 3: 285, with fig. & photo. 2016; X. T. Ma et al. Wild Flow. Xizang 25, photo 48. 2016. Described from Lapland.

99a. var. **alpinum**

多年生小草本植物，全部无毛。叶 4–5 或更多，均基生，为 2 回羽状三出复叶；叶片长 1.5–4 cm；小叶薄革质，通常无柄，圆菱形、菱状宽倒卵形或倒卵形，长和宽均为 3–5 mm，基部圆形或宽楔形，3 浅裂，全缘，脉不明显；叶柄长 1.5–3.5 cm。花葶 1–2 条，高 6–20 cm，不分枝；总状花序长 2.2–9 cm；苞片狭卵形，长 1–2.8（–4）mm；花梗向下弧状弯曲，长 1–12 mm。花：萼片 4，早落，椭圆形或长圆形，2–2.2×0.8–1.2 mm，顶端钝或微尖，每一萼片具（1–）2–3 条基生脉（中脉和 1 侧生脉不分枝，另 1 侧生脉 1 回二歧状分枝）。雄蕊（5–）7–10；花丝丝形，长 2.8–3.5 mm；花药狭长圆形或长圆形，1.2–2×0.5–0.7 mm，顶端具长 0.1 mm 的短尖头。心皮 3–5；子房椭圆形，约 0.8×0.4 mm；花柱长约 0.6 mm；柱头宽箭头状三角形，约 0.6×0.7 mm，有 2 宽侧生翅。瘦果无柄，近长圆形，约 2×0.8 mm，每侧有 3 条纵肋；宿存柱头长约 1 mm。6–8 月开花。

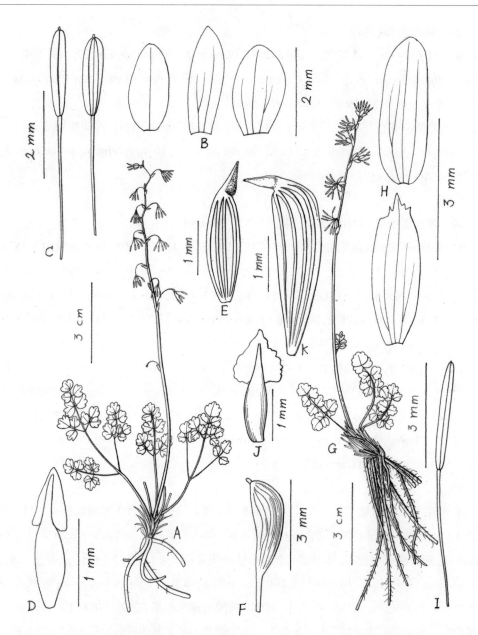

图版99. A-E. **高山唐松草** A. 开花植株，B. 三萼片，C. 二雄蕊，D. 心皮（据陈舜礼112），E. 瘦果（据王忠涛等165）。 F. **柄果高山唐松草** 瘦果（据青藏队12355）。 G-K. **直梗高山唐松草** G. 开花植株，H. 二萼片，I. 雄蕊，J. 心皮（据蓝顺彬424），K.瘦果（据谢磊，毛建丰130）。

Plate 99. **Thalictrum alpinum** A-E. var. **alpinum** A. flowering plant, B. three sepals, C. two stamens, D. carpel (from S. L. Chen 112), E. achene (from Z. T. Wang et al. 165). F. var. **microphyllum** achene (from Qinghai-Xizang Exped. 12355). G-K. var. **elatum** G. flowering plant, H. two sepals, I. carpel (from S. B. Lan 424), K. achene (from L. Xie & J. F. Mao 130).

Small perennial herbs, totally glabrous. Leaves 4–5 or more, all basal, 2-pinnat-ternate; blades 1.5–4 cm long; leaflets subcoriaceous, usually sessile, orbicular-rhombic, broadly rhombic-obovate or ovate, 3–5×3–5 mm, at base rounded or broadly cuneate, 3-lobed, entire, nerves inconspicuous; petioles 1.5–3.5 cm long. Scapes 1–2, 6–20 cm long, simple; racemes 2.2–9 cm long; bracts narrow-ovate, 1–2.8(-4)mm long; pedicels downwards arcuate, 1–12 mm long. Flower: Sepals 4, caducous, elliptic or oblong, 2–2.2×0.8–1.2 mm, at apex obtuse or slightly acute, each sepal with (1–)2–3 basal nerves (midrib and 1 lateral nerve simple, and another lateral nerve once dichotomizing).Stamens (5–)7–10; filaments filiform, 2.8–3.5 mm long; anthers narrow-oblong or oblong, 1.2–2×0.5–0.7 mm, apex with mucrones 0.1 mm long. Carpels 3–5; ovary elliptic, ca. 0.8×0.4 mm; style ca. 0.6 mm long; stigma broadly sagittate-triangular. ca. 0.6×0.7 mm, with 2 broad lateral wings. Achene sessile, suboblong, ca. 2×0.8 mm, on each side longitudinally 3-ribbed; persistent stigma ca. 1 mm long.

分布于中国云南西部、西藏、青海、甘肃、内蒙古、新疆；以及亚洲北部和西部、欧洲、北美洲。生高山草地或灌丛中，海拔 2500–4800 m。

根和全草有清热燥湿、解毒、凉血之效，用于小儿肺炎、泄泻痢疾、头痛目赤等症。根含高山唐松草碱二酮（thalpindione）、厚果唐松草次碱（thalidasine）等生物碱。(《中国药用植物志》，2016）

标本登录：云南西部，1930，G. Forrest 28556。

西藏：芒康，青藏队 76-11846；林芝，西藏中草药队 3114；改则，郎楷永 10278；仲巴，青藏队 75-6772；土门格拉，杨金祥 1894；拉孜，陈书坤，杜庆 75-420；帕里，青藏补点队 75-344；定日，张永田、郎楷永 3796；聂拉木，张永田，郎楷永 3866，4320；吉隆，青藏队 75-6227；札达，青藏队 76-7828，76-12826。

青海：可可西里，岗齐曲，郭柯 K-45。

甘肃：玛曲，郎木寺，白水江队 1597。

新疆：青河，阿尔泰山，秦仁昌 1134；吐鲁番，李安仁，朱家柟 5691；乌鲁木齐，天山，边田文弘，杨戈 102；石河子，陈舜礼 112；尼勒克，关克俭 4031；察布查尔，王忠涛 165；且末，新疆队 90-9408；策勒，郭柯，郑度 12095；塔什库尔干，新疆队 78-1421。

99b. 柄果高山唐松草（《中国植物志》，变种）

var. **microphyllum** (Royle) Hand.-Mazz. Symb. Sin. 7: 311. 1931; W. T. Wang & S. H. Wang in Fl. Reip. Pop. Sin. 27: 589. 1979; W. T. Wang in Fl. Xizang. 2: 75. 1985; in Vasc. Pl. Hengduan Mount. 1: 505. 1993; in Fl. Yunnan. 11: 177. 2000; et in High. Pl. China 3: 480.

2000; D. Z. Fu & G. Zhu in Fl. China 6: 286. 2001. —*T. microphyllum* Royle, Ill. Bot. Himal. 51. 1831. Type specimen collected from Kedarkanta, Himalaya at an elevation of between 12000 and 13000 feet.

T. alpinum L. var. *stipitatum* auct. non Yabe: Boivin in Rhodora 46: 358. 1944, p. p. quoad syn. *T. alpinum* var. *microphyllum* (Royle) Hand.-Mazz.

与高山唐松草的区别在于本变种的瘦果具柄，每侧有 4-5 条纵肋。花葶高 9-17 cm，不分枝。萼片具 3 条不分枝的基生脉。

This variety differs from var. *alpinum* in its stipitate achene with 4–5 longitudinal ribs on each side. Scapes 9–17 cm tall, simple. Sepal with 3 simple basal nerves.

分布于中国西藏、云南西北部、四川南部；以及印度北部。生高山草地；在西藏海拔 4000-5100 m，在云南西北和四川南部海拔 2700-4200 m。

标本登录：西藏：双湖，郎楷永 10074；墨竹工卡，傅国勋 142；错那，吴征镒等 75-874；札达，青藏队 76-8018。

云南：丽江，王启无 71028, 71118，俞德浚 15626，赵锡光 30309；中甸，俞德浚 12268, 12328, 12691；维西，王启无 64658；Mekong-Salwin divide, Sawalonba, 俞德浚 22657。

四川：盐源，青藏队 83-12355。

99c. **直梗高山唐松草**（《中国植物志》，变种）

var. **elatum** Ulbr. in Notizbl. Bot. Gart. Berl. 10: 877. 1929; Pl. Tsinling. 1(2): 247. 1974; W. T. Wang & S. H. Wang in Fl. Reip. Pop. Sin. 27: 591, pl. 149: 1-4. 1979; W. T. Wang in Fl. Xizang. 2: 76. 1985; J. W. Wang in Fl. Hebei 1: 445. 1986; Y. K. Li in Fl. Guizhou. 3: 73, pl. 30: 6–9. 1990; X. Y. Yu et al. in Fl. Shanxi. 1: 607, fig. 386. 1992; W. T. Wang in Acta Phytotax. Sin. 31(3): 216. 1993; et in Vasc. Pl. Hengduan Mount. 1: 504. 1993; L. H. Zhou in Fl. Qinghai. 1: 336. 1997; W. T. Wang in Fl. Yunnan. 11: 175. 2000; et in High. Pl. China 3: 481. 2000; D. Z. Fu & G. Zhu in Fl. China 6: 286. 2001, p.p. excl. syn.; L. Q. Li et al. Pl. Baishuijiang St. Nat. Reserve Gansu Prov. 70. 2014; W. T. Wang in High. Pl. China in Colour 3: 394. 2016; L. Xie in Med. Fl. China 3: 286. 2016. Holotype：四川：松潘, alt. 3100 m, 1922-07-06, H. Smith 2575(not seen).

T. esquirolii Lévl. & Van. in Bull. Acad. Geogr. Bot. 17: n. 210-211, 11. 1907; Hand.-Mazz. Symb. Sin. 7: 311. 1931. Holotype：贵州：Tsin Tchen, Gan-pin, 1906-05-02, Esquirol s.n.（E, not seen; photo, PE）.

T. tofieldioides Diels in Not. Bot. Gard. Edinbury. 5: 263. 1912. Syntypes：云南：丽江，

玉龙山，1906-05，G. Forrest 2149(not seen); 昆明，Tchong-chan, 1905-06-25, Ducloux 3578(not seen).

T. nudum Lévl. & Van. ex Hand.-Mazz. Symb. Sin. 7: 311. 1931, pro syn.

T. alpinum auct. non L.: Rehd. & Kobuski in J. Arn. Arb. 14: 13. 1933; Hand.-Mazz. in Acta Hort. Gotob. 13: 172. 1939.

T. alpinum L. var. *stipttatum* auct. non Yabe: Boivin in Rhodora 46: 358. 1944, p.p. quoad syn. *T. alpinum* var. *elatum* Ulbr. et spesim. sichuanensia et yunnanensia.

99ci. f. **elatum**

与高山唐松草的区别在于其花梗直，向上开展，花药条形和每侧具 4 条纵肋的瘦果。花葶通常高 7-15 cm，不分枝，有些产云南北部和四川西部的植株的花葶高达 25-38 cm，有 1 条分枝。花梗长 1.8-9 mm。萼片具 3 条基生脉（中脉和 1 条侧脉不分枝，另外 1 条侧生脉 1 回二歧状分支；或 3 条脉均 1-2 回二歧状分枝。）

This variety differs from var. *alpinum* in its straight, upwards spreading pedicels, linear anthers and achenes longitudinally 4-ribbed on each side. Scapes usually 7-15 cm tall, simple, but in some plants found in northern Yunnan and western Sichuan, their scapes up to 25-38 cm tall and 1-branched. Pedicels 1.5-9 mm long. Sepal with 3 basal nerves (midrib and 1 lateral nerve simple, and another 1 laternal nerve once dichotomizing; or 3 nerves once or twice dichotomizing).

分布于中国西藏东部、云南北部、贵州、四川西部、青海南部和东部、甘肃南部、宁夏、陕西南部、河北西北部；以及印度北部。生高山草地，海拔 2400-4600 m。

根和全草有清热燥湿、凉血解毒之效，用于小儿惊风，痢疾、肠炎等症。本变种的干燥全草在四川等地作"惊风草"药材使用。(《中国药用植物志》，2016)

标本登录：西藏：左贡，青藏队 76-12109；比如，王金亭 3452，陶德定 11370；藏北加林地区，甘肃农大队 102。

云南：鹤庆，汤宗孝，田效文 421；宁蒗，姜恕 6085；丽江，王启无 71118，秦仁昌 20805，青藏队 81-230；中甸，俞德浚 11378，青藏队 81-1242，周士顺 6187；德钦，俞德浚 8506，8626，横断山植被组 4526，谢磊，毛建丰 180，谢磊 155；东川，蓝顺彬 424。

四川：木里，俞德浚 14645，田效文，汤宗孝 108；得荣，青藏队 81-2314；乡城，D. E. Boufford, B. Bartholomew et. al. 28521；稻城，D. E. Boufford, B. Bartholomew et al. 28052；昭觉，田效文，汤宗孝 153；康定，杨竞生 91-244；乾宁，四川植物队 5414；九龙，黄治平 390；巴塘，四川植物队 1665；德格，四川植物队 7210；甘孜，赵清盛 111147；色达，

虞泽荪 6586；绰斯甲，第八森林队 5416；金川，李馨 77937，汤宗孝，田效文 1036；刷经寺，张泽云，周洪富 22586；茂汶，罗毅波 1053；松潘，汤宗孝 709；红原，姜恕 7252；阿坝，汤宗孝 1067；若尔盖，姜恕，金存礼 1379。

青海：玉树，杨永昌 854；玛沁，何廷农 585；德令哈，郭本兆 11609；天俊，青甘队 59–1067；海晏，钟补求 8318；大通，张志和 4349；门源，青甘队 60–2431；祁连，青甘队 60–2799；格尔木，青甘队 59–474。

甘肃：玛曲，朱格麟 2574，4134；迭部，白龙江队 1749；卓尼，J. F. Rock 12246；岷县，王作宾 14042，洮河队 3052；天祝，？308；山丹，何业祺 4037。

陕西：太白山，钟补求 2973，杨金祥 1040。

山西：五台山，？118；五寨，管山，？81；宁武，刘继孟 1793；繁峙，段长虹 573。

河北：小五台山，王启无 62038，吴韫珍，杨承元 37074，段俊喜 541。

99cii. **毛叶高山唐松草**（《中国植物志》，变型）

f. puberulum W. T. Wang & S. H. Wang in Fl. Reip. Pop. Sin. 27: 591, 621, 1979; W. T. Wang in Fl. Yunnan. 11: 175. 2000, Holotype: 云南：中甸，哈巴山，草坡，1939–05–27，冯国楣 1094 (PE). Paratypes: 云南：丽江，1937–04–26，俞德浚 5073（PE）。四川：木里，1960–06–21，应俊生 4126（PE）；康定，新都桥，1963–07–08，王文采等 676（PE）；金川，1958–06–25，李馨 78736（PE）。

与直梗高山唐松草的区别在于本变型的小叶下面被短柔毛。

This form differs from f. *elatum* in its leaflets abaxially puberulous.

分布于云南西北部和四川西部。生山地草坡，海拔 2500–4600 m。

99ciii. **毛脉高山唐松草**（《东北林学院植物研究室汇刊》，变型）

f. setulosinerve (Hara) W. T. Wang in Bull. Bot. Lab. N.-E. Forest. Inst. 8: 36. 1980; et in Fl. Xizang. 2: 76. 1985. —— *T. alpinum* L. var. *acutilobum* Hara in Fl. East. Himal. 3: 40. 1975. —— *T. setulosinerve* Hara in J. Jap. Bot. 51(1): 7. 1976. Type: Nepal: Late, Kali Gandaki, alt. 18 000 ft., 1954–06–03, Stainton, Sykes & Williams 1044 (holotype, BM, not seen). Bhutan : Shingbe, Me La, 1944–06–29, Ludlow, Sherriff & Hiks 20419 (paratype, BM, not seen).

与直梗高山唐松草的区别在于本变型的小叶下面被短柔毛，在脉上被微硬毛。

This form differs from f. *elatum* in its leafles abaxially puberulous and on nerves hirtellous.

分布于中国西藏东南部；以及不丹、尼泊尔。生高山草甸上，海拔 4100 m。

标本登录：西藏：波密，古乡，alt. 4100 m，1965–08–15，应俊生，洪德元 1167。

亚属 2. 单性唐松草亚属　美洲唐松草亚属(《中国植物志》)

Subgen. **Lecoyerium** B. Boivin in Rhodora 46: 391. 1944; Tamura in Sci. Rep. Osaka Univ. 17: 52. 1968; et in Hiepko, Nat. Pflanzenfam., Zwei Aufl., 17a Ⅳ : 488. 1995; Emura in J. Fac. Sci. Univ. Tokyo, Sect. 3, Bot. 9: 98. 1972; W. T. Wang & S. H. Wang in Fl. Reip. Pop. Sin. 27: 591. 1979. Type: *T. dioicum* L.

植物杂性，雌雄同株或雌雄异株，具单性花，有时具两性花。萼片4(-6)。雄蕊常下垂；花丝丝形，稀棒状；花药顶端具短尖头。心皮比萼片长，具一近鞭形长柱头。

Plants polygamous, monoecious or dioecious, with unisexual flowers, sometimes with bisexual flowers. Sepals 4(–6). Stamens usually pendulous; filaments filiform, rarely clavate; anther apex usually mucronate. Carpel longer than sepals, with a long subflagelliform stigma.

约70种，多数分布于北美洲南部和南美洲北部，少数分布于非洲和欧洲南部，只2种分布于亚洲（中国西南部）。

100. 鞭柱唐松草(《中国植物志》)　图版100

Thalictrum smithii B. Boivin in J. Arn. Arb. 26: 114, pl. 1: 20–22. 1945, p.p. excl. paratypis omnibus et descr. Fl.; W. T. Wang & S. H. Wang in Fl. Reip. Pop. Sin. 27: 592, pl. 140: 14–18. 1979; W. T. Wang in Fl. Xizang. 2: 76. 1985; in Vasc. Pl. Hengduan Mount. 1: 505. 1993; in Fl. Yunnan. 11: 177. 2000; et in High. Pl. China 3: 481, fig. 769. 2000; D. Z. Fu & G. Zhu in Fl. China 6: 298. 2001; Y. Yang, X. Mo & B. Liu, color Fl. Panzhihua Cycas Nat. Reserve 48. 2011; W. T. Wang in High. Pl. China in Colour 3: 396. 2016; L. Xie in Med. Fl. China 3: 287, with fig. 2016; X. T. Ma et al. Wild Flow. Xizang 28, photo 54. 2016. Type: 四川：木里，洼耳寨，沙湾，alt. 3000 m, 1937-10-11, 俞德浚 14487 (holotype, GH, photo, PE, isotype. PE)。

多年生草本植物，茎高30-90 cm，无毛，不分枝或上部分枝。下部茎生叶为3回羽状复叶；叶片长10-20 cm；小叶草质，宽菱形、卵形或宽卵形，长和宽均为0.6-1.5 cm，顶端急尖，基部圆形或浅心形，3浅裂（裂片有1-2牙齿），上面无毛，下面脉上有稀疏稍硬毛,（毛长0.2-0.6 mm）或变无毛，脉上面稍凹陷，下面稍隆起；小叶柄长0.5-9 mm；叶柄长1.5-5 cm，无毛；托叶狭，边缘不规则开裂。聚伞圆锥花序长10-40 cm，有多数花，两性，雄性或雌性。两性花：萼片4，宽卵形，约2×1.2 mm，无毛，每一萼片具3条基生脉（中脉1-2回二歧状分枝，2侧生脉1回二歧状分枝）；雄蕊约7枚，无毛，

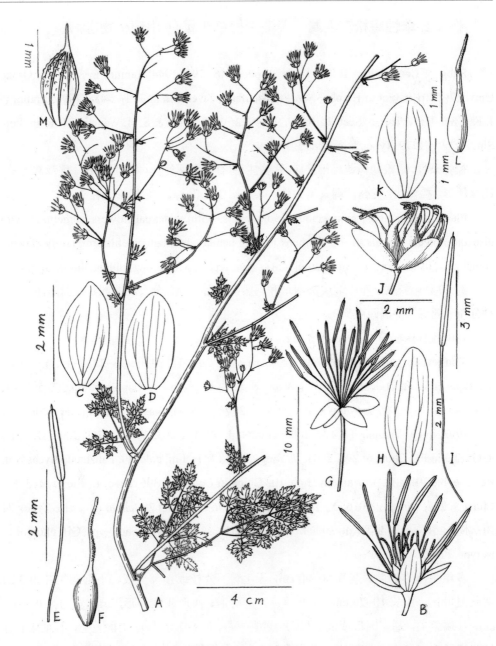

图版100. 鞭柱唐松草　A.具两性花的开花茎顶部，B.两性花，C、D.二萼片，E.雄蕊，F.心皮（据李馨70700）G.雄花，H.萼片，I.雄蕊（据郎楷永，李良千，费勇637），J.雌花，K.萼片，L.心皮（据俞德俊6698），M.瘦果（据isotype）。

Plate 100. **Thalictrum smithii**　A. apical part of flowering stem with bisexual flowers, B. bisexual flower, C,D. two sepals, E. stamen, F. carpel (from X. Li 70700), G. staminate flower, H. sepal, I. stamen (from K. Y. Lang, L. Q. Li & Y. Fei 637), J. pistillate flower, K. sepal, L. carpel (from T. T. Yu 6698), M. achene (from isotype).

花丝丝形，长 1.5-2.5 mm，花药狭长圆形，1.8-2.5×0.4-0.5 mm，顶端具短尖头；心皮 6-9，子房长 0.8-1.2 mm，柱头近鞭形，长 1.2-2 mm。雄花：萼片 4，长圆状卵形，约 3.2× 1.2 mm，无毛，每一萼片具 3 条基生脉（中脉不分枝，2 侧生脉均 2 回二歧状分枝）；雄蕊约 17 枚，花丝丝形，长约 5.5 mm，花药条形，约 3×0.4 mm，顶端具短尖头。雌花：萼片 4，宽卵形，约 1.5×1 mm，无毛，每一萼片具 3 条基生脉（中脉不分枝，2 侧生脉约 1 回二歧状分枝）；心皮约 11 枚，子房长约 1 mm，柱头近鞭形，长约 0.9 mm。瘦果卵球形或椭圆体形，1.5-2.2×0.6-1 mm，有 6 或 8 条钝纵肋，被长 0.1-0.6 mm 的短柔毛，常变无毛；宿存柱头长 0.5-1.5 mm，5-7 月开花。

Perennial herbs. Stems 30-90 cm tall, glabrous, simple or above branched. Lower cauline leaves 3-pinnate; blades 10-20 cm long; leaflets herbaceous, broad-rhombic, ovate or broad-ovate, 0.6-1.5×0.6-1.5 cm, at apex acute, at base rounded or subcordate 3-lobed (lobes 1-2-dentate), adaxially glabrous, abaxially on nerves with sparse slightly stiff hairs 0.2-0.6 mm long or glabrescent, nerves adaxially slightly impressed, abaxially slightly prominent; petiolules 0.5-9 mm long; petioles 1.5-5 cm long, glabrous; stipules narrow, irregularly divided. Thyrses 10-40 cm long, many-flowered, bisexual, staminate or pistillate. Bisexual flower : Sepals 4, broad-ovate, ca. 2×1.2 mm, glabrous, each sepal with 3 basal nerves (midrib once or twice dichotomizing, and 2 lateral nerves once dichotomizing); stamens ca. 7, glabrous, filaments filiform, 1.5-2.5 mm long, anthers narrow-oblong, 1.8-2.5×0.4-0.5 mm, apex mucronate; carpels 6-9, ovary 0.8-1.2 mm long, stigma subflagelliform, 1.2-2 mm long. Staminate flower: Sepals 4, oblong-ovate, ca. 3.2×1.2 mm, glabrous, each sepal with 3 basal nerves (midrib simple, and 2 lateral nerves all twice dichotomizing); stamens ca. 17, filaments filiform, ca. 5.5 mm long, anthers linear, ca. 3×0.4 mm, apex mucronate. Pistillate flower: Sepals 4, broad-ovate, ca. 1.5×1 mm, glabrous, each sepal with 3 basal nerves (midrib simple, and 2 lateral nerves all once dichotomizing); carpels ca. 11, ovary ca. 1 mm long, stigma subflagelliform, ca. 0.9 mm long. Achenes ovoid or ellipsoidal, 1.5-2.2×0.6-1 mm, longitudinally obtusely 6- or 8-ribbed, minutely puberulous (hairs 0.1-0.6 mm long), often glabrescent; persistent stigmas 0.5-1.5 mm long.

分布于西藏东南部（林芝，隆子）、云南西部和中北部、四川西部。生山地林边、草坡或田边，海拔 1820-3300 m。

B. Boivin 指定 *Thalictrum smithii* Boivin 的 holotype 是采自四川木里的 T. T. Yü（俞德

浚）14487号标本，此外，指定3号标本为paratypes：2号是采自云南福贡（Shangpa）的H. T. Tsai（蔡希陶）54683和58383号，另一号H. T. Tsai 57280号标本缺野外记录。在PE，有isotype 1张，每一paratype均有2张标本。经过我们对7张标本研究之后了解到：2号标本H. T. Tsai 54683和58383的植株高大，全部无毛，聚伞圆锥花序也大，分枝繁多，具多数两性花，花的花丝丝形，心皮柱头有2侧生狭翅，这2号标本代表的植物乃是多叶唐松草 *T. foliolosum* DC.，在中国西南部分布较广。H. T. Tsai 57280也为具两性花的标本，植株全部无毛，茎高40-45 cm，其特征是自基部分枝，有2-4条长分枝，花序为具5-10花的伞房状复单歧聚伞花序，花丝丝形，心皮花柱狭长圆形，此标本代表了一新种，在本书中发表，被置于帚枝唐松草 *T. virgatum* Hook. f. & Thoms. 附近。holotype 和 isotype T. T. Yu 14487号是一结果标本，瘦果卵球形或椭圆球形，长1.5-2 mm，有8条钝纵肋，小叶下面脉上疏被长约0.4 mm的稍硬毛，在20世纪根据上述瘦果和小叶的形态特征从采自四川的标本中鉴定出 *T. smithii* 多号具花和具果的标本，了解到此种的花两性或单性，并根据其心皮花柱细长，呈鞭形，了解到此种实为隶属单性唐松草亚属 Subgen. *Lecoyerium* 的植物。还了解到Boivin根据其对3号paratypes的错误鉴定，以及对花构造的错误认识，将 *T. smithii* 误置于唐松草亚属叉枝唐松草组 Subgen. *Thalictrum* sect. *Leptostigma* 中，而不知道此种是隶属 Subgen. *Lecoyerium* 的植物。

全草有清热明目之效，用于腹痛下痢等症。（《中国药用植物志》，2016）

标本登录：云南：腾冲，alt. 1820 m, Gaoligong Shan Biodiv. Survey 29568（♂, GH）；维西，alt. 3100 m，植物所横断山队81-1331（♀）；中甸，虎跳峡，植物所横断山队81-2779（fr.）；贡山，俞德浚20881（fr.）；德钦，青藏队81-2962（♀）；禄劝，毛品一1324（fl. buds）。

四川：木里，武素功3535（fr.），青藏队83-14580（fr.）；德荣，D. E. Boufford et al. 30996（♀）；乡城，D. E. Boufford et al. 30476（☿）；稻城，四川植物队2573（fr.），D. E. Boufford, B. Bartholomew et al. 28205（♂）；康定，郎楷永，李良千，费勇637（♂, ♀）；雅江，杨竞生91-161（fl. buds）；理塘，D. E. Boufford, B.Bartholomew et al. 36247（♂）；巴塘，D. E. Boufford, B. Bartholomew 35511, 35614（♀）；德格，四川植物队7035, 7259（♂）；道孚，伍煜庭，高宝莼11659（♂），张彩飞2001（♂）；金川，李馨77374（fl. buds），第八森林队4689（♂）；绰斯甲，第八森林队2328, 2378（♀）；松岗，李馨70614, 70891（♂）；马尔康，李馨70659（♀），70700（♂），四川植物队22936, 23121（fr.），郎楷永，李良千，费勇2187（fl. buds）。

西藏：左贡，alt. 4000 m，青藏队 76-12097（♀），昌都，D. E. Boufford et al. 31154，32429（fr.）；察雅，青藏队 76-12324（♂）；隆子，青藏补点队 75-0453（♂），75-0555（fr.）；林芝，西藏中草药队 72-3232（♀）；米林，alt. 3000 m，倪志诚等 2825（fr.）。

101. 定结唐松草　图版 101

Thalictrum dingjieense W. T. Wang, sp. nov. Holotype：西藏（Xizang）：定结县，那当至日屋镇途中（Dingjie Xian, on the way from Nadang to Riwu Zhen），alt. 3913 m，林中，草本，花白色（in forest, herbs, fls. white），2013-07-14，PE- 西藏队（PE-Xizang Exped.）3334（PE）.

Species nova haec est arcte affinis *T. smithii* B. Boivin, a qua foliolis minoribus subtus minute glanduloso-puberulis, staminum filamentis longioribus (8 mm longis) differt. In *T. smithii*, foliola majora usque ad 15 mm longa et lata subtus ad nervos hirtella, filamenta breviora 1.5-5.5 mm longa sunt.

Perennial herb. Stem ca. 80 cm tall, glabrous, with short sterile branches. Basal leaves not seen. Cauline leaves 3-ternate, shortly petiolate or sessile; blades 4-5.5×3-4.6 cm; leaflets thinly papery, broad-obovate, pentagonol, suborbicular or obliquely elliptic, 3-9×2-9 mm, at apex obtuse, at base rounded, subtruncate or broadly cuneate, 2-3-lobed (lobes 1-2-rotund-denticulate or entire), rarely undivided, adaxially glabrous, abaxially minutely glandular-puberulous (hairs 0.1-0.25 mm long), 3-5-nerved, nerves adaxially flat or slightly impressed, abaxially prominent; petiolules 0.2-3 mm long, with blade branches glabrous; petioles 0-2 cm long, glabrous, at base narrowly vaginate. Thyrse terminal, ca. 23 cm long, sparse-flowered; rachis and thyrse branches glabrous; bracts foliaceous, 1-5 cm long; pedicels filiform, 14-23 mm long, glabrous. Flower bisexual: Sepals 4(?), white, obovate, ca. 4×2.2 mm, glabrous, at apex rounded, each sepal with 3 basal nerves (midrib and 1 lateral nerve above once dichotomizing, and the another lateral nerve twice dichotomizing). Stamens ca. 7, glabrous; filaments filiform, 8 mm long; anthers narrow-oblong, ca. 2×0.5 mm, at apex mucronate. Carpels 4-7; ovary narrow-obovate, ca. 2×0.8 mm, glabrous; style ca. 2.6 mm long; stigma subflagelliform, ca. 2.4 mm long.

多年生草本植物。茎高约 80 cm，无毛，有不育的短分枝。基生叶未见。茎生叶为 3 回三出复叶，具短柄或无柄；叶片 4-5.5×3-4.6 cm；小叶薄纸质，宽倒卵形、五角形、近圆形或斜椭圆形，3-9×2-9 mm，顶端钝，基部圆形、近截形或宽楔形，2-3 浅裂（浅裂

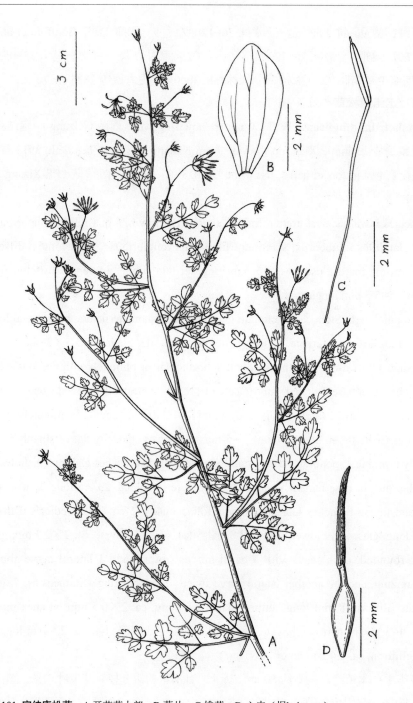

图版101. **定结唐松草** A.开花茎上部,B.萼片,C.雄蕊,D.心皮(据holotype)
Plate 101. **Thalictrum dingjieense** A. upper part of flowering stem, B. sepal, C. stamen, D. carpel (from holotype).

片有 1-2 小圆齿或全缘），稀不分裂，上面无毛，下面被小腺毛（毛长 0.1-0.25 mm），具 3 或 5 出脉，脉上面平或稍下陷，下面隆起；小叶柄长 0.2-3 mm，与叶片分枝均无毛；叶柄长 0-2 cm，无毛，在基部具狭鞘。聚伞圆锥花序顶生，长约 23 cm，具稀疏的花；花序轴和花序分枝无毛；苞片叶状，长 1-5 cm；花梗丝形，长 14-23 mm，无毛。花两性：萼片 4（？），白色，倒卵形，约 4×2.2 mm，无毛，顶端圆形，每一萼片具 3 条基生脉（中脉与 1 条侧脉在上部 1 回二歧状分枝，另一侧脉 2 回二歧状分枝）。雄蕊约 7，无毛；花丝丝形，长 8 mm；花药狭长圆形，约 2×0.5 mm，顶端具短尖头。心皮 4-7；子房狭倒卵形，约 2×0.8 mm，无毛；花柱长约 2.6 mm；柱头近鞭形，长约 2.4 mm。7 月开花。

特产西藏南部定结县。生山地林中，海拔 3913 m。

本种在亲缘关系方面与鞭柱唐松草 *T. smithii* 甚为相近，区别特征为本种的小叶较小，下面被很短的腺毛，雄蕊花丝较长（长 8 mm）。在鞭柱唐松草，小叶较大，长和宽均达 15 mm，下面被微硬毛，花丝较短，长 1.5-5.5 mm。

不了解的种：

1. Thalictrum oshimae *Masamune* in J. Soc. Trop. Agr. 6: 569. 1934; T. S. Liu & C. F. Hsieh in Fl. Taiwan 2: 512. 1976. Holotype: 中国台湾：新竹，Izawayamasita, 1934-07-04, G. Masamune s.n. (not seen)

2. Thalictrum ussuriense *Luferov* in Biull. Moskovak. Obshch. Isp. Prir. Otd. Biol. 94 (5): 103. 1989.

中名索引

（按笔画顺序排列）

二 画

二翅唐松草　27, 42, **66**, 291
二翅唐松草组　33, **59**, 291

三 画

大叶唐松草　19, 23, 29, 38, 48, 52, 58, **62**, **63**, 122
大关唐松草　41, 42, **62**, **94**, 95
大花唐松草　22, 23, 38, 52, **67**
大花台湾唐松草　22, **64**, 149
小叶唐松草　291
小花唐松草　23, 31, 34, 41, 42, **62**, 84
小果唐松草　22, 38, 52, **66**, 225
小金唐松草　19, 33, **59**, **66**, 234, 235
小金唐松草系　33, **59**, 234
小喙唐松草　22, 27, 30, **63**, 129
小喙唐松草系　28, 30, 32, **58**, 128
川鄂唐松草　19, 22, 38, 42, **68**, 295, 300, 301, 304
川鄂唐松草系　33, **59**, 295
马尾连　22, 53, 54, 170, 265, 308
叉枝唐松草　23, 28, 29, 32, 34, 42, **62**, 81
叉枝唐松草系　28, 31, 32, **58**, 81
叉枝唐松草组　23, 25, 28, 29, 32, 44, **58**, **81**, 162, 364

四 画

心基唐松草　**68**, 271, 272
六脉萼唐松草　22, 41, **66**, 238
爪哇唐松草　19, 35, 37, 44, 52, **63**, 109
云南唐松草　41, 42, **63**, **106**, 108
五果绢毛唐松草　42, **62**, **88**, 89
长柱唐松草　41, **66**, 243, 244
长柱贝加尔唐松草　167
长柄唐松草　19, 22, 23, 27, 50, 52, **65**, 199, 201
长梗亚欧唐松草　**70**, 330, 334,
长喙唐松草　24, 38, 52, **64**, **142**, 143
贝加尔唐松草　27, 38, 39, 42, 44, 51, 52, **65**, **170**, 172
贝加尔唐松草系　28, 30, 32, 33, **58**, 167, 170
毛叶网脉唐松草　**65**, 182
毛叶高山唐松草　**70**, 360
毛脉高山唐松草　**70**, 360
毛发唐松草　19, **295**
毛蕊唐松草　23, 41, **65**, 197

五 画

玉山唐松草　**63**, 116, 117
石砾唐松草　20, 40, 45, **70**, 351
石砾唐松草系　29, 34, **59**, 351
东亚唐松草　2, 19, 23, 35, **70**, 329

白茎唐松草 19, 23, 25, 27, 31, 38, 45, **66**, 247
白茎唐松草系 33, **59**, 245
台湾唐松草 22, 32, **64**, 147, 149
丝叶唐松草 22, 38, 47, **67**, 70, 256

六　画

兴山唐松草 41, 64, **65**, 188
吉隆唐松草 41, 45, **66**, 245
西南唐松草 28, 33, 38, 44, **58**, **65**, 184
西南唐松草系 28, 30, 33, **58**, 184
亚东唐松草 41, 46, **68**, 302, 303
亚欧唐松草 38, 52, **70**, **326**, 328, 334, 335
札达唐松草 41, **69**, **312**, 313
尖叶唐松草 16, 27, 31, 38, 48, **66**, 213
尖叶唐松草系 29, 30, 33, **58**, **206**, 228
尖叶唐松草组 16, 25, 32, 48, **58**, 205
光果唐松草 **65**, 174
网脉唐松草 38, 48, 52, **65**, **180**, 182
网脉唐松草系 25, 33, 48, **58**, 179
多叶唐松草 19, 51, 52, **69**, **316**, 364
多枝唐松草 20, 23, 38, 42, 48, 49, **63**, 99
华东唐松草 20, 28, 38, 48, 49, 52, **63**, 96, 99
阴地唐松草 23, 27, 38, **66**, 207
阴阳和 251, 253

七　画

花唐松草 19, 22, 23, **66**, **223**
纺锤唐松草 41, **63**, 135

芸香叶唐松草 27, 52, **66**, **239**, 240
芸香叶唐松草系 29, 33, **59**, 239
芸香叶唐松草组 25, 33, 44, 45, **59**, 234
希陶唐松草 41, **67**, 254
丽江唐松草 23, 27, 38, 42, **65**, 190
角药偏翅唐松草 **68**, 290

八　画

岳西唐松草 41, **66**, 211
河南唐松草 19, 38, **68**, 295
定结唐松草 41, 46, 47, **70**, 365
直梗高山唐松草 23, 38, **70**, **358**, 360
武夷唐松草 23, 41, 48, **65**, 182
金丝马尾连 22, 38, 52, **67**, 251, **265**, 268
单性唐松草亚属 16, 19, 29, 31, 34, 35, 42, 46, **59**, **361**, 364
细柱唐松草 19, 27, 32, 38, 44, **58**, 64, 65, **167**
细柱唐松草系 32, 33, **58**, 167
细唐松草 38, **69**, 335
陕西唐松草 23, 41, **68**, 304
帚枝唐松草 22, 42, 51, 52, **67**, **251**, 364
帚枝唐松草系 29, 33, 44, 52, **59**, 250
帚枝唐松草组 33, **59**, 250

九　画

思茅唐松草 23, 32, 41, **62**, 150
思茅唐松草系 32, **58**, 149
南湖唐松草 20, **64**, 145
柔柱微毛唐松草 **63**, 115

弯柱唐松草 42, **64**, 139
扁果唐松草 **69**, 326
美丽唐松草 19, 23, 27, 46, 52, **67**, 280
美花唐松草 19, 22, 41, 52, **68**, 278
美洲唐松草亚属 29, 361
柄果高山唐松草 **70**, 357
星毛唐松草 38, **69**, 70, 309
昭通唐松草 51, 52, **67**, 268
狭序唐松草 22, 27, 38, 44, 51, **64**
狭药绢毛唐松草 42, **62**, 88
狭裂瓣蕊唐松草 **64**, 179
盾叶唐松草 22, 35, 51, 52, **65**, 70, 217
钩柱唐松草 22, 27, 28, 32, 38, 44, **58**, **64**, 154, 157
钩柱唐松草系 28, **58**, 154

十 画

圆叶唐松草 20, 22, 31, 34, **67**, 70, 269
圆叶唐松草系 33, **59**, 268
钻柱唐松草 22, 25, 32, 41, **162**
钻柱唐松草组 32, **58**, 162
高山唐松草 20, 23, 38, 42, 45, **70**, 355, 358, 359
高山唐松草系 29, 34, **59**, 355
高原唐松草 19, 23, 27, 46, 51, 52, **69**, 319
宽柱唐松草 25, 32, 41, **64**, 165
宽柱唐松草组 32, **58**, 164
宽萼偏翅唐松草 53, **68**, 289
唐松草 3, 16, 38, 42, 51, **66**
唐松草组 16, 20, 23, 25, 28, 31, 33, 44, **59**, 230
唐松草亚属 16, 19, 23, 28, 29, 32, 35, **58**, **80**, 364
唐松草属 1, 2, 3, 4, 6, 7, 9, 14, 15, 16, 17, 19, 20, 22, 23, 25, 27, 28, 29, 31, 32, 34, 35, 42, 43, 44, 46, 48, 51, 53, 54, **57**, **164**
珠芽唐松草 19, 22, 52, **68**, 276
珠芽华东唐松草 **63**, 99
峨眉唐松草 **62**, 118
展枝唐松草 19, 38, 51, **69**, 347
绢毛唐松草 23, 31, 34, 38, 42, **62**, **86**, 88, **90**, 149

十 一 画

密叶唐松草 **62**, 104
粗壮唐松草 19, 38, 48, **52**, **62**, **63**, 125
渐尖偏翅唐松草 **68**, 289
淡红唐松草 145
深山唐松草 48, 52, **66**, 221
菲律宾唐松草 32, **64**, 131
黄唐松草 19, 52, **70**, 345
偏翅唐松草 19, 25, 27, 38, 42, 51, 52, 53, **68**, 250, 273, 285, 289, 290
偏翅唐松草系 22, 29, 46, 52, **59**, 273

十 二 画

散花唐松草 22, 23, 27, 38, 41, 50, **65**, 203
紫堇叶唐松草 **69**, 337
短梗箭头唐松草 35, 52, **69**, 343
短蕊唐松草 23, 32, 41, **62**, 152
稀蕊唐松草 22, 27, 38, 42, **66**, 209

十 三 画

错那唐松草　19, 41, **67**, **273**, 274,
矮唐松草　41, **65**, **194**
察瓦龙唐松草　41, 58, 64, **65**, **67**, 184, **192**
察隅唐松草　41, **63**, **133**
新宁唐松草　23, 41, **63**, **102**
滇川唐松草　38, 51, 52, **68**, **295**, 308
滇南唐松草　**63**, **108**
腺毛唐松草　19, 22, 38, 42, 52, **69**, **322**
腺毛唐松草系　29, 45, **59**, **295**, 309
腺毛唐松草组　16, 19, 23, 25, 29, 31, 33, 46, **59**, **294**
腺毛箭头唐松草　23, **69**, **344**
微毛唐松草　25, 38, 44, 52, **63**, **112**

十四画以上

澜沧唐松草　27, 41, **64**, **137**
藏南唐松草　19, 41, **69**, **70**, **314**
鹤庆唐松草　25, 31, 34, 42, 42, 44, **63**, **92**,
箭头唐松草　19, 38, 51, 52, **69**, **339**, 342, 343
糙叶唐松草　20, 25, 31, 34, 41, 42, 44, **63**, **90**
糙叶唐松草系　28, 31, 32, 48, **58**, **90**, 128
螺柱唐松草　22, 41, **64**, **160**
鞭柱唐松草　38, 42, 46, 51, 52, **70**, **361**, 367
瓣蕊唐松草　20, 23, 27, 35, 38, 58, **64**, **167**, 175
瓣蕊唐松草系　28, 33, **58**, **174**
瓣蕊唐松草组　33, 44, **58**, **167**

拉丁学名索引

(按字母顺序排列，正体字为正名，斜体字为异名)

Isopyrum 219
multipeltatum 219
  trichophyllum 258
Schlagintweitella 20, 29, 30, 31, 34, 45, 353
  fumarioides 353
  glareosa 353
Thalictrum 1, 28, **29**, **56**, **58**, **59**, **60**, **61**, 230
  Sect. *Actaeifolia* 90
  Sect. Baicalensia **170**
  Sect. Dipterium **59**, **291**
  Sect. Dolichostylus **58**, **162**
  Sect. Erythrandra 9, 60
  Sect. Euthalictrum 3
  Sect. *Genuina* 7, 309
  Sect. *Homothalictrum* 7, 294
  Sect. Leptostigma **28**, **58**, **81**
  Sect. *Microgynes* 6, 81
  Sect. Omalophysa **58**, **196**
  Sect. Physocarpum **58**, **205**
  Sect. Platystachys **59**, **250**
  Sect. Platystylus **58**, **164**
  Sect. Rutifolia **59**, **234**
  Sect. Schlagintweitella **29**, **59**, **351**
  Sect. Thalictrum **29**, **59**
  Sect. Tripterium **28**, **59**, **230**
  Ser. Alpina **29**, **59**, **355**
  Ser. Baicalensia **28**, **58**, **170**

  Ser. Clavata **29**, **59**, **61**, **206**
  Ser. Fargesiana **28**, **58**, **184**
  Ser. *Flexuosa* 29
  Ser. Indivisa **59**, **268**
  Ser. Leuconota **59**, **245**
  Ser. Megalostigmata **58**, **167**
  Ser. Osmundifolia **59**, **295**
  Ser. Petaloidea **28**, **58**, **174**
  Ser. Rutifolia **29**, **59**, **239**
  Ser. Saniculiformia **81**
  Ser. Scabrifolia **28**, **58**, **90**
  Ser. Simaoensia **58**, **149**
  Ser. Sinomacrostigmata **59**, **242**
  Ser. Sparsiflora **196**
  Ser. Squamifera **29**, **59**, **351**
  Ser. Thalictrum **59**, **309**
  Ser. *Tripterea* 230
  Ser. Uncata **28**, **58**, **154**
  Ser. Violacea **29**, **59**, **61**, **273**
  Ser. Virgata **29**, **59**, **250**
  Ser. Xiaojinensia **59**, **234**
  Subgen. *Euthalictrum* 81
  Subgen. Lecoyerium **29**, **59**, **361**
  Subgen. Thalictrum **28**, **58**, **80**
  Subgen. *Tripterium* 230
  Subsect. *Actaeifolia* 90
  Subsect. *Alpina* 355

Subsect. *Baicalensia*　170
Subsect. *Claviformes*　205
Subsect. *Flexuos*a　294
Subsect. *Genuium*　294
Subsect. *Indivisa*　9, 34, 268
Subsect. *Omalophysa*　196
Subsect. *Petaloidea*　174
Subsect. *Rutifolia*　234
Subsect. *Saniculiformia*　81
Subsect. *Sparsiflora*　196
Subsect. *Squamifera*　351
Subsect. *Thalictrum*　7, 309
Subsect. *Violacea*　273
　　var. *clematidifolium*　127
actaeifolium　10, 90, 127
acutifolium　**66, 75, 213**
acutifolium　12, 16, 21, 31, 38, 206
affine　79, 342
alpinum　**70, 80, 355**
　　var. *acutilobum*　360
　　var. elatum　**70, 80, 358**
　　　　f. puberulum　**70, 80, 360**
　　　　f. setulosinerve　**70, 80, 360**
　　var. microphyllum　**70, 80, 357**
　　var. *stipitatum*　45, 251, 358
alpinum　1, 4, 6, 16, 20, 38, 42, 45
amplissimum　331
angustifolium　342
aquilegifolium　**66, 75, 230**
　　ssp. *asiaticum*　231

　　var. *asiaticum*　231
　　var. *sibiricum*　1, 27, 35
aquilegifolium　5, 12, 19, 27, 31, 35, 230
argyi　109
atriplex　**64, 73, 157**
austrotibeticum　**69, 78, 314**
baicalense　**65, 74, 170**
　　f. levicarpum　**65, 74, 174**
brachyandrum　**62, 71, 152**
brevisericeum　**62, 71, 86**
　　var. angustiantherum　**62, 71, 88**
　　var. pentagynum　**62, 71, 88**
callianthum　**68, 77, 278**
chaotungense　52, 268
chayuense　**63, 73, 133**
chelidonii　**68, 77, 276**
　　var. *reniforme*　280
chiaonis　215
chinense　331
cirrhosum　**69, 78**
　　var. *acutifolium*　1 213
　　var. *cavaleriei*　213
　　var. *filamentosum*　223
clematidifolium　127
contortum　51, 231
coreanum　16, 51, 217, 219
cultratum　320
　　ssp. *tsangense*　353
cuonaense　**67, 77, 274**
cysticarpum　276

daguanense 62, 71, **94**

deciternatum 305, 306

declinatum 213

delavayi **68, 77, 285**, 287

   var. acuminatum **68, 289**

      f. *appendiculatum* 289

   var. decorum **68, 77, 289**

   var. mucronatum **68, 78, 290**

   var. *parviflorum* 285

delavayi 1, 10, 19, 22, 25, 27, 38, 42, 51, 52, 53, 273, 285, 289, 290

dingjieense **70, 80, 365**

dipterocarpum 53, 285, 290

   var. *mucronatum* 290, 374

diffusiflorum **68, 77, 282, 283**

duclouxii 287

elegans **66, 76, 291**

englerianum 251

esquirolii 358

faberi **62, 63, 71, 72**

falcatum 127

fargesii **65, 74, 184**

fauriei 147

filamentosum **66, 75, 223**

finetii **68, 78, 306**

flavum **70, 79, 346**

flavum 2, 11, 19, 52

foeniculaceum **67, 76, 256**

foetidum **69, 78, 79, 322**

   ssp. *glabrescens* 326

   var. glabrescens **69, 79, 326**

   var. glandulosissimum **67, 77, 265**

foliolosum **69, 78, 316**

fortunei **63, 72, 96, 98**

   var. bulbiliferum **63, 72, 99**

fusiforme **63, 73, 135**

giraldii 171

glandulosissimum **67, 77, 265**

   var. chaotungense **67, 268**

glareosum 353

grandidentatum **63, 72, 120**

grandiflorum **67, 76, 259**

hamatum 154

hayatanum 147

honanense **68, 78, 297**

ichangense **65, 74, 75, 217**

   race *coreanum* 219

   var. *coreanum* 219

isopyroides **69, 79, 337**

javanicum **63, 72, 109**

   var. *puberulum* 44, 114, 116

javanicum 3, 4, 7, 10, 19, 35, 44, 52, 114, 116

jilongense **66, 76, 245**

kemense 334

   var. *stipellatum* 334

lancangense **64, 73, 137**

lasiogynum **65, 74, 197**

latistylum **73, 165**

laxum 127

lecoyeri **63, 72, 114**
    var. debilistylum **63, 72, 115**
leuconotum **66, 76, 247**
leuconotum 1, 10, 19, 23, 25, 27, 31, 45, 243
leve **63**
macrophyllum 124
macrorhynchum **64, 73, 142**
macrostigma 1, 241, 243
mairei 234
megalostigma **58, 65, 74, 167**
micrandrum 147, 358
microphyllum 45, 358
minus **70, 79, 326**
    var. elatum **70, 80, 358**
    var. *foetidum* 322
    var. hypoleucum **70, 79, 329**
    var. *kemense* 334
    var. rotundidens **70, 79, 334**
    var. stipellatum **70, 79, 330**
minus 4, 11, 19, 23, 35, 38, 52, 294, 322
minutiflorum **62, 71, 84**
morii 147
multipeltatum 219
myriophyllum **62, 71, 104**
neurocarpum 280
nudum 359
oligandrum **66, 75, 209**
omeiense 2, 10, 29, 38, 41, 51
oshimae 367
osmundifolium **68, 78, 300**

pallidum 185
panzhihuaense **67, 776, 261**
petaloideum **64, 74, 175**
    var. *latifoliolatum* 175
    var. supradecompositum **64, 74**
philippinense **64, 73, 131**
przewalskii **65, 75, 199**
pseudoichangense **65, 74, 228**
pumilum **74, 194**
punduanum **68, 271**
    var. hirtellum **65, 68, 75, 78, 182, 271**
purdomii 331
radiatum 82
ramosum **63, 72, 99**
reniforme **67, 77, 280**
reticulatum **65, 75, 180**
    var. hirtellum **65, 68, 75, 78, 271**
robustum **62, 71, 72, 125**
rockii 199
rostellatum **63, 72, 128**
rotundifolium **67, 76, 269**
rubescens **64, 73, 145**
rutifolium **66, 76, 239**
samariferum 293
saniculiforme **62, 71, 81**
scabrifolium **63, 72, 90**
    var. *leve* 1, 92
scabrifolium 1, 10, 20, 25, 31, 34, 41, 42, 44, 90, 92, 106
scaposum 225

sessile 60, 61, 63, 71, 72, 74
setulosinerve 360
sexnervisepalum 66, 273, 237
shensiense 68, 76, 304
silvestrii 185
simaoense 40, 62, 75, 150
simplex 69, 39, 341
    var. affine 69, 70, 342
    var. brevipes 69, 79, 343
    var. glandulosum 69, 79, 344
simplex 11, 19, 23, 35, 38, 51, 52, 342, 343, 345
sinomacrostigma 66, 76, 243
smithii 70, 80, 361
smithii 1, 13, 16, 38, 42, 46, 51, 52, 254, 316, 318, 363, 364, 365, 367
sparsiflorum 65, 75, 203
spiristylum 64, 73, 160
squamiferum 70, 80, 351
squarrosum 69, 79, 347
supradecompositum 51, 179
tenii 295
tenue 69, 79, 335
tenuisubulatum 64, 73, 162
thibeticum 7, 285
thunbergii 331
    var. *majus* 331
trichopus 68, 78
trigynum 349

tripeltatum 217
tsaii 67, 76, 254
tsawarungense 65, 67, 74, 77, 192
tuberiferum 66, 75, 221
umbricola 66, 75, 206
uncatum 64, 73, 154
    var. angustialatum 64, 73, 157
uncinulatum 139
unguiculatum 213
urbainii 64, 73, 147
    var. majus 64, 73, 149
ussuriense 367
verticillatum 251
virgatum 67, 76, 251
    var. *stipitatum* 251
viscosum 67, 77, 263
wangii 65, 190
wangii 11, 27, 38, 42, 192
wuyishanicum 75, 182
xiaojinense 66, 76, 235
xingshanicum 74, 188
xinningense 63, 72, 102
yadongense 68, 78, 302
yui 319
yuoxiense 66, 75
yunnanense 63, 72, 106
    var. austroyunnanense 63, 72, 108
zhadaense 69, 78, 312